About Island Press

Since 1984, the nonprofit organization Island Press has been stim-
ulating, shaping, and communicating ideas that are essential for
solving environmental problems worldwide. With more than 1,000
titles in print and some 30 new releases each year, we are the
nation's leading publisher on environmental issues. We identify
innovative thinkers and emerging trends in the environmental
field. We work with world-renowned experts and authors to
develop cross-disciplinary solutions to environmental challenges.

Island Press designs and executes educational campaigns, in
conjunction with our authors, to communicate their critical mes-
sages in print, in person, and online using the latest technologies,
innovative programs, and the media. Our goal is to reach targeted
audiences—scientists, policy makers, environmental advocates,
urban planners, the media, and concerned citizens—with information that can be used to create the framework for long-term
ecological health and human well-being.

Island Press gratefully acknowledges major support from The
Bobolink Foundation, Caldera Foundation, The Curtis and Edith
Munson Foundation, The Forrest C. and Frances H. Lattner Foundation,
The JPB Foundation, The Kresge Foundation, The Summit Charitable
Foundation, Inc., and many other generous organizations and indi-
viduals.

The opinions expressed in this book are those of the author(s)
and do not necessarily reflect the views of our supporters.

Praise for *Corridor Ecology*, second edition

"Humanity readily gobbles up and impacts both land and sea, interrupting the life histories of the plants and animals upon which we depend. How do we identify and protect corridors to help our fellow travelers keep on keeping on? *Corridor Ecology* has long been the definitive guide and resource for scientists and policy makers. This new edition is a welcome and necessary update, accessible to the general reader."

—Mary Ellen Hannibal, author of *The Spine of the Continent* and *Citizen Scientist*

"The ultimate survival of nature depends on the critical concept of 'corridor ecology,' as this superb book describes in timely, absorbing detail. The worldwide destruction and fragmentation of habitats, coupled with the deadly effects of climate change, already affects all plant and animal life. Only corridors connecting habitats, coupled with restoration of critical areas, can offer survival to species."

—George B. Schaller, Senior Conservationist, Wildlife Conservation Society

"When I think about the flow of life across the landscape, I think about grizzly bear (*smxeyčn*), deer (*sxʷllesšn*) and muskrat (*ččlexʷ*) dying on the roads, and the fact that it is humans' responsibility to make this world safe for all of us, people and animals. Providing linkages between protected areas allows for the flow of life to remain intact and constant. Providing for future generations is our responsibility."

—Whisper Camel-Means, Tribal Wildlife Biologist, Confederated Salish and Kootenai Tribes

"Billions of animals and plants have been lost across the planet in the last century. Protecting habitat and maintaining or restoring connectivity is essential for nature and our civilization to survive. We are fortunate to have a radically updated edition of the most comprehensive book on corridors for ecologists, managers, politicians, and laypersons."

—Paul R. Ehrlich, coauthor of *The Annihilation of Nature*

"A transforming planet, from global climate change and other human impacts, increases the urgency for conservation practitioners and managers to consider impacts to species movements and habitat dynamics

across landscapes and seascapes. *Corridor Ecology* provides the tools necessary for sustainable conservation outcomes."
—Stacy Jupiter, Regional Director, Melanesia Program,
Wildlife Conservation Society

"Axiomatic in ecology is that everything is connected, but in nature conservation we have regularly failed to build connectivity into conservation solutions. Newly updated, *Corridor Ecology* is an outstanding contribution to nature conservation. In a fragmented world with a rapidly changing climate, building ecological connectivity is essential to living in harmony with nature."
—Stephen Woodley, Vice-Chair for Science and Biodiversity,
World Commission on Protected Areas, IUCN

"Growing up in the Y2Y region, I have seen firsthand its amazing ecological value. Now, current work connecting wild areas and sharing the science substantially increases our effectiveness in conservation efforts."
—Scott Niedermayer, former NHL hockey player, four-time Stanley
Cup Champion, and champion for nature.

"Fragmentation of wild natural areas is a growing problem in every corner of the world. With rigor and didactic prose, Hilty and colleagues offer key ecological principles to connect networks of protected areas in human landscapes. When planning development, societies cannot afford to lose natural areas; corridors can provide a solution for coexistence with nature."
—Avecita Chicchón, Program Director, Andes-Amazon Initiative,
Gordon and Betty Moore Foundation

"I recommend *Corridor Ecology* to terrestrial and aquatic conservation and development practitioners, including in Africa, where seeking a sustainable relationship between development and conservation is ever more urgent. Maintaining ecological connectivity and preventing habitat loss and fragmentation is critical to the long-term conservation of wildlife and wildlands, and benefits the well-being of humans."
—Philip Muruthi, Vice President, Species Conservation,
African Wildlife Foundation

Corridor Ecology
Linking Landscapes for Biodiversity Conservation and Climate Adaptation

SECOND EDITION

Jodi A. Hilty, Annika T. H. Keeley,
William Z. Lidicker Jr., and Adina M. Merenlender

Foreword by Hugh Possingham

ISLANDPRESS

Washington | Covelo | London

Library of Congress Control Number: 2019931846

Keywords: conservation corridor, wildlife corridor, ecological corridor, stepping-stones, riparian corridors, ecological networks, movement pathways, green infrastructure, greenway, linkage, connectivity, structural connectivity, functional connectivity, climate-wise connectivity, marine connectivity, landscape, seascape, conservation, conservation planning, wildlife, climate change, ecology, landscape fragmentation, habitat loss, island biogeography, dispersal, metapopulations, connectivity modeling, scale, implementation, case studies

CONTENTS

Foreword   *xi*
Preface   *xiii*

Introduction   1

Chapter 1. Background: Habitat Loss, Fragmentation, and
Climate Change   6
    Human-Induced Change and Habitat Loss   7
    Climate Change Overview   9
    Limitations to Protected Areas   13
    Reconnecting Our Planet   14
    Growth of Connectivity Science and Practice   16

Chapter 2. The Ecological Framework   19
    Island Biogeography   19
    Metapopulation Theory: Conceptual History   24
    Metapopulation Processes   30
    Dispersal   30
    The Demography of Extinction   36
    Genetic Structuring   40
    A Longer-Term Perspective   44
    Metacommunity Theory   47
    Beyond Metacommunities: Landscape and Ecoscape Concepts   52

Chapter 3. Understanding Fragmentation   55
    Natural versus Human-Induced Fragmentation   55
    Speed and Pattern of Change   56
    Consequences of Human-Induced Fragmentation   58
    Changes in Species Composition of Patches   63
    Genetic Considerations Affecting Species Extinction   66
    Role of the Matrix   67
    Edges and Edge Effects   73

Chapter 4. Approaches to Achieving Habitat Connectivity   90
    What Is a Corridor?   90
    Types of Corridors   93

Riparian Areas   101
Corridors for Individual Species Conservation   102
Corridor Complexities   104
Biological Benefits   108
Benefits to Humans   112

Chapter 5. Corridor Design Objectives   116
Focal Species Considerations   117
Habitat Requirements   120
Dispersal Considerations   122
Generalist versus Specialist   123
Behavioral Factors   123
Sensitivity to Human Activity   124
Physical Limitations   125
Topography and Microclimate for Climate-Wise Connectivity   127
Corridor Quality: Continuity, Composition, and Dimension   129
Continuous Corridors   129
Stepping-Stone Connectivity   132
Habitat Quality   134
Corridor Dimensions   136
Landscape Configuration   139
Riparian Corridors   141
Hydrologic Habitat Connectivity: Structural, Functional,
  and Ecological   142
Ecological Networks for Conservation   144

Chapter 6. Potential Pitfalls or Disadvantages of Linking Landscapes 146
Impacts of Edge Effects   148
Corridors as Biotic Filters   148
Facilitation of Invasions   151
Invasions of Deleterious Native Species   153
Demographic Impacts   155
Social Behavior   157
Negative Genetic Effects   158
Conflicting Ecological Objectives   160
Economic Considerations   161

Chapter 7. Identifying, Prioritizing, and Assessing Habitat
Connectivity   164
Establishing Collaborations   164
Addressing Scale   165
Identifying Terrestrial Corridors for Conservation and Restoration   167
Prioritization   178

Climate Resilience Benefits   185
Assessing Corridors   186
Caveats   193

Chapter 8. Climate-Wise Connectivity   195
Principles of Climate Space   196
Designing Climate-Wise Connectivity   205
Including Refugia in Climate-Wise Connectivity Design   210
Estimating Range Shifts Using Species Distribution Modeling   210
Recommendations   214

Chapter 9. Ecological Connectivity in the Ocean   216
Mark H. Carr and Elliott L. Hazen
What Constitutes Pelagic Connectivity and Corridors?   217
Where Are the Major Pelagic Marine Corridors?   219
Threats to Pelagic Corridors and Potential Conservation Approaches   222
What Constitutes Connectivity and Corridors in the Coastal Ocean?   226
Threats to Coastal Species, Ecosystems, and their Connectivity   233
Implications of Coastal Corridors for Species and
    Biodiversity Conservation   234

Chapter 10. Protecting and Restoring Corridors   238
Opportunities and Challenges   239
Law and Policy Mechanisms   243
Stewardship of Working Lands   245
Private Land Conservation   248
Types of Agreements   248
Restoring Land   251
Lessons from Corridor Projects   253

Conclusion   274

References   *283*
About the Authors   *337*
Index   *339*

FOREWORD

This updated edition of *Corridor Ecology* could not be more timely. For decades, corridors have been employed across fragmented landscapes in the hope that plants as well as animals can continue to move about the landscape as they have for millennia. Corridors of every scale, from narrow urban riparian strips to the Yellowstone to Yukon megacorridor, have a role to play in nature conservation. But now many species, indeed, entire ecosystems, are moving in response to climate change. Assembling the evidence for corridor effectiveness and practical rules about conservation interventions that can be used to protect, enhance, and construct corridors has never been more important. This book is a crucial contribution to that effort.

A range of intriguing corridor issues will require our attention at varying spatial, temporal, and institutional scales. For example, over a geographic area are small or large-scale corridors the issue? Over time, are we interested in short-term or century-scale outcomes? Are the actors involved in decision making large governments or small local communities, or both, or everything in between? This book adeptly moves across all scales and draws on a huge variety of sources, both empirical and theoretical.

Although the theory of corridors emerged largely from thinking about terrestrial systems, corridors are important in any realm, and this book covers them all. Given the linear nature of their focal ecosystem, riparian conservation scientists have long understood the importance of corridors. The impact of weirs and dams on streams and rivers are dramatic, not just because they can stop species from moving but because of the huge biophysical changes they impose. In marine systems on the other hand, the role of corridors is not well understood, and the field of marine connectivity is still new. The chapter on marine corridors, written by marine biologists, adeptly introduces the concept of marine connectivity and the practical implications for conservation actions. And only last week I became aware of the idea of corridors in the air, where international policy issues are especially important. How do we facilitate safe passage for propagules and individuals moving on wind currents when wind farms, night lights, and human-controlled flying objects clutter the sky?

In all systems—terrestrial, marine, aquatic, and air—I worry about the balancing act required to meet the needs of not only a few iconic species but all of them. Actions that deliver the most appropriate level of connectivity for one species are unlikely to deliver the optimal level of connectivity for all the other species. This is especially true in marine environments, where species life-histories can be remarkably diverse. What is good for the whale may have little meaning for a small reef fish.

In the past few years, I have reflected on conservation management dichotomies. Should we invest more in action A or action B? For example, should we create a single large reserve or several small ones? The short answer is, "it depends." I feel we need to provide advice about how our conservation investment should be divided between alternative strategies. For example, should we "restore or protect," "expand or manage," or "represent or connect"?

In the study of corridors and a world of constrained conservation resources, I see many interesting conservation dichotomies. For example, should we restore habitat to connect patches or make one of the patches bigger? Should we focus more of our effort on connectivity within a habitat or between habitats? Or should we increase connectivity along mountain ridges and riparian habitats at the expense of representing habitats more equitably?

Fundamental knowledge about the costs and benefits of corridors will be essential to resolving these conundrums. Hilty, Keeley, Lidicker, and Merenlender do a remarkable job of exploring the rich topic of corridor ecology and conservation. They are concise yet comprehensive. The book is founded in basic theory, and compelling case studies are woven through every chapter. In line with modern conservation science, this newly updated comprehensive synthesis covers not only ecology, but also relevant components of social science, policy, and related disciplines. This newest edition of *Corridor Ecology* comes at just the right time and will be an outstanding resource for practitioners, researchers, teachers, and students.

Hugh Possingham, FNAS
Chief Scientist, The Nature Conservancy
Brisbane, Queensland, AU, and Arlington, Virginia, USA
July 2018

PREFACE

Conservation is humanity caring for the future.
—*Nancy Newhall (1908–1974), Editor and photography critic*

Since the first edition of *Corridor Ecology* was published, science has advanced significantly, and the implementation of corridors is vastly more widespread. A proliferation of studies focusing on habitat connectivity originate from all over the world; over half of these studies coming from the United States, followed by Spain, China, and Australia, respectively (Correa Ayram et al. 2016).

Over 180 new papers have been published on connectivity and climate change alone, an area barely discussed in 2005 when we submitted the first book for publication (Keeley, Ackerly et al. 2018). Likewise, the growth and sophistication of connectivity modeling approaches exploded in recent years. Connectivity in the marine environment has recently emerged as a field of research, and we are pleased that Mark Carr and Elliott Hazen have contributed a chapter on the topic. There is even some discussion around aero-corridors (e.g., Hale et al. 2015), although air space conservation is still in its infancy and terrestrial stepping-stone nonaerial corridors continue to be the focus of connectivity for most flying species.

The focus of this new edition continues to be terrestrial but in key places we discuss connectivity as related to aquatic systems. Terrestrial and aquatic realms are often interconnected requiring attention to both. Therefore, in this new edition, we provide examples of connectivity across the realms of land, water, and sea. Corridors often serve to link different types of habitats because many species utilize one type of habitat for one part of their life cycle and a different type of habitat for another part of their life cycle. For example, amphibians may require wetlands or rivers for early life stages and upland habitat during the adult stage; salmon spawn in rivers but spend most of their lives in oceans. Where one community begins and another community ends really is dependent on the species involved. For marine turtles, the beaches where they lay their eggs are as fundamentally important as the ocean environment. In other words, while we humans attempt to put clear labels and distinguishing lines on maps and create typologies of the world, many processes and organisms move across boundaries. This

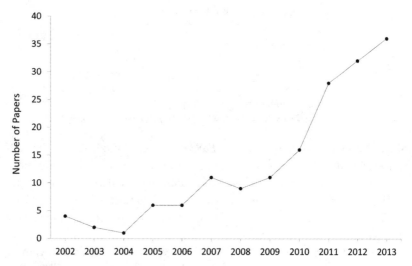

FIGURE P.1 Number of articles published on habitat connectivity by year. (Based on data from fig. 1 in Correa Ayram et al. 2016.)

means that conserving biodiversity in all its forms requires understanding and thinking about connectivity not only within terrestrial, marine, or aquatic systems but also across systems.

Similar to 2005 when we began to work on the first edition of this book, there is a pressing need to distill this large body of new work into a synthesis that can be accessed by students, scientists, and practitioners, and serve as a launching point to further advance the science and practice of corridor conservation. Likewise, we believe that the strength of this book comes from the different perspectives that each of us brings to the task, both because of our unique expertise and because we are at different stages in our professions. We are very fortunate to have had the opportunity to include Annika Keeley in the author team for this revision as she brings fresh, new perspectives on the topic and a deep understanding of existing scholarship especially in the realm of climate-wise connectivity. Bill Likicker has more than sixty years of academic experience in basic science departments, which helped ensure a sense of historical depth, as well as a chronology of theory and research. Adina Merenlender is advanced in her career as a conservation biologist, conducting research on the interactions between land use and biodiversity, and she founded the California Naturalist program to foster social learning and community engagement in nature conservation. Jodi

Hilty is midcareer as a leader in the nonprofit conservation community and works to ensure that science is used to provide the best decisions, is incorporated into policy, and guides the implementation of projects.

As with the first edition, we committed to writing this book with one voice. This meant that while we each took initial leadership on updating existing and writing new chapters, through the process of consolidating sections and multiple reviews, we all had substantial input on all chapters. This book would not have been possible without all of us, and therefore the authors are listed in alphabetical order.

We owe thanks to the many people who helped us develop as scientists and conservationists, ultimately leading us to this book. These include our many mentors, from parents to teachers and from advisors to students. You know who you are. We also want to thank the many folks who contributed directly to this book, providing advice, information, figures, or review of chapters either in the first or second version of this book. These include Alexei Andreev, Misti Arias, Luciano Bani, Reginald H. Barrett, Jon Beckmann, Paul Beier, Steve Beissinger, G. S. (Hans) Bekker, Joel Berger, Sudeepto Bhattacharya, Hein D. van Bohemen, Douglas Bolger, Brent Brock, Colin Brooks, Brian Brost, Doug Burchill, Jeff Burrell, Whisper Camel, Sue Carnevale, Cheryl Chetkiewicz, Juliet Christian-Smith, Tony Clevenger, Charlie Cooke, Caitlin Cornwall, Tim Crawford, Laury Cullen, Richard Dale, Brent J. Danielson, Heather Dempsey, Joe DiDonato, Brock Dolman, Tim Duane, Paul Elsen, Sally Fairfax, Doris Fischer, Kathleen Fitzgerald, Karen Gaffney, Joshua Ginsberg, Dennis Glick, Morgan Gray, Craig Groves, Amanda Hardy, John Harte, Kerry Heise, Emma Hilty, Maya Hilty, David H. House, Marcel Huijser, Bob and Kris Inman, Jochen Jaeger, Aaron Jones, Brian Keeley, Ronja Keeley, Karen Klitz, Claire Kremen, Lorraine LaPlante, William F. Laurance, Raymond-Pierre Lebeau, Domingos Macedo, Andrea Mackenzie, Philip Maruthi, Dale McCullough, Seonaid Melville, David Moffet, Katia Morgado, Karl Musser, Jeff Opperman, Richard S. Ostfeld, Laura Pavliscak, Kristeen Penrod, Scot Pipkin, Hubert Potočnik, Justina C. Ray, Kent H. Redford, Karen Richardson, Adena Rissman, Jane Rohrbough, Eric Sanderson, Graça Saraiva, Jessica Shepherd, Amber Shrum, Peter B. Stacey, Robert C. Stebbins, Paul Sutton, Gary Tabor, Ben Teton, Andra Toivola, Patricia Townsend, Oscar Venter, Severine Vuilleumier, Sonam Wangdi, Bill Weber, Kerrie Wilson, and Steve Zack.

We much appreciate the resources provided by the University of California at Berkeley. We also thank the Yellowstone to Yukon Conservation

Initiative and, in particular, Bill Weber and Colleen Brennan, representing the Board, in supporting Jodi on this project. Mark Carr was supported by NSF Award No. 1260693. Thanks also to the Wilburforce Foundation. At Island Press, editors Erin Johnson and Sharis Simonian provided important feedback and encouragement throughout the process.

Last and most important, we thank our immediate families for support, especially given the many weekend and evening hours needed to complete this project. This includes Jodi's husband and best friend, David House, and their two girls, Jesse and Remi; Annika's family, Brian, Ronja, and To-via; Bill's wife, Louise; and Adina's family, Kerry Heise, Noah, and Ariella.

Introduction

Looking at images of Earth from space we are confronted with how fragmented our blue planet is. Buildings, roads, and shipping lanes divide our landscapes and seascapes, isolating natural communities in the last remaining open spaces. These images also point to melting polar ice caps and rising sea levels that are upon us with the climate changing faster than life on Earth has ever experienced. In order to adapt to these changes, species, including our own, are on the move. Researchers are documenting and modeling how plants and animals move through landscapes, including areas modified by humans. These movements range from daily patterns to long-term shifts in species distributions. With the need for organisms to move among habitat patches more critical than ever, how do we protect the ability for species to roam?

Animals and plants move whether we like it or not. In 2017, Grizzly Bear 148 lumbered across a rugby field in Banff National Park, Canada, where more than one hundred kids were practicing, and she bluff charged or chased several people and their dogs in Banff and then beyond the park near the town of Canmore. She never hurt anyone, but her lack of fear of and aggression toward humans concerned officials. Despite the outcry from over nine thousand people who wanted the authorities to let her stay, Grizzly Bear 148 was moved far north and eventually killed by a hunter only days before a ban on grizzly bear hunting went into effect in British Columbia.

The problem spans urban and rural areas. In Los Angeles, some mountain lions are as famous as the Hollywood movie stars who live in their habitat. Perhaps the most well-known mountain lion in Southern California is P-22, who currently lives in Griffith Park after crossing eight-lane highways to get there. Many others are not that lucky and get killed trying to navigate through the urban landscape. This was the case for lioness P-23, who was struck by a vehicle west of Los Angeles in Malibu. In rural areas of Africa, elephants leave protected areas and move through farms, often raiding the farmers' crops. In India, tigers are so densely packed into some protected areas that they wander outside, coming into conflict with humans, even killing them in some instances.

Bears, mountain lions, elephants, and tigers are some of the big animals that require large landscapes to roam and we inevitably run into them, but maintaining and restoring habitat connectivity is more than just facilitating the safe passage for large mammals. To confront the ecological problems associated with habitat fragmentation and forecasted changes in species distribution in response to climate change, conservation science has focused on protecting and enhancing connectivity by maintaining and restoring landscape linkages in the form of corridors (fig. I.1). At the same time, many community groups have come to recognize the importance of securing safe passage for wildlife. We have had the privilege of working with the Sonoma Ecology Center and the Sonoma Land Trust for the past twenty years as they realized the Sonoma Valley Corridor vision that Christy Vreeland first brought to their attention. Christy worked at a public facility on a piece of land that she referred to as the last dark area in the valley at night—a corridor of natural vegetation joining two adjacent mountain ranges that she hoped animals could move through. That land is now conserved and surrounded by private conservation lands that enhance the permeability of the Sonoma Valley Corridor (see chap. 10). Christy died after a long fight with cancer a few years after she got the ball rolling, but nature is benefiting from the actions she inspired.

The Sonoma Valley Corridor is one of many stories that demonstrate the extensive community efforts focused on protecting and restoring natural or seminatural passageways to maintain and enhance connectivity. Actions vary in purpose and scale from building highway underpasses to reduce amphibian roadkill to linking vast continental ranges for migrating herds of ungulates. Yet the utility and effectiveness of these efforts for increasing species' persistence are still on the frontier of conservation.

This book focuses on connectivity as a measure of the extent to which plants and animals can move between habitat patches. This is required both for their persistence in today's fragmented landscapes and into the future under the threat of continued human development and as the primary adaptation strategy in a changing climate. We refer to landscape features such as corridors, a permeable matrix, and ecological networks as potential means for achieving connectivity. Our goal is to provide guidelines that

FIGURE I.1   UC California naturalists removing barbed wire to increase permeability for wildlife on Tejon Ranch Conservancy land near Los Angeles (a), where
lions, like this one, and other animals are free to roam (b). (Photos by Scot Pipkin
and Tejon Ranch Conservancy.)

combine conservation science and practical experience for maintaining, enhancing, and creating connectivity among areas of remaining natural habitat for biodiversity conservation and increased resilience to climate change
across landscapes and seascapes. This book will serve scientists, landscape
planners, land managers, and all those studying and working on land and
wildlife conservation. We seek to engage a readership from a variety of disciplines including the social, biological, and physical sciences, a spectrum that
encompasses those concerned with both natural and human-dominated
systems. We are especially keen to promote appreciation for the importance
of connecting humans and natural systems in land planning efforts.

To increase our understanding of how to address impacts of fragmentation on species, investigators address different aspects of connectivity. Topics of research include the effects of distance between fragments; habitat
quality, both in remaining patches and surrounding these patches; and how
variations in species' life-history attributes affect species movement between
fragments. Likewise, researchers have begun to explore how dispersal of
organisms, home range dynamics, and migration are affected by habitat
discontinuities. Some efforts focus on single-species movement, while others examine ways of maintaining community coherence. Some applied re-

search addresses natural habitat connectivity by examining the suitability of corridors as compared to adjacent land-use types for movement by animals and plants.

While researchers continue to measure movement patterns and the influence of the built environment on movement behavior, model and map connectivity for corridor planning, and assess how well corridors function to facilitate movement, conservation planners and land managers have wholeheartedly adopted the corridor concept. Some conservation organizations have been formed explicitly to implement connectivity as a way of conserving biodiversity. Other organizations interested in conserving or restoring land consider connectivity in setting priorities for their expenditures. These practical conservation decisions benefit from reference to the scientific findings pertaining to connectivity and prior conservation projects. In an effort to extend science and conservation practice, this book examines the science behind the common assumptions about habitat connectivity and the utility of corridors for different ecological objectives and in various contexts, as well as provides guidelines for implementation.

Unfortunately, as we see in the case of Grizzly Bear 148, planning efforts are challenged to account for the requirements of human and natural systems to coexist. This leaves individuals and organizations attempting to protect and enhance connectivity without guidelines that incorporate the relevant human context in which corridors must function. In this book we review and assess the scientific concepts and evidence that support the use of corridors as a conservation strategy. Then we provide guiding principles as well as cautionary notes for those implementing corridor projects. The book begins by documenting the need for such guidelines and the scientific evidence that supports the current conceptual and practical directions of this field. In the first chapter we set the context of land-use change that leads to the need for restoring connectivity. Chapter 2 reviews the ecological principles that constitute the scientific underpinnings for our understanding of the patterns, processes, and consequences of fragmentation. Chapter 3 provides evidence for the consequences of habitat fragmentation for biodiversity conservation and explains the importance of the landscape surrounding core habitat areas, often referred to as the *matrix* by conservation biologists. Chapter 4 surveys the various approaches to facilitating connectivity across the landscape, including greenbelts, wildlife passageways, hedgerows, and others. The benefits of these types of landscape features for wildlife and people are discussed.

The middle of the book details the methods used to achieve connectivity, and its potential benefits, as well as the pitfalls for biodiversity conservation. Chapter 5 focuses on how different types of habitat and species requirements influence the type and configuration of desirable corridors. It is widely assumed that there are environmental benefits of corridors and similar landscape features, but the evidence for that is sometimes equivocal. In chapter 6 we critically review the diverse problems that can be associated with indiscriminate reliance on corridors and other linkages for a large number of conservation objectives.

The final part of this book provides information on corridor mapping and modeling methods to improve local and regional planning, and ways to conserve habitat links over the long term. The focus of chapter 7 is on planning for connectivity primarily through identifying where habitat corridors currently exist, could be established, or could be restored. It also reviews the various methods to help identify, prioritize, and assess specific sites that may meet connectivity objectives. Climate change has added a new dimension to connectivity conservation. Chapter 8 delves into principles of climate space that influence how species will move in response to changes in climate and describes the two primary approaches to designing climate-wise connectivity. Wholly different from terrestrial connectivity discussed throughout the book, is marine connectivity conservation. Chapter 9 reflects research on predicting species ranges and patterns of migration in the oceans, larval and gamete dispersal, and how marine species are responding to a changing global climate. It discusses how to best include connectivity in marine coastal and pelagic conservation efforts. These efforts include the establishment of marine protected area networks and corridor projects. The final chapter discusses considerations for implementing corridor projects and provides case studies of conservation efforts from around the world. The brief conclusion recaps critical points presented throughout the book and suggests possible future research directions.

Chapter 1

Background: Habitat Loss, Fragmentation, and Climate Change

Protected areas have been increasing in number and expanse globally, yet biodiversity continues to be imperiled. This is in part because, while protected areas are critical for the conservation of biodiversity, they are not sufficient. Conserving biodiversity, especially given unprecedented rates of climate change, requires conservation planning and implementation across large areas. This means that protected areas need to be framed in the context of a larger landscape or seascape and in relation to other protected areas. Accordingly, protected areas are significantly more effective in conserving biodiversity if they are part of an ecological network. For this to occur, areas that facilitate connectivity between protected areas must be managed to maintain or restore this connectivity. Such an approach to conservation is especially important during this time of climate change.

Global, national, and regional policy is increasingly calling for improvements to habitat connectivity. At the global level, the call for connectivity can be found in many major conventions and documents such as the Aichi Biodiversity Targets, the Convention for Biological Diversity guidance documents, and the World Business Council for Sustainable Development's Call to Action for Landscape Connectivity 2017. Here we see that connectivity conservation to address the biodiversity crisis is a priority for conservation globally.

This chapter provides a coarse overview of rapid changes occurring across our planet that contribute to the need for wildlife corridors to conserve connectivity, including habitat loss and fragmentation as well as climate change. We also discuss the importance of connectivity as a component of ecological networks for conservation and to support the global shift toward large-scale conservation. Finally, we provide basic definitions related to connectivity and corridors.

Human-Induced Change and Habitat Loss

Habitat loss and associated species loss are primarily a result of the acceleration of human-induced changes that occurred over the past century. The human population has increased sixfold since the 1800s, and the earth has been transformed to accommodate human habitation and rising consumption (Ehrlich and Ehrlich 2004). A human footprint is detectable across 75 percent of the land area in the world, an increase of 9 percent between 1993 and 2009 (Sanderson et al. 2002; Venter et al. 2016; fig. 1.1). Likewise, human influence in the marine environment is significantly impacting the chemistry and pollution of the ocean and is having a negative impact across marine ecosystems (Doney 2010). Only one-third of the world's 177 largest rivers are free flowing (https://www.internationalrivers.org/resources/where-rivers-run-free-1670).

Attempts to map large undeveloped or wilderness areas (greater than 4,000 square kilometers, or about 1,500 square miles) globally estimated that such areas only constitute somewhere between 16 and 25 percent of the land on Earth outside the polar regions (McCloskey and Spalding 1989; Venter et al. 2016). The paucity of wilderness on Earth restricts the space suitable for persistence of some species, especially those that require intact ecosystems or have large ranges. In more human-impacted areas, mammal movement has been reduced by as much as half compared to equivalent areas with a much smaller human footprint (Tucker et al. 2018). This demonstrates that human presence and activities impact many species' ability to move. Where wilderness does remain, it is often in isolated patches. These isolated patches of natural habitat rarely contain the biodiversity that existed in the region prior to fragmentation. Smaller patches simply do not provide sufficient habitat for some species (Laurance et al. 2002).

Widespread habitat loss and fragmentation due to human activities clearly threatens species survival. As a result, extinction rates are 1,000 times higher than the historic background rates documented in the fossil record (Millenium Ecosystem Assessment 2005). If the current rate of biodiversity loss continues, we will experience the most extreme extinction event in the past 65 million years (Wilson 1988; Ripple et al. 2017). These losses will be devastating for humans, given that we depend on the goods and services that intact ecosystems offer. Many of our medicines and fibers, and all of our food, the basis for our economies and our survival, are derived from the nonhuman species with which we share this planet. Complex ecosystems are also responsible for many of the natural processes on which we depend, such as maintaining air quality, soil production, chemical and

FIGURE 1.1 The global human footprint map for 2009 using a 0–50 cool to hot color scale (Venter et al. 2016).

nutrient cycling; moderating climate; providing freshwater, fish and game, and pollination services; breaking down pollutants and waste; and controlling parasites and diseases (for more in-depth discussion, see Jacobs et al. 2013). An interdisciplinary field of study has arisen around trying to understand ecological resilience, which is defined as the capacity of a system to withstand changes to the processes that control its structures (Holling and Gunderson 2002). One of the primary observations is that disturbing ecosystems can reduce their resilience and result in dramatic shifts to less desirable states that weaken the ecosystem's capacity to provide goods and services (Folke et al. 2004).

Some scientists have quantified ecosystem services in financial terms (Daily and Ellison 2002). Studies estimate that Earth's biosphere provides up to US$72 trillion worth of goods and services per year that we currently do not pay for (Costanza et al. 1997; Corporate Ecoforum and The Nature Conservancy 2012). While quantifying ecosystem services may help enlighten people to the importance of natural systems in their daily life, it is difficult to quantify the value to humans of each of the 10 to 30 million or so species inhabiting Earth. The ecological roles of most species are unknown to us, but we do know the key roles that some species play in the normal functioning of biotic communities. For example, beavers are ecosystem engineers that create wetlands, which increases species richness at the landscape scale (Wright et al. 2002). A study of the functional role of species in ecological communities reveals that while frequently multiple species contribute to ecosystem process and function in the same way, referred to as *redundancy* (Walker 1992), ecosystem resilience may depend on that redundancy. In other words, even removing redundant species could have long-term consequences (Naeem 1998).

While the cost of species extinction to ecosystems and the goods and services that they provide should be one of today's primary concerns, what appears to motivate many people to conserve nature is the intrinsic value of biodiversity (Rolston 1988; Vucetich et al. 2015). With the loss of biodiversity, we are losing important opportunities for personal inspiration and cultural enrichment, whether by bird watching, fishing, or enjoying a natural scenic view. There is increasing evidence that spending time in nature has significant health benefits for young and old alike. To ignore the emotional and, for some, spiritual connections to nature and focus solely on the goods and services people rely on is a mistake. We must acknowledge our ethical and moral responsibility to prevent irreversible change to Earth's systems so that we do not harm other species and our own future generations.

Climate Change Overview

Global annual average surface air temperature has increased by about 1.0°C (1.8°F) from 1901 to 2016. Earth is currently experiencing the warmest period in the history of modern civilization. Many of us have experienced the implications of this such as record-breaking, climate-related weather extremes causing flooding and fires. As we move into the middle of this century, annual average temperatures are expected to rise, such as another 1.5°C for the United States, relative to the recent past under all plausible future climate scenarios (Wuebbles et al. 2017). It is increasingly unlikely that the global average temperature rise will remain below a 2.0°C change, a global temperature target, by the end of the century (IPCC 2014). This target was recommended as a level that would have global impacts but allow stabilization of natural systems.

It is clear that climate change is impacting the conservation of global biodiversity and will radically alter ecosystems and result in species' extinctions (Watson et al. 2012; Jantz et al. 2015). Today's climate change models forecast that the global climate and sea level are on a trajectory to change through the end of the century and beyond even if we successfully curbed greenhouse gas emissions in the immediate future (Ripple et al. 2017). While climate change is not unprecedented in Earth's history, today's rates of climate change far exceed anything previously experienced.

During the Quaternary (the last 2.6 million years), there were many periods of cooling interspersed by times when the climate warmed, bringing about expansions and contractions of ecosystems. Species have three mechanisms for adapting to climate change. A species might adapt within its current range through phenotypic plasticity (observable characteristics of an individual resulting from the interaction of its genotype and its environment); adaptive evolution (evolutionary changes that are adaptive to a given environment); or a species' range might shift in elevation, latitude, longitude, or aspect (Hilty et al. 2012; Pecl et al. 2017). Paleoecological studies have documented that during the Quaternary, in response to changing environmental conditions, populations increased or decreased in abundance, species shifted their ranges, and new species evolved while only a few species went extinct. Different types of range shifts have been documented. The entire range of some taxa, for example, spruce (*Picea* ssp.) in eastern North America, has shifted to higher latitudes. Other taxa, for example, pine trees (*Pinus* ssp.) in Nevada, shifted their ranges up in elevation, but the amount of shift and even the direction have varied. Contraction to or expansion from refugial populations are other patterns that have been observed (Davis and Shaw 2001). Because species respond individualistically to climate change, shifting ranges at different rates and in different directions, community turnover historically was common, and frequently resulted in novel associations (Gill et al. 2015).

In current times, synergistic effects between habitat loss, habitat fragmentation, and global climate warming (Ripple et al. 2017) are compounding the effects of habitat loss on biodiversity. Some species may not be able to shift their distributions or evolve new adaptations fast enough to accommodate global climate change. Fragmented landscapes decrease the opportunity for movements that could result in range shifts. Numerous studies document the historical movement of species as the climate changed in the past (e.g., DeChaine and Martin 2004), but the unprecedented speed of modern climate change combined with habitat loss could make historical processes of adaptation less applicable today (Steffen et al. 2015).

Models can predict the effects of climate change on vegetation communities. While predictions depend on the model specifications, the climate models, and greenhouse gas concentration trajectories used, the general message is that the natural vegetation in its current state is at risk from climate stress. In Europe, models predict that up to 80 percent of the vegetation types will be replaced by a different vegetation type by the end of the century if carbon emissions are not drastically curbed (Hickler et al. 2012). Without effective carbon mitigation in the Mediterranean Basin, increased temperatures and water deficits will cause the desertification of large parts of the landscape (Guiot and Cramer 2016). In western North America, models predict a widespread shift from tree-dominated landscapes to shrub- and grass-dominated landscapes (Jiang et al. 2013; Thorne et al. 2017). However, at high elevations, forests are simulated to move upward causing the contraction of alpine grass and shrub vegetation (Shafer et al. 2015).

Climate change research suggests that a critical factor for species and system adaptation is resilience. Terrestrial and aquatic ecosystems tend to be more resilient if they are conserved in large and unfragmented areas such that the ecological processes that sustain these systems can continue unhindered (Walker and Salt 2006). Climate change can directly affect species populations and communities through, for example, higher average or extreme temperatures, increased water deficit, or more severe flooding. Other stressors can exacerbate negative effects of climate change, and vice versa (Staudt et al. 2013). An increase in disturbances such as fires, flooding, windstorms, or insect outbreaks can decrease system stability (Buma and Wessman 2011). Resource extraction, such as timber harvest, can also decrease ecosystem resilience making the forest more susceptible to climate change (Staudt et al. 2013). Higher temperatures can magnify the adverse environmental effects of pollutants by increasing their availability in the environment and amplifying their negative effects. Biological disturbances can reach catastrophic dimensions, for example through the invasion of nonnative species that have a competitive advantage, or the emergence of pest species or pathogens that have expanded their range or are more numerous due to milder winters. Since the late 1990s, mountain pine beetles (*Dendroctonus ponderosae*) have caused pine mortality throughout millions of acres in North America, and the outbreak has been attributed to warmer temperatures, increased drought stress, and generally unhealthy growing conditions (West et al. 2014).

Lessons learned from previous climate changes in Earth's history, observations of the effects in the first decades of human-induced climate changes on species and ecosystems, and predictions of the magnitude of and biological responses to climate change give us a foundation for thoughtful adjustments to conservation planning and action, including corridor conservation. Climate change is literally forcing us to reorient our approach to conservation as we are no longer managing to maintain a historical reference point, but rather are managing for change (Hilty et al. 2012). The fields of protected-area design, connectivity modeling, restoration, risk assessment, and assisted migration have evolved a good deal over the past fifteen years. This includes even the abandonment of long-existing principles, such as the requirement to source seeds for restoration from local

FIGURE 1.2   Russian River corridor through North Coast California vineyard landscape. (Photo by Adina Merenlender.)

populations (Havens et al. 2015). The enormous scale at which climate change impacts are happening reinforces the importance of planning conservation at larger spatial scales. It is only by achieving conservation at such scales that we have a chance of conserving otherwise increasingly imperiled biodiversity around the world. Integrated landscape conservation needs to design and implement conservation networks connected by corridors and embedded in a permeable landscape. Planning and managing ecosystems in these protected areas for resilience is essential to reduce stressors that exacerbate the effects of climate change. Species will be able to persist in landscapes with large, connected natural areas where pressures from overharvesting, invasive species, and pollutants are low (fig. 1.2).

In corridor conservation, we need to shift from planning habitat connectivity for a few charismatic focal species to an ecosystem approach where the entire biota can move through the landscape in response to climate change. This can be accomplished by applying coarse-filter strategies designed to conserve the majority of biodiversity (Lawler, Watson, et al. 2015; Anderson et al. 2016). Complementary fine-filter approaches may be necessary to focus on conserving individual species that slip through the coarse filter and require specific conservation action (Hunter 2005). We will discuss these different approaches in more detail when we consider climate-wise connectivity in chapter 8. Assisted migration, translocating species to locations where the climate is suitable, is an example of a fine-filter approach. Originally this strategy was mostly rejected based on the great uncertainty of the ecological effects of moving a species into formerly unoccupied habitat. Now the concept has gained greater acceptance especially among the forestry industry with numerous models predicting the need for arid-adapted seed stock and relocation of species that cannot adapt to changing local conditions (Lawler, Ackerly, et al. 2015; Thomas 2011; Seddon et al. 2014).

Limitations to Protected Areas

Approximately 12.5 percent of the land and 3 percent of marine environments have been designated as protected (World Bank Group 2017). Unfortunately, for several reasons even the composite of these protected areas is not sufficient to conserve global biodiversity, and many have argued for increasing the extent of protected areas (Watson et al. 2014; Wilson 2016). Existing protected areas are often too small to conserve long-term viable populations of species or fail to incorporate natural processes (e.g., floods and fire) that may be necessary to retain all elements over time. Furthermore, illegal activities like hunting and logging are impairing what is within some protected area boundaries, calling into question how protected some areas really are (Watson et al. 2014).

Today's protected areas are more likely to be found in certain types of ecosystems—less-productive, high-elevation sites—leaving more-populated areas with high levels of richness and endemism at risk (Jenkins et al. 2015). Much of the more-productive land on Earth has been prioritized for and converted into agricultural and other human uses, and often is privately owned. As a result, many species are imperiled. In the United States, over 90 percent of all federally threatened and endangered flora and fauna can be found on nonfederal land (Wilcove et al. 1996). Globally, only 28 percent of the Important Bird Areas and 22 percent of sites harboring more than 95 percent of the population of a species identified by the Alliance for Zero Extinction are within existing protected areas (Watson et al. 2014).

Low-intensity land uses in landscapes surrounding protected areas can buffer parks and provide additional habitat for some species less sensitive to human activities, contributing critically to biodiversity conservation in the protected areas (Kremen et al. 1999). However, often what happens in areas adjacent to protected areas can affect species and ecosystems within the protected areas, further impairing them. Human activities can spill into protected areas, altering ecological processes, such as by introducing species or changing fire regimes, or impacting animals that need to move beyond the protected area boundaries (Newmark 1987). The world's very first national park, Yellowstone, is the largest national park in the contiguous United States, yet it is too small to support viable populations of some large carnivores for the long term, and survival of such animals outside the park is lower. Some of these large carnivores, such as grizzly bear (*Ursus arctos*), struggle to exist on lands occupied by humans even at densities as low as one house per square mile. Most grizzly bear deaths are human caused, resulting from human–wildlife conflicts (Primm and Clark 1996; Haroldson and Frey 2002; Schwartz et al. 2012). Likewise, Gir Forest National Park in India is a classic example of a patch of habitat surrounded by very high human densities and is, in fact, regularly encroached on by human activities. It is also the only place in the world where Asiatic lions (*Panthera leo leo*) still occur, although their numbers are very low due to limited habitat (Khan 1995; Oza 1983). Human and lion conflicts are an ongoing problem in and adjacent to this park. Because there are no other large islands of natural habitat in the nearby region and thus no opportunity to restore connectivity, the government has made a decision to translocate lions to another region of India as a means of helping the Asiatic lion (IUCN Red List of Threatened Species 2017). Ultimately, those species that naturally occur at low densities and require large home ranges are potentially most impacted by habitat loss. Human-wildlife conflict continues to be one of India's largest conservation challenges with approximately one person killed by wildlife each day due to the large number of people and the limited amount of protected area for large animals such as lions, tigers, and elephants (George 2017).

Reconnecting Our Planet

Given that many protected areas are too small to allow for the persistence of viable populations of species and ecological processes needed to maintain the protected area, one solution is more and larger protected areas. However, new or expanded national parks may not be a viable option in many places and may not solve the problem entirely. Another tool is to connect existing protected areas through corridors, which may increase movement between patches by approximately 50 percent, compared to disconnected patches, and increase persistence of many species in protected areas (Bennett 2003; Gilbert-Norton 2010). The need to recover endangered species and rare habitat types has driven much of the demand for habitat connectivity. The most common approach is to maintain and restore habitat that will provide connections between protected natural areas for wildlife movement.

Ecological networks composed of connected protected areas possess a variety of traits that make them inherently more resilient than individual protected areas to a range of threats, including climate change. Many large-scale conservation efforts are actively seeking to identify, restore, and create protected areas and corridors. Examples are Australia's Great Eastern Ranges corridor; Two Countries One Forest in North America's Eastern Appalachian region; Europe's Natura 2000 conservation network; the Yellowstone to Yukon Conservation Initiative; Baja California to Bering Sea in the Pacific Ocean; the Vatu-i-Ra Seascape in Fiji; among many more large landscape and seascape initiatives (Hilty et al. 2012). Such networks can benefit a spectrum of ecosystem types, including terrestrial, freshwater, and marine, and also protect stop-over sites, such as on flyway routes. In reality, many species flow or move between land, sea, and freshwater. Species may use different habitats during different times of the year or during different life-history stages such that consideration of corridors between terrestrial, freshwater, and marine ecosystems, as well as within them, is important.

It is not always clear that connecting wildlands through the conservation of remnant habitat or restoration of movement pathways across a disturbed landscape will enhance a species' persistence within protected areas. A few issues are worth considering. First, limited conservation dollars may mean making choices between protected areas and corridors, although ideally over the long term, conservation action results in a broad ecological network with both adequate protected areas and connectivity between them. Second, the configuration of existing and potential future protected areas may be important to consider when deciding whether to invest in corridor conservation. Models suggest that more clustered patches may decrease extinction risk for species and that distance between patches beyond 1.25 times a species' maximum dispersal distance could increase a species extinction rate in a given region (Kitzes and Merenlender 2013). In some cases, it is premature to suggest methods for enhancing connectivity when not enough is known about the requirements of focal species, and whether increased connectivity will result in boosting their persistence (see chap.

6). However, connectivity is one of the top recommendations for building ecosystem resilience around climate change (Heller and Zavaleta 2009), and is often the only option given developed surrounding landscapes. Further, failure to act to conserve corridors could foreclose future opportunities given the pace of human development globally.

Growth of Connectivity Science and Practice

The need for ecological connectivity has become an accepted norm in the scientific and conservation practitioner communities. It is increasingly well founded in scientific theory and research (Worboys et al. 2010; Trombulak and Baldwin 2010; Hilty et al. 2006; Crooks and Sanjayan 2006; Worboys et al. 2015). Guidelines for creation, implementation, and management are increasing (e.g., Bennett 2003; Beier et al. 2011; Hilty et al. 2012; Olds et al. 2016) and, as will be discussed further in chapter 7, methods for measuring, modeling, and mapping connectivity have grown exponentially over the past decade.

We use the term *ecological connectivity* to mean a measure of the ability of organisms, gametes, and propagules to move among separated patches of suitable habitat (Rudnick et al. 2012; see chap. 4 for more discussion). Abiotic flows such as air, water, and nutrients are often integral processes, but the focus here is on ecological connectivity for the world's biota. Ecological connectivity can be viewed at various spatial scales. Temporal and spatial scales are relevant in the discussion of ecological connectivity and conservation networks. Time proceeds in a one-way linear dimension, while spatial scale refers to the two or three dimensions of an object or process and is characterized by both grain and extent. Ideally, corridors serve to facilitate connectivity over time and can operate at a range of spatial scales. Temporal scales are particularly relevant during this time of climate change due to the need for species to adapt by shifting their range locations, and what serves as a corridor today might not continue to function for the original focal taxa in the future. New corridors may also emerge over time as the land cover changes due to natural phenomena such as river bank realignment or melting ice, or changes in land use. Also, connectivity itself may have a temporal component, such as where movement of organisms occurs on a seasonal, annual, or multiyear cycle.

Corridors can be relevant across a range of spatial extents. These could range from a road underpass such as for pygmy possums (*Burramys parvus*) in Australia (Asher 2016) to much larger-scale continental efforts such as

Paseo Pantera across Latin America (https://www.panthera.org/initiative /jaguar-corridor-initiative). In addition, connectivity may need to be considered in a three-dimensional context including the vertical dimension, which could include ocean or freshwater depth, canopy height, or underground components.

We concentrate on connectivity between patches of suitable habitat that are potentially relevant to the experiences and welfare of individual organisms in the near term as well as over the next century. Because organisms vary tremendously in their abilities to travel and in their motivation to leave their birthplace, the degree of connectivity in a given mosaic of biotic communities will vary greatly according to the perspective of each species. Methods for measuring, modeling, and mapping connectivity have grown exponentially over the past decade (see chap. 7) and new guidelines for implementation and management are emerging.

This brings us to a consideration of what is meant by corridor. As you will discover throughout this book (see in-depth discussion in chap. 4), the term corridor is used in a variety of ways. For this book, we define it as any space that facilitates connectivity, thereby improving the ability of organisms, gametes, and propagules to move among patches of their habitat. It is important to recognize that what serves as a corridor for one species may not be a corridor or may even be a barrier to another. Corridors can be natural features of a landscape, are sometimes an emergent property of the built environment, or can be created through habitat restoration. Because many related terms have arisen related to the subdiscipline of corridor ecology, we provide box 1.1 on page 18 with terms used in this book and commonly throughout the scientific literature and in the practitioner community.

As Albert Einstein wisely noted in another context, "The problems that exist in the world today cannot be solved by the level of thinking that created them." The conservation of biodiversity depends on landscape and seascape scale conservation that facilitates connectivity among protected areas (Noss et al. 2012). Maintaining connectivity can link otherwise disconnected populations, maintain genetic exchange, facilitate seasonal movements, provide for multigenerational movement, and enable movement over time in response to climate change. Corridors are a critical tool for long-term conservation.

BOX 1.1

CORRIDOR-RELATED TERMINOLOGY

Note that organisms can move at different life-history stages: as individuals, gametes, propagules, or plant parts capable of vegetative reproduction.

Ecological, Habitat, Landscapes, or Seascape Connectivity

A measure of the ability of organisms to move among separated patches of suitable habitat that may be variously arranged.

Structural Connectivity
A measure of habitat permeability based on the physical features and arrangements of habitat patches, disturbances, and other landscape elements presumed to be important for organisms to move through their environment.

Functional Connectivity
The degree to which evidence indicates that landscapes or seascapes facilitate or impede the movement of organisms.

Corridor

Any space that facilitates connectivity over time among habitat patches.

Linkage

Although the term is frequently used synonymously with corridor, "linkage" technically refers to broader regions of connectivity important to facilitate the movement of multiple species and maintain ecological processes.

Specific Corridors

Seasonal Migration Corridor
Used by wildlife for annual migratory movements between source areas (e.g., winter and summer habitats).

Dispersal Corridor
Used for one-way movements of individuals or populations from one resource area to another.

Commuting Corridor
Linked habitat elements that support daily movements including breeding, resting, and foraging.

Landscape Permeability

The degree to which regional landscapes, often encompassing a variety of natural, seminatural and developed land cover types, are conducive to wildlife movement and that sustain ecological processes.

Matrix

A component of the landscape, often altered from its original state by human land use, which may vary in attributes from human-dominated to natural, and in which corridors and habitat patches are embedded.

Chapter 2

The Ecological Framework

Organisms on planet Earth are not evenly distributed across its surface. The tremendous variety of conditions found from the deepest oceans to the tops of the highest mountains, from the poles to the equator, and around various latitudes ensures that the various kinds of organisms will be discontinuously distributed. On smaller than global spatial scales, species are generally limited to certain continents or biotic provinces within continents and to places that suit their adaptations. It is sometimes the case, moreover, that what constitutes suitable habitat will vary with stage of development or season. A frog, for example, may need a pond while it is a tadpole and a forest as an adult. Many sandpipers will use Arctic tundra in the summer for breeding and a tropical mudflat in the winter.

Organisms have always had to deal with discontinuous habitats, as well as their own changing needs over both short and long terms. What is new is the accelerating rate of habitat fragmentation that has been occurring over the last few hundred years in response to human population growth, technology advances, and the recent unprecedented speed of climate change. Increasingly, organisms are confronted with disappearing and severely fragmented habitat, as well as exotic predators, diseases, and competitors, making it difficult for them to cope effectively with inevitable catastrophes. In this chapter, we review the ecological concepts that address these challenging issues of spatial discontinuity to which organisms are being forced to adapt (Tilman and Kareiva 1997).

Island Biogeography

The first major effort to theorize about how organisms deal with disconnected patches of habitat was the theory of island biogeography. This perspective was formalized in a now classic book by R. A. MacArthur and E. O. Wilson, *The Theory of Island Biogeography*, published in 1967, which triggered a flood of interest among ecologists in this topic and energized much research effort. The theory in its original form dealt with real islands, bits of land surrounded by water. The focus was not so much on how individual organisms might move among the islands and the mainland, but on what influenced the diversity or number of species on islands. The particular species involved were not of direct concern; rather, the number of species or species richness was.

To address this interesting and important question, four simple propositions were made:

1. Larger islands will host more species than smaller ones because large islands are likely to have greater habitat diversity. Greater topographic diversity will lead to more microclimates and to more soil types, so that a greater diversity of plants and microorganisms can live there and will thus support a larger variety of animals, as well. Furthermore, large islands will present larger targets for potential colonizers to find.
2. Islands close to the mainland will be more diverse than more distant islands. The mainland is presumed to be the source of colonizers to the islands, and so the closer the island is to the source of immigrants, the more likely it is to be reached by them.
3. Small islands will suffer higher rates of species extinctions than large islands. This is because small islands will support generally smaller numbers of individuals for each species present, and residents will therefore be continuously at a greater risk of having their numbers decline to zero. Moreover, species that compete with each other (have negative impacts on each other) will be less likely on small islands to find some microhabitat or spatial refuge where they can avoid each other.
4. Islands close to the mainland will experience lower extinction rates regardless of size because they will benefit from a larger input of new colonists, including individuals of species already present on the island. Thus, even species persisting precariously on a near island will frequently have their numbers enhanced by new arrivals from the mainland. Species on distal islands will only rarely benefit by the arrival of new recruits to their ranks.

The interactions among these four basic principles can be shown in two graphs (fig. 2.1). The resulting species richness on any given island can

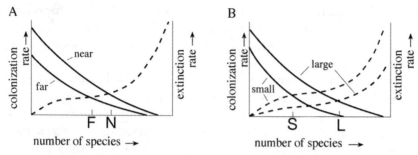

FIGURE 2.1 Graphic depiction of species richness on islands as determined by distance from the island and island size. Solid lines represent the rate of colonization by new species; dashed lines are species extinction rates. Except on large islands, extinction rates increase rapidly at very low species richness because communities will be simple at that stage and may lack elements that new colonists need. Graph A shows the number of species on islands of a given size as a function of distance from source populations. *F* and *N* are equilibrium values for far and near islands, respectively. Graph B shows the number of species on islands at a given distance from mainland sources as a function of island size. *S* and *L* are equilibrium values for small and large islands, respectively.

be seen as an equilibrium between the rate of new species becoming established on an island and the rate of colonizers going extinct. Species number (richness) is therefore a dynamic concept and one that is sensitive to even small changes in colonization and extinction rates. Species richness equilibria will be greatest on large islands close to mainland sources, whereas small or distant islands will have the lowest equilibria. Although the principles involved here are simple and intuitive, just how island distance and size play off each other depends on the details of any given situation. One application of this theory that has been generally neglected is that of aquatic "islands" in a "sea" of land. By analogy, large lakes close to other lakes should have richer biotas than small lakes well isolated from others.

Because of its primary focus on species richness, the theory of island biogeography ignores a number of interesting and important issues related to colonization and survival on islands. The first example is the fact that island biota will be biased toward organisms with particular morphological or life-history traits. Specifically, island colonists are likely to be species with good vagility: that is, those that can fly, swim long distances, or float easily in water or air currents. Colonization will also be favored by species in which the smallest number of individuals can start a new population. This might be a single seed, spore, or pregnant female. If at least one male and one female are necessary for establishment, the chances of that happening would be the probability of a male arriving multiplied by that of a female arriving at about the same time, a much smaller number than for the arrival of a single colonist. The second issue is that colonization by one or just a few individuals may not provide sufficient genetic variability so that ensuing offspring can successfully adapt to their new environment. If genetic variability is important for success, a large pool of immigrants would be needed. Third, the least likely species to succeed would be those that require large social groups to be successful. For them, enough individuals would have to arrive together, so that their advantageous social behavior could be established. The fifth issue is that island life may perpetrate a physiological bias. For example, Wilcox (1978) reported that organisms with high metabolic rates, such as terrestrial mammals, suffer higher rates of extinction than do reptiles and amphibians. Flight in birds and bats overcomes this metabolic disadvantage to some extent. Sixth, large species characteristically have higher extinction rates on islands than small species. This is simply because a given island is effectively smaller for large species than for small ones, given that large body size is generally correlated with large home range size.

Another major general category of shortcomings of island biogeographic theory is that it does not consider evolutionary changes subsequent to colonization and does not allow for speciation either on an island or within an archipelago of islands. The initial colonists, given sufficient time, can evolve into many species. The Hawaiian and Galápagos Islands are great examples of speciation happening in spectacular fashion.

Still another limitation to the theory is that it ignores species-specific demographic behaviors and especially does not consider that these behaviors may have to adjust to a different species composition on the island than in the ancestral home on the mainland. Not only may the mix of competitors that a species encounters be different, but there will likely be different predators and parasites to challenge it, as well. Conversely, the absence of familiar predators or parasites may offer different challenges and opportunities. A colonist that finds itself without its usual enemies may instead face the unfamiliar challenges of high-density stresses and food shortages. Lastly, it should be mentioned that the focus of island biogeography on island size and isolation ignores the reality that particular air and water currents may well enhance or diminish the probability of potential colonists reaching a particular island, relative to others.

A general complication of island theory is that strictly speaking it applies only to oceanic islands, that is, to those that emerge from the ocean devoid of terrestrial life. Then the processes of colonization and extinction can proceed as expected by theory. Many islands, however, are land-bridge islands. This means that they were once part of a mainland but became islands through earth movements or rising sea levels. Such islands start life with a sample of the mainland biota already in place. Because these islands tend to be fairly close to a mainland, they will also experience colonization events, and therefore a portion of their biota will behave as prescribed by classical island biogeographic theory. However, these land-bridge islands will also support organisms that cannot disperse over water and so will not be supplemented by colonization events. The continuing presence of such species on the island will be determined mainly by the probability of extinction. Extinction rates will of course be influenced by island size, topographic diversity, and the chance occurrence of other species by colonization. The richness of this non-colonizing element on land-bridge islands will therefore gradually diminish over time as inevitable extinctions take their course. This process of community relaxation is illustrated in figure 2.2 by species diversity of lizards on land-bridge islands off Baja California as a function of time elapsed since island isolation (Gonzalez 2000). Island theory projects an equilibrium species richness for islands, but the decrease in species diversity over time, caused by zero to minimal colonization, introduces a nonequilibrium element into the situation. Understanding the rate at which species will be lost due to island isolation is important because, as we will discuss later, similar phenomena can occur in terrestrial community fragments, as, for example, those produced by forest cutting.

Further problems were encountered when attempts were made to generalize island biogeographic theory to terrestrial arrays of fragmented habitats. The theory was applied to so-called sky islands in the Great Basin of North America (Brown 1978; Brown and Lomolino 1998); these were isolated patches of boreal or subalpine habitats on the tops of the higher mountains in the region. Desert and various semiarid habitats effectively isolated these cooler and moister fragments. Except for birds and other strong-flying creatures, this isolation was generally found to be complete. That is, the colonization rate was zero or close to it, at least for the groups of organisms analyzed in detail, such as terrestrial mammals. These sky island biotas were not in equilibrium between colonization and extinction rates, as prescribed by the theory. They were, in fact, remnants of more widespread species distributions that occurred during cooler and moister periods in the Pleistocene. What we have left is remnant or refugial populations influenced primarily by factors affecting extinction rates, not colonization rates, which were effectively zero. Bird diversity on these sky islands fits the classic

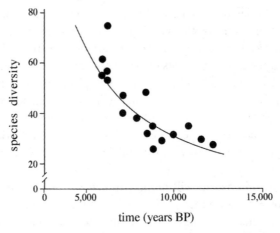

FIGURE 2.2 Lizard species data from land-bridge islands off of Baja California, Mexico, showing a decline of species richness in the absence of colonization over time since island isolation (data from Wilcox 1978).

island biogeographic model better because colonization rates are usually above zero.

A related insight is that the nature of the habitats that separate "islands," be they aquatic or terrestrial, is an important variable. The intervening areas can vary from completely inhospitable to seasonably passable or can even be just low-quality habitat that is suitable for temporary occupancy.

In spite of all of the inadequacies in classical biogeographic theory, it was nevertheless important in getting ecologists and conservation biologists to begin thinking critically about spatial and temporal variations in the distribution of organisms. Moreover, biogeographic theory continues to provide useful insights that guide us into the development of a more sophisticated understanding of fragmented habitats. Therefore, rather than discard these insightful developments, we should instead appreciate their historical importance, and simply incorporate island biogeography into more complex and more generally applicable formulations.

Metapopulation Theory: Conceptual History

The metapopulation concept was a major advance in both spatial ecology and conservation biology. It is a framework for understanding, and hence predicting the behavior of species in fragmented landscapes. A *metapopulation* is simply a population of populations, or a system of local populations (demes) variously connected by movements of individuals (dispersal) among the population units. The concept thus recognizes that species are usually arranged in variously disconnected patches across the species' distribution. This fact has been recognized for a long time by naturalists and field ecologists. As habitat fragmentation accelerated in recent decades, the existence and importance of metapopulation structuring became critical to conservation science.

Even before World War II, Soviet ecologists concerned with rodent pest control were emphasizing the disjunct nature of species distributions and the relevance that it had for demographic behavior (Lidicker 1985, 1994). Then in 1970, P. K. Anderson applied those insights to his research on house mice (*Mus musculus*). In 1969 and 1970, Richard Levins was motivated by the epidemiological problem of how to control parasites and diseases that were living in a mosaic of habitat patches, namely the bodies of the various individuals that constituted their host species. He published two critically important papers in which he modeled metapopulations mathematically, and in the second paper, he introduced the term "metapopulation." His insight was to treat each population of pathogens living in an individual host as analogous to an individual in a population. Thus, his populations (demes) had birth and death functions just as individuals did. Because of his modeling approach, his proposal of the new term "metapopulation," and especially because the time was propitious, Levins's papers represented a defining moment in ecology.

It is important to understand the basics of Levins's model because it requires a number of important assumptions that need to be evaluated, and because it was used as a starting point for development of many more complex and realistic models. Levins's fundamental equation

$$dp/dt = mp\,(1-p) - ep$$

expresses how the proportion of habitat patches or hosts (p) occupied by the species in question changes over time (dp/dt). This proportion can vary from zero (extinction) to full occupancy ($p = 1$). In this equation, m is the dispersal rate (migration) of individuals from an occupied patch to one that is unoccupied; mp is therefore the total amount of successful dispersal from all occupied patches and is like a reproductive or birth rate. So the first element in the growth equation expresses the rate of establishment of new colonies in available empty patches ($1 - p$). The second element is the rate (probability) of extinction of a single occupied patch (e) multiplied by the proportion of occupied patches. This is the death rate (mortality) of the array of occupied patches. Obviously, if the birth of new patches exceeds the death (extinction) of occupied patches, the proportion of occupied patches (p) will increase. If births and deaths are equal, p is a constant, and if the death rate exceeds the birth rate, p will decline toward zero. So, just as with a single population, it is the ratio of losses to additions that determines the growth trajectory of the metapopulation.

A number of critical assumptions should now be apparent for this simple but insightful model:

1. Extinction rate (e) is a constant; that is, all demes have the same probability of going extinct, and that rate does not change over time.
2. Dispersal rate (m) is a constant; that is, all demes produce successful dispersers to unoccupied patches of habitat at the same rate.
3. All unoccupied patches ($1 - p$) are equally accessible to dispersers; that is, their actual spatial arrangement is irrelevant.
4. The matrix between the habitat patches is uniform everywhere and its nature is specified only to the extent that it influences m; that is, if the matrix were completely inhospitable to dispersers, m would be zero even if occupied patches were sending out a steady stream of emigrants.
5. All habitat patches are equivalent; that is, they are all of the same size, they can all support the same size population, and their explicit spatial position in the metapopulation array is irrelevant.
6. When a disperser arrives at an unoccupied patch, the patch immediately becomes fully inhabited and begins to produce emigrants. That means that there is no local population dynamic, and patches exist in only two states: + or − (i.e., occupied or unoccupied).
7. There is no dispersal among occupied patches.

Note that metapopulations of this type always have some empty patches, unless $e = 0$. Patches will "blink on and off" as demes go extinct and the habitat patch subsequently becomes recolonized. If m is not less than e, the long-term fate of the metapopulation will depend on the probability of all occupied patches becoming extinct at the same time (e^n), where n is the number of occupied patches. Therefore, continued existence of metapopulations not only will be sensitive to changing values of the parameters m and e, but also will be dependent on the number of patches, particularly the number of those occupied. Because all of the assumptions implicit in

Levins's model would rarely seem to be true, this type of metapopulation is probably rare in the real world. Nevertheless, the model represents a useful abstraction of spatially structured populations. An example of the oversimplification inherent in this minimalist model is given by Hastings (2003), who shows that if the probability of demic extinction is not a constant but increases with patch age (because of succession or impacts of the focal species itself), then colonization rates needed to sustain the metapopulation may be twice as much as those determined by the Levins's approach (Dobson 2003). Therefore, a dispersal rate that can sustain a metapopulation composed only of young patches will become inadequate as the patches age or become variable in age structure.

Because of the abstract nature of Levins's metapopulation model, it is important that metapopulations not be defined in terms of that model. Rather, a general definition is needed, as we provided previously (a system of variously connected demes of a single species), to accommodate the variety of spatially structured populations of organisms that actually exists. It would be a serious mistake if we restricted our investigation of spatial structuring to a limited subset of what organisms do. Our understanding would inevitably be confined in such a situation to the defined subset of patterns. We would then have to come up with new definitions for spatial patterns that are excluded from our definition, or remain ignorant of those aspects of the living world. Our approach here is to start with a general definition and then determine if useful subpatterns can be defined. It was in this spirit that Harrison (1991) proposed recognizing four categories of metapopulation (fig. 2.3):

1. *Patchy*. A demic structure is present, but the demes are well connected, and dispersal is common among the populations. If and when a deme becomes extinct, the patch it occupied is quickly recolonized (rescue effect). This is a common pattern and is the one most resistant to metapopulation extinction.
2. *Core-satellite*. This pattern is also called "mainland-island" or "source-sink." It consists of one or more large extinction-resistant populations (large habitat patch) plus one or more separated, usually peripheral, smaller patches of habitat. Dispersal is mostly in the direction of mainland to satellite. The smaller populations experience occasional extinctions, but eventually those patches are recolonized from the core population. This kind of metapopulation is also relatively extinction resistant because of the relative security of the core or source populations.

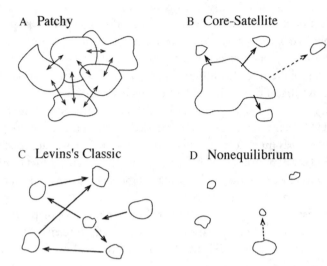

FIGURE 2.3 Four types of metapopulations (Harrison 1991). Dispersal movements among habitat patches are shown by arrows, and extremely rare dispersal events are depicted as dashed arrows. In Levins's classic model, movements are only from occupied to unoccupied patches, and the spatial arrangement of the patches is not specified.

3. *Levins's Classic.* This is the mathematical modeler's favorite pattern, and its features have already been given. The long-term prospects for such a metapopulation depend critically on the number of patches and the level of connectivity among them.

4. *Nonequilibrium.* This arrangement has the highest risk of overall extinction, because connectivity among the demes is weak or absent, and there is no large secure deme among them. With minimal dispersal among populations, the rescue effect is unreliable or nonexistent. As demes suffer extinction, there is little prospect of these habitat patches being recolonized, at least over reasonable time spans. Nonequilibrium metapopulations are in a sense moribund. Unfortunately, this pattern is also common and becoming more so with increased habitat fragmentation. Such situations are the ones to which most conservation efforts are directed.

Of course, many species will be composed of an array of metapopulations consisting of a mixture of types. This reality is nicely illustrated by the Florida scrub jay (*Aphelocoma coerulescens*), an endangered species confined to oak scrub habitats on sandy soils in the central part of peninsular

Florida (Stith et al. 1996). As of 1996, the entire species was distributed into forty-two metapopulations. The boundaries of these population arrays could be objectively defined because the dispersal distances for the birds are accurately known. The frequency distribution of the number of adult pairs of jays in these metapopulations is shown in figure 2.4. Note that half of the metapopulations consisted of ten or fewer pairs. Only four metapopulations contained the relatively safe number of three hundred or more pairs. Assigning the forty-two metapopulations to Harrison's categories is extremely revealing. There are three patchy populations, although they are all too small to be secure (fifteen to twenty-six pairs of jays). Fifteen arrays can be classified as core-satellite, although only six of those have a large enough mainland (core) population to be reasonably extinction resistant. Three metapopulations are of the classic Levins type, and therefore questionably secure, and twenty-one are in the vulnerable nonequilibrium state. So if one looked only at the total species numbers, about four thousand pairs of adults, one might conclude that there were no grounds for alarm. Placing the birds in these spatially structured metapopulations, on the other hand, reveals that only six metapopulations are reasonably extinction resistant. Moreover, a clear vision is provided to conservationists as to how to develop a species survival plan. This is a wonderful example of the power of the metapopulation approach.

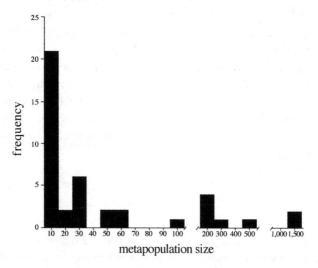

FIGURE 2.4 Frequency distribution of the number of adult pairs of Florida scrub jays (*Aphelocoma coerulescens*) in the forty-two metapopulations that constituted that endangered species in 1996 (Stith et al. 1996).

Metapopulation Processes

Besides parts, metapopulations possess characteristic processes. Of central significance is dispersal among the patches. The quantity, quality, and timing of these movements are what give the metapopulation its all-important connectivity. Some movements in and out of metapopulations are also possible, although these must be rare or the metapopulation boundaries would have to be enlarged to accommodate them. The second major process is that of the demography of the target species within the patches. Demographic processes will be influenced by the species' life-history features, its morphology, the quality of the patch, stochastic events, and all of the metapopulation properties extending beyond the boundaries of a particular patch. One aspect of demography that will be of particular importance is the synchronicity of demographic behavior across the various demes. Especially relevant in this regard are the timing of population density changes, the rate of emigrant production, and the responses of demes to extrinsic factors such as catastrophes. Finally, metapopulations develop genetic structuring. Variations in genetic makeup across the metapopulation will likely influence the performance of particular demes, their probability of extinction, their rate of production and chances for emigrant success, and potential for adaptability to changing conditions. All of these metapopulation parts and processes will be discussed in more detail in this and other chapters. Excellent sources for further exploring modern approaches to metapopulation biology include Taylor (1991), Hanski and Gilpin (1997), Hanski (1999, 2001), and Mestre et al. (2017). Box 2.1 summarizes the parts and processes that characterize metapopulations. First there are the habitat patches, which have various properties important for the constituent species; second, between the habitat fragments there is the matrix, which will be discussed in detail in chapter 3; third, there is the focal species, which lives in some or all of the patches of habitat and survives to varying degrees in the matrix as well.

Dispersal

Absolutely critical to the maintenance and long-term prospects of any metapopulation is the dispersal of individuals or propagules among the habitat patches. This is what determines the connectivity among the constituent demes and hence is the most important of the vital processes characterizing metapopulations (box 2.1). Dispersal is rarely a simple process and so needs

BOX 2.1

METAPOPULATION PARTS AND PROCESSES

PARTS (with their components)

1. Patches of habitat suitable for a focal species
   a. Spatial arrangement
   b. Size and shape
   c. Quality, including shifts over time and space, such as due to climate change
2. Matrix
   a. Size and shape
   b. Quality for supporting focal species dispersal
   c. Spatial variation, including placement and nature of corridors
   d. Changes in frequency of disturbance events such as fires
   e. Increased frequency of severe weather events
3. Populations of focal species
   a. Inhabiting some or all of the habitat patches
   b. Dispersing among habitat patches

PROCESSES (with important influencing factors)

1. Dispersal
   a. Timing of dispersal
   b. Dispersal distances
   c. Quality of dispersers
   d. Emigration and immigration rates
   e. Successful colonization rate
2. Demographic processes including demic extinction
   a. Local demic behavior
   b. Production of emigrants
   c. Acceptance of immigrants
   d. Synchronicity among demes
3. Processes affecting genetic structuring
   a. Number of demes (occupied patches)
   b. Size of demes
   c. Rates of emigration and immigration
   d. Demic extinction rates
   e. Susceptibility to stochastic influences (resilience), including genetic drift
   f. Differential selective pressures among demes

to be explored in more detail. Corridors can be an important ingredient in this story, because it is essential to understand what conditions will favor dispersal movements by a focal species and thereby improve the connectivity of a metapopulation. Conversely, we may want to know how to control a pest or exotic species, in which case we will need to know how to disrupt metapopulation connectivity.

What Is Dispersal?

The subject of dispersal biology is troubled by much semantic confusion. So we will start with some basic definitions of terms we use in this book (see Stenseth and Lidicker [1992] for further discussion).

Dispersal—Process of individuals leaving the place where they are resident (home) and looking for a new place to live. This behavior can occur both within and between habitat patches.

Disperser—Individual in the process of dispersal. If the search for a new home is successful, such an individual would be a successful disperser.

Excursion—An exploratory trip away from home including a return home. Such movements may be preliminary to dispersal, and if an excursion ends with the death of the individual, we cannot distinguish it from failed dispersal.

Disseminule—A life-history stage adapted for dispersal, for example, planktonic larvae, winged seeds, spores, or spiderlings floating on a spun thread.

Migration—Term often loosely used to refer to any movements of individuals or genes (that is, gene flow), including dispersal. We will restrict the term, as do vertebrate biologists, to mean seasonal movements between breeding and nonbreeding ranges.

Emigrant—A disperser that leaves its home population (deme) and thus represents a loss to that population. Emigration rate is the number of emigrants produced by a deme per unit of time.

Immigrant—A disperser that enters and becomes established in a new population. Immigration rate is the number of new arrivals per unit of time.

Colonist—A disperser that takes up residence in an unoccupied habitat patch. Colonization rate is the number of new colonies established per unit of time.

Phoresy—Dispersal aided by some other species, for example, a deer carrying a seed that sticks to its fur and is dropped in some new location, or a bird eating a fruit and later defecating viable seeds.

Dispersal distance—The distance moved by a disperser from its current home to a new one. The frequency distribution of dispersal distances exhibited by a population (fig. 2.5) is a useful statistic.

Who Disperses?

The first essential point about the dispersal process is that dispersers are not likely to be a random sample of the source population. Dispersal is often biased by sex, age, genetic makeup, or health. Although there are many exceptions, mammals tend to have male-biased dispersal (Wolff 1999) and birds, female biased. Moreover, the sex that dominates among dispersers often tends to travel farther. This implies that if both sexes are needed for successful colonization of empty habitat patches, the process will be constrained by the least dispersive sex. In the extreme case where one or both sexes do not disperse at all, the species will be able to spread only by propagule or gamete dispersal, or the incremental addition of new home ranges to the periphery of preexisting ones. For species with this kind of life history, habitat fragmentation poses an almost insurmountable barrier to long-term survival. Every metapopulation would quickly become the nonequilibrium type and be destined for extinction unless connectivity were reestablished.

Dispersal is also often age biased. Typically, it is young individuals that are most likely to become dispersers. Pre-reproductives such as recently weaned mammals or recently fledged birds often readily disperse. Sometimes there is a life-history stage specifically adapted for dispersal; such stages are called *disseminules*. Among plants, these are seeds or spores. Pollen provides a mechanism for movement of genetic material, but pollen alone cannot establish new colonies. Adults of non-sedentary species also can disperse and often do, but typically that happens in response to environmental cues, such as deteriorating food supplies or the beginning of the breeding season, and not primarily to particular developmental stages.

Another cause of dispersal bias is that dispersers may have significant differences in their genetic composition compared to those that stay at home (Cockburn 1992). This possibility comes from the discovery that some organisms display genetic variation in the tendency to disperse. In such cases, colonizers of empty habitat patches will on average carry more genes encouraging dispersal than other individuals of the same species that stay at home (philopatry).

The potential evolutionary consequences of fragmentation may well be something that conservationists will be well advised to consider in their planning. As fragmentation begins, we could anticipate selection for

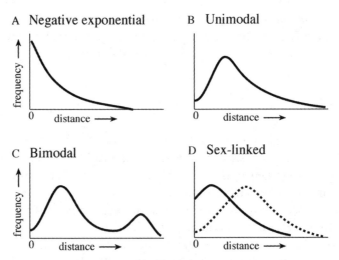

FIGURE 2.5 Four possible patterns of dispersal distance distributions. The negative exponential pattern (A) is characteristic of seed dispersal from a parent plant (seed shadow) and other passively dispersed propagules. This pattern is the most frequently used by modelers. Unimodal distributions (B) are usually strongly skewed toward long distance and represent a commonly found pattern in nature. Bimodal patterns (C) are found when there is a genetic polymorphism for dispersal behavior, or where phoresy occurs and can be influenced by the pattern of habitat fragmentation. Sex-biased dispersal (D) is a commonly found pattern in mammals and can include cases in which one sex is strongly philopatric.

increasing rates of dispersal in the metapopulation. This is because small fragments are likely to quickly reach carrying capacity, and so dispersers that can find empty or low-density patches would be more successful in passing on their genes. If strong cohesive social forces initially inhibited emigration, they would likely be weakened as fragmentation proceeds. However, if fragmentation is progressive, there will be a point where successful dispersal becomes so unlikely that selection will favor those that opt for philopatry and not for accepting the risks of dispersal. Tucker et al. (2018) used GPS tracking collars on 7 species of mammals, and documented that movements decreased markedly with increasing evidence of human presence.

An often-overlooked life-history trait that can strongly influence a species' dispersal behavior is its social system, which can do that in a number of ways. One of these has already been mentioned, namely that successful colonization of empty patches may require the nearly simultaneous arrival of a large number of conspecifics so that they can establish the necessary social environment for their survival. A corollary to this is the requirement that such coordinated dispersal may mean that groups of individuals will have to emigrate together. The centripetal forces of strong social bonds may make that difficult to organize. Moreover, species that profit from tight social interactions will likely find that dispersal by isolated individuals will be especially precarious. An example of this was reported by Laurance (1995), who found that the highly social lemur-like ringtail opossum (*Hemibelideus lemuroides*) almost never left large patches of tropical rain forest (in Queensland, Australia) to traverse corridors and colonize empty forest patches. Thus, while it remained common in undisturbed rain forest, it was completely absent in fragmented areas. Cockburn (2003) makes the more general case that cooperatively breeding birds have a reduced capacity for colonization compared to those with the usual single-pair mating system. Finally, social behavior may also discourage immigration into local populations by making it difficult for prospective immigrants to integrate into existing social groups.

Why Leave Home?

A really intriguing aspect of dispersal biology is the issue of why an individual would want to leave home in the first place. This is a fascinating subject that cuts across the disciplines of physiology, behavior, and ecology, and it is discussed in more detail in Lidicker and Stenseth (1992). The traditional view of dispersal is that it happens when conditions at home become intolerable, thus motivating exodus. Deteriorating conditions can be caused by resource depletion, climate change, social intolerance, accumulation of toxic materials, or high levels of predation and/or parasitism. A corollary of this scenario is that those who leave as dispersers are those who tolerate the deteriorating conditions least well. They may be young or old, the sick, or social subordinates that compete least well for mates or resources. Moreover, these difficulties generally occur at times of the year when conditions are most harsh. It follows that the hapless dispersers will face the challenges of travel at times when conditions are poor and they themselves are not in optimal condition for coping with these difficulties. Consequently, they will have little chance of success. This traditional type of dispersal is called *saturation dispersal* (Lidicker 1975).

Saturation dispersal certainly does occur, but now we recognize that much dispersal is of a different sort, namely *presaturation dispersal* (Lidicker 1975). In this type, individuals leave home for a variety of reasons other than an unhappy home life. In the case of disseminules, there is a morphological imperative. For juveniles, there may be a physiological imperative as they approach sexual maturation. There may be motivation to find unrelated mates so as to avoid inbreeding. At any age, there may be opportunities to colonize empty patches of habitat and have numerous successful progeny. Decisions to move of this type will generally be independent of population density and will most likely occur when environmental conditions are favorable, greatly improving the chances for success. The time course for both kinds of dispersal in relation to density is shown schematically in figure 2.6.

Whatever the proximal motivation for dispersal, each prospective disperser faces an implicit balance sheet of pros and cons that will influence the likelihood of successful movement. Natural selection will favor those individuals who make the most advantageous "choices" most of the time. Table 2.1 summarizes the potential advantages and disadvantages that may be relevant to any dispersal "decision."

The Demography of Extinction

The second category of metapopulation processes to be considered is that of demographic behavior (box 2.1). Specifically, we are interested in the probability that the number of individuals in a deme will fall to zero, that is, local extinction. The factors that influence that probability are many and complex (Beissinger and McCullough 2002; Turchin 2003), and this book cannot adequately treat them all. For a succinct overview of the basics, see Lidicker (2002). What we can do is provide a brief outline of the factors that are important in the demography of small populations. We will emphasize those aspects that are most relevant to connectivity among demes, including the role of corridors.

We have emphasized the critical role that dispersal plays in the viability of metapopulations (see also Stacey et al. 1997). At the same time, we have shown that dispersal can be anything but a simple process, making it imperative that conservation planners consider the particulars of dispersal behavior for target species. Moreover, there are other demographic implications of dispersal, on both the local population and metapopulation levels. Movements in and out of populations represent two of the four vital rates that constitute the population growth equation:

$$dN/dt = (b + i) - (d + e),$$

where b, i, d, and e are birth, immigration, death, and emigration rates, respectively. Dispersal movements thus have the potential for significantly

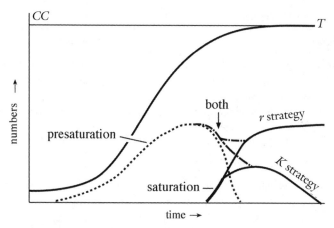

FIGURE 2.6 The time course of saturation and presaturation dispersal as a function of population density (solid S-shaped line) (Lidicker 1975). *CC* is estimated carrying capacity for the population; *r* strategy is a life-history mode characterized by high reproductive rates, one corollary of which is that slowing of population growth rates usually occurs by increasing mortality; *K* strategy is a contrasting mode, in which slow growth is associated with reduced reproduction rather than increased mortality.

affecting demic growth rates (dN/dt). For example, a high rate of emigration could periodically or chronically depress growth in particular habitat patches. Dale (2001) argues that female-biased dispersal in birds will lead to heavily unbalanced sex ratios in small habitat fragments. This in turn will strongly suppress reproduction in those patches, and push the deme toward extinction.

The other aspect of demography that is important here is the synchronicity of demographic behavior among the demes in a metapopulation. If there is much asynchrony among demes, it is unlikely that all of them will go extinct at the same time, and metapopulation persistence is thus encouraged. Much synchrony, on the other hand, could result in a situation where all demes simultaneously suffer catastrophic declines, resulting in simultaneous extinctions and metapopulation disappearance. So it is important to consider what factors encourage asynchrony. At least three things are important in influencing demic synchrony. The first is geographic extent of the metapopulation and amount of variation among the habitat patches. The more spacious a metapopulation and the more variation it encompasses, the less likely it is that a single event, catastrophic or not, will have a similar impact on all of the component demes. Conversely, spatially limited

TABLE 2.1. The Dispersal Balance Sheet

Types of factors	Potential advantages	Potential disadvantages
Environmental	Escape from unfavorable conditions (economic, physical, social) Reduced exposure to predators and competitors Reduced exposure to population crashes	Uncertainty of finding food, shelter, appropriate social arrangements Greater exposure to predators and competitors
Quantitative Genetic[a]	May find uninhabited or incompletely filled habitat patch Promiscuity (both sexes have multiple mates in any given breeding season) Frequency-dependent selection may favor rare phenotypes	May not find any suitable habitat Uncertainty of finding a mate Strange phenotypes may be avoided
Qualitative Genetic[b]	Heterosis and avoidance of inbreeding Greater chance for new and advantageous recombinations	Less viable offspring may be produced (breakdown of coadapted systems; disadvantageous recombinations)

Source: Modified from Stenseth and Lidicker 1992, table 1.2.

[a]Quantitative genetic factors are those that influence the quantity of future reproduction by the disperser.

[b]Qualitative genetic factors influence the fitness of future offspring.

metapopulations with relatively homogeneous conditions throughout will more likely respond to changes in a uniform manner. The second factor influencing synchronization is dispersal. The more the demes are connected by dispersal, the more likely they are to share similar demographic patterns, and hence suffer simultaneous declines. Ironically, while dispersal is essential for metapopulation survival, too much dispersal can be disadvantageous unless metapopulations are large and diverse. The third factor is the situation in which synchronicity results from strong seasonal forcing of population dynamics (Grenfell et al. 1995). Fontaine and Gonzalez (2005) describe an experimental study of captive rotifer metapopulations in which the strongest synchrony occurred when high levels of dispersal were combined with periodic environmental fluctuations.

Why is it that small populations are subject to very high extinction risks? There are possible genetic reasons for this that will be discussed. But there are also nongenetic factors that may play important roles. These can be random (stochastic) influences or more insidious nonrandom (deterministic) forces. Demographic stochasticity is the process whereby the demographic structure of a small population can be radically altered by random birth, death, emigration, or immigration events. Such alterations may make it difficult or impossible for a small population to survive. For example, a population might suddenly find itself consisting only of males, or only of post-reproductive individuals, or subject to some environmental perturbation that temporarily caused all reproduction to fail. Because such changes can happen at random, they are unpredictable in their timing and in their impact. They can leave a small deme either increasingly vulnerable or on the path to extinction. Dai et al. (2012) discuss some signals that portray a small population's loss of resilience and hence imminent collapse.

Obviously, the size and quality of habitat patches will influence the number of the target species that can be supported in the patch. For a given quality, larger patches will usually support larger populations. This is not inevitably the case, however, as larger patches may also support predator and parasite populations that are not present in small patches. Generally, demes occupying larger patches are less likely to go extinct than those living in smaller ones. However, even common species may undergo strong seasonal or multiannual fluctuations in abundance. At the low points of these fluctuations, such populations would be subject to a much higher risk of extinction. Planning for species' survival therefore demands that attention be given to low points in the changes in numbers, and not merely to a population's behavior at moderate to high densities. For example, fish populations living in streams subject to Mediterranean climate need especially to be monitored during drought years, when numbers may be at unusually low levels. Related to this last issue is the question of how species respond to stochastic catastrophes. By definition, the occurrence and intensity of such events are unrelated to the state of the local populations. Instead, their causes are to be found outside the boundaries of the subject populations. This means that populations must be able to cope with these catastrophes at any stage in their demographic cycles. If the catastrophe is severe enough to cause localized extinctions, survival of the metapopulation will depend on it being large enough and/or diverse enough that portions of it will escape the catastrophic impact.

Another class of risks faced by small populations is that of minimum threshold densities caused by anti-regulating influences (Lidicker 1978,

2002, 2010). This phenomenon is sometimes known as the Allee effect. Anti-regulating forces (sometimes called *inverse density dependent*) are non-random in that their influence is predictably related to population density (fig. 2.7). These effects are the opposite of those caused by the regulating factors normally involved in encouraging population growth at low densities and stopping it at high densities. In fact, they stimulate growth at high densities and discourage or stop it at low densities. By these actions, they are destabilizing influences on population growth. Such forces are especially well developed in social species but are minimally present in any sexually reproducing species.

Two classic examples of this phenomenon are (1) Darling (1938) demonstrated that some species of colonial seabirds cannot breed successfully if group numbers drop below a critical threshold; and (2) the inability of the musk ox (*Ovibos moschatus*), an Arctic ungulate, to defend its young against wolf predation if the numbers in a group get too small. Such anti-regulating forces also occur in plants as it has been demonstrated in a variety of plant species that small patches fail to attract pollinators, leading to reproductive failure and demic extinction (Bawa 1990; Lamont et al. 1993; Roll et al. 1997; Groom 1998). In effect, these thresholds make survival of small demes more precarious by increasing the minimum numbers needed for a deme to persist.

A final point about small populations is that human managers can influence metapopulation survival by manipulating connectivity among patches, by controlling deleterious predators or parasites, and by making strategic introductions to either supplement existing demes or establish colonies in empty habitat patches. A good example of a metapopulation in which successful management of this type is a realistic possibility is that of tule elk (*Cervus elaphus nannodes*) in California (McCullough et al. 1996). This subspecies almost became extinct, but currently exists in twenty-two, mostly isolated, populations. Demes are relatively easily monitored demographically and genetically, providing an excellent opportunity for competent management.

Genetic Structuring

The last metapopulation process to be briefly discussed is that of genetic structuring (box 2.1). Again, we can provide only an outline for this huge topic, emphasizing some general considerations. Comments will be organized into three topics: (1) the genetics of small populations, (2) genes in a metapopulation context, and (3) evolutionary trends. For more detailed discussion, see McCauley (1993), Hedrick (1996), Whiteley et al. (2015), and Frankham et al. (2017).

The Genetics of Small Populations

It is to be expected that metapopulations will be composed partly or entirely of populations of relatively small size compared to those in nonfragmented landscapes. Small size is associated with a number of genetic consequences. These include loss of alleles, increased influence of stochasticity (random changes), and risk of inbreeding depression (discussed further on). If we imagine a population of diploid organisms consisting of ten individuals, it is easy to see why this is the case. Such a population would possess only twenty genes at any particular locus. For those loci that are polymorphic, many alleles present in a large population could easily be missing from this small subsample of genes. From this sampling effect alone, small populations will have reduced genetic variation. The situation could quickly get worse, however, as small populations are subject to significant random changes as one goes from generation to generation. Because of uneven contributions among the ten individuals to subsequent generations, some alleles initially present could readily be lost; in fact, it is quite possible that a single allele will be fixed at a given locus. A single mutation occurring in this population would immediately represent 5 percent of the genes at that locus. Such a mutation could become fixed through random processes even if it were somewhat deleterious relative to the other alleles. If the mutation were significantly deleterious, it would substantially reduce the viability of the deme, since the single individual possessing it would represent an extraordinary 10 percent of the total population.

Actually, the genetic situation could be a whole lot worse than described in this hypothetical population of ten. That is because the fate of genes from one generation to the next depends not only on stochastic perturbations, but also on what is called the effective population size (N_e). This number is generally smaller than the actual population size (N) and in fact represents the size of a population that would behave genetically the same way as the subject population is actually behaving. N_e is equal to N only if all members of the deme participate equally in reproductive activities for a given breeding season and the population size remains constant. Equal participation requires that all individuals produce the same number of successful progeny, and that the sex ratio be equal. These conditions are all

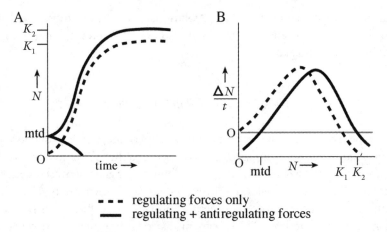

- - - regulating forces only
—— regulating + antiregulating forces

FIGURE 2.7 The effects of antiregulating factors on population growth rates and equilibrium densities. N equals population size (numbers). Mtd is minimum threshold density, and K_i are equilibrium densities. Graph A shows S-shaped growth with and without antiregulating forces. Graph B shows population growth rate as a function of population size with and without antiregulating influences.

unlikely to be met in any real small population, and to the extent that they are not true, N_e is reduced accordingly. For starters, all nonreproductives, the young, the very old, and those in between that cannot find mates are not counted in N_e. If the sex ratio is not equal among those that do participate in reproduction, N_e is reduced still further. For example, if our population of ten has eight individuals that do the reproducing, but only two of them are males, N_e would be six instead of ten. Beyond this, if there are variations in the reproductive successes among participating adults, N_e is also reduced. Such variations could be due to age and fitness differences, or to chance. Finally, if we are calculating the effective population size over a longer time period than one generation, we must take into account differences in numbers from one generation to another. N_e will be closer to the low points in density fluctuations than to the high values. Of course, all of these factors could be acting simultaneously, so that N_e may be very much less than N.

Last, it needs to be mentioned that small populations risk inbreeding depression, which could further debilitate the deme. Inbreeding depression arises from two causes: (1) increasing chances for deleterious alleles to occur in homozygous form so that the full negative effect of these alleles will be expressed, and (2) loss of fitness caused by the disappearance of heterosis, which is the added fitness (vigor) that often accompanies individuals carrying high levels of heterozygosity in their genomes. Progressive loss of alleles and increased frequency of breeding with closely related individuals both increase the overall homozygosity of small populations. A case in point is reduced lifespan associated with genetic drift (Lohr et al. 2014). Inbreeding will also lower N_e. As a deme becomes more and more homozygous, it becomes, from a population genetics perspective, increasingly like a haploid organism (one set of chromosomes instead of two and therefore half the total number of genes). That is, N_e approaches $N/2$ from this cause alone.

An interesting and unexpected benefit to genetic diversity within a deme is illustrated by honeybee colonies. Jones (2004) has shown that genetically diverse colonies are better able to maintain stable brood nest temperatures than less diverse colonies. Maintaining optimal nest temperatures over a broad range of ambient temperatures clearly translates into a distinct reproductive advantage. Just how many individuals are needed in a population to maintain long-term fitness is an intensively debated topic. Reed (2005) analyzed data from the literature on plant populations and concluded that at least two thousand individuals are required for long-term persistence. Dixo et al. (2009) report a case of reduced genetic diversity and connectivity associated with habitat fragmentation in a species of toad endemic to the Brazilian Atlantic Forest.

Genes in a Metapopulation Context

When we consider population genetic processes in an array of semi-isolated demes, several important properties are manifest. First of all, the total number of individuals in a metapopulation, the sum of all the demes combined, is ordinarily going to be much less than the size of the population in an unfragmented state. This is simply because fragmentation involves a loss of habitat available to a species since portions of the original habitat are converted to non-habitat.

The second important fact about metapopulation genetics is that the various demes will tend to become different from one another. This tendency to differentiate is caused by two things: random changes in genetic composition and progressive changes caused by differential selective pressures on the various demes. The greater the spatial extent of the metapopulation and the greater the variation in local conditions across the demic array, the stronger will be the tendency for selective pressures to vary, and hence to influence genetic differentiation within the metapopulation. Random changes (genetic drift) will become increasingly important as deme sizes decrease since small demes experience increased demographic stochasticity.

Countering the tendency for demes to differentiate is the movement of genetic material among demes (gene flow). Mostly, genes are carried by dispersing individuals, but in some cases gametes can also serve this function, as in the case of pollen. It has been estimated that the arrival of one unrelated and successful immigrant per generation is sufficient to counter the forces of genetic drift, except perhaps in very small populations (Mills and Allendorf 1996; Wang 2004). This emphasizes yet another reason conservation managers need to have detailed information on the dispersal behavior of focal populations.

A long-term research program on butterfly metapopulations led by Ilkka Hanski illustrates the comprehensive approach to understanding metapopulation demographic and genetic dynamics that deserves emulation (summarized in Baguette et al. 2017). It integrates patch features, short- and long-term variations in species numbers, demic variations in genetic attributes, variations in demographic behavior among demes and in trajectories modified by natural selection, patch occupancy as a function of patch size and isolation, variations in patch colonization rates with possible genetic drivers, genetic correlates of metapopulation size and persistence, and variations in dispersal behaviors in response to a changing environment.

A Longer-Term Perspective

Conservation efforts are confronted with such an array of urgent problems that it is not surprising that longer-term genetic changes are often neglected. To the extent that such trends can be anticipated, our effectiveness will be significantly improved. Some of these trends have already been mentioned, such as effects on dispersal and body size, but there are some additional processes to consider. For example, Tilman (1990, 1994) argues that in plants fragmentation not only selects for improved dispersal, but simultaneously reduces competitive ability. Schmidt and Jensen (2003) studied changes in body size in twenty-five species of Danish mammals over a period of 175 years of increasing habitat fragmentation. They found that small-bodied species tended to get larger, and larger species got smaller. This is the same direction of changes as found on real islands (Van Valen 1973). If rodents can experience evolutionary changes in seventeen decades, it seems likely that insects and other creatures with higher potential rates of evolutionary change could respond much more rapidly to fragmentation.

There also may be a long-term trend for continuing loss of genetic variation in the metapopulation as a whole. This is because as demes become periodically extinct, there is a loss of all the genetic variation formerly possessed by that population, except as that variation may still be present in successful dispersers from that deme and their progeny. The empty habitat patch will then typically be recolonized by a few founders who carry only a limited sample of the genetic information found in the metapopulation as a whole. Still another trend to be expected is that selection will favor the evolution of stronger inbreeding avoidance mechanisms in order to counter the deleterious effects of too much inbreeding. If this happens, it will help to reduce losses of genetic variation, unless the deme is so small that inbreeding can be avoided only by not breeding at all. Other metapopulation features that would help to reduce genetic losses are large sizes of habitat patches, large numbers of patches, and adequate connectivity among patches. Good connectivity will result in quantitatively more gene flow and also improved quality if there are multiple sources of immigrants to a patch. Spatial heterogeneity in patch conditions will encourage differential selective pressures and hence increase genetic diversity. This is in addition to its beneficial role of discouraging dangerous demographic synchronicity among the demes.

One warning, however, is that if selective pressures diversify the various demes in a metapopulation too much, there is the risk of outbreeding depression occurring. This can happen when individuals who are quite distantly related or adapted to rather different microhabitats mate and produce offspring. These offspring may turn out to be not well adapted to either of the parental habitats, or they may exhibit some other genetic breakdown. Such breakdowns or disruptions of coadapted gene complexes may include balanced polymorphisms, linkage groups, or epistatic complexes (interactions among loci). Sometimes the deleterious effects of outbreeding do not become manifest until the second generation after the initial mating (Edmands 1999). The results of outbreeding may be analogous to what happens in some hybrid zones between species or distinctive subspecies. Decreasing individual fitness, increased parasite loads, and increasing mutation rates have been described in some hybrid zones. If there are a lot of such immigrants they may impede or even reverse a deme's adaptation to local conditions, and thus increase the risk of extinction. Well-meaning wildlife managers have on occasion caused extinction by introducing distantly related individuals to an endangered population and inadvertently hastening extinction (Riley et al. 2003). In Pacific salmon, different spawning runs in some river systems represent temporally (instead of spatially) isolated demes. Gharrett and Smoker (1991) have shown that hybridization between even- and odd-year classes of pink salmon (*Onchorhynchus gorbuscha*) leads to poor return rates for the hybrids and their progeny.

Adding complexity, we must contemplate the real possibility that these various genetic risks that are inherent to small populations will interact synergistically with each other and with various environmental stressors. If this should happen, it will be even more difficult to predict the net effects on population viability, as net effects might be less than, equal to, or greater than the sum of the factors acting separately. In an ingenious experiment using gray-tailed voles (*Microtus canicaudus*) living in large outdoor enclosures, Peterson (1996) showed that inbreeding and insecticide application to alfalfa (*Medicago sativa*) had additive negative effects on the voles. Separately, inbreeding and insecticide each negatively affected vole population growth. When both were present, the combined negative effects were approximately equal to the sum of the two effects acting separately: that is, they were additive.

Finally, it is essential to face the prospect that the reduced overall genetic variation often associated with metapopulation structure will likely make it more difficult to adapt to changing conditions in the future. Roff (2003) suggests, based on the findings of Hoffmann (2003), that tropical rain forest species may be especially at risk in this regard. This impaired capability for evolutionary adaptation may be especially urgent with the current trend toward rapid global climate change, which is magnified, of course, by other human assaults on the environment with which organisms must cope, such as pollution. Thus, genetic factors must be considered both for their short-term impacts on demic extinction rates and for the longer-term effects on continuing evolutionary adaptations. Demic connectivity clearly relates to the many challenges posed by these trends, and which humanity must at some point confront.

Two papers by I. Hanski illustrate how the complexities of metapopulation analysis can be profitably modeled. This research builds on classical metapopulation theory to predict metapopulation persistence across the landscape using spatially explicit models (Hanski 1998, 1999). These models include measures of habitat quality and the effects of allelic variation and dispersal dynamics on metapopulation size and dynamics. They were tested using twenty-two years of data on butterfly presence and absence across thousands of habitat patches. However, as insightful and important as metapopulations are, we need to be reminded that they focus on a single species and therefore are at the population level rather than the community level of biological organization. We now advance to the next higher level of biological complexity.

Metacommunity Theory

A metacommunity is an array of patches of a particular type of community variously connected by dispersers. This concept is a much newer concept than that of *metapopulation*, the term apparently first being used by David S. Wilson (1992). Its roots, however, can be traced to the end of the nineteenth century (Hansson 1995; Lidicker 1995b). Because the units that comprise a metacommunity are a particular kind of community, this requires a classification system for communities. There is much controversy about what constitutes the most useful or perhaps "natural" definition for community types, but there is no doubt that classifications of communities are possible and useful. The metacommunity concept thus raises us to a new level of biological complexity and displays a new array of emergent properties (see box 2.2). The focus is now on biological systems at the community level of complexity, namely the assembly of two or more living species in a particular place over a specified time interval. The classification of a community of interest to a type should not be confused with the actual community (biocenosis) itself as they often are. The real community is a specified fragment of Earth's biota, and it is assigned a "type," which is a label that informs us about its characteristics. It is also plausible that different classification systems may be used for different purposes. In practice, we often use "habitat" to be loosely equivalent to some community type. This is justified because often numerous species of organisms find a particular community type to be approximately equivalent to their habitat. Because species often vary in their abilities to track changes in their environment, new species assemblages will often form in response to long-term changes such as climate change (Schloss et al. 2012). New species interactions will then likely affect many properties of the community (Gilman et al. 2010). Guichard et al. (2004) discuss the metacommunity concept as applied to marine communities.

It may be helpful to note that we can recognize four types of metacommunities based on the underlying primary reason for the disjointed distribution of a community type over space; only one of these is based on human modification of the environment:

- *Heterogeneous physical environment.* The physical environment is quite variable over the earth, and organisms respond to this by living only in those places with characteristics conducive to their survival and reproduction. Variables such as latitude, altitude, and topography (slope, aspect) help to define the mix of creatures living in a particular spot. Also, the substrate, or local physical context, is critically important. In terrestrial environments, this can be soil type and depth. In aquatic environments, water depth, salinity, flow rate, pH, oxygen content, temperature, and nutrient content are all important, as is the nature of the bottom. Disjunct distributions of these variables result in a fragmented distribution of community types, however they might be delimited.

- *Biotic heterogeneity.* Even in a given physical environmental setting, the organisms present may vary because the extent of the suitable site may be too small for some species to live there, or it may be so isolated that some species cannot reach the site, or there may be barriers like oceans, mountain ranges, or large rivers that constrain the mix of species present on the site. Finally, physically similar sites may have different communities inhabiting them because some may be recovering through succession from natural catastrophes.

- *Island archipelagos.* As discussed earlier, oceanic and continental islands may hold different communities because of differences in their age, size, distance from sources of colonization, and chance.

- *Anthropogenic fragmentation.* An increasingly large number of metacommunities are the direct result of humans destroying one community type and establishing another in its place, for example, agricultural crops, suburbia, clear-cut forests, polluted rivers.

Next, we can ask if metacommunities have characteristic parts, as do metapopulations (box 2.1). In fact, they do, and they are much the same as for metapopulations. There is the physical place of the patch with its various attributes, and there is the matrix of different community types that make up the intervening space between the fragments. However, the living component is not merely a focal species but the entire assembly of species that give the community the properties of its community type. Not all of the community fragments will have an identical list of component species, but they will all have a similar collection of species and particularly will have those species that help to define the community type. These keystone species will be those with high dominance (influence on the nature of the community) or that are particularly abundant or that are diagnostic for some reason for this particular type of community. The mix of species may include nonnatives or exotics, as well as species that are part of the community only seasonally.

Metacommunity processes are summarized in box 2.2. This list does not include the many processes carried out by communities that are independent of their location in a metacommunity, such as primary productivity. Just as with metapopulations, patches are variously connected by movement of individuals, they suffer extinctions, and new patches can be established. And, of course, the entire metacommunity can decline to extinction or suddenly be eliminated by a catastrophe, anthropogenic or natural. In an experimental study of simple metacommunities involving only a species of bacteria, a flagellate, and a ciliate, it was demonstrated that smaller and more fragmented communities went extinct faster (Burkey 1997). The species richness of patches may be highest at intermediate levels of connectivity (Kneitel and Miller 2003), as strongly isolated patches will suffer higher extinction rates whereas extensive connectivity will likely ensure that dominant competitors and predators will reach all patches and thereby possibly suppress diversity.

Loss of a trophic level in a patch will usually change the demographic behavior of the remaining species, so that a cascade of further extinctions may result (Estes et al. 2011; Ripple et al. 2014). The number of trophic levels that a patch can support depends on the interaction of patch quality and size (fig. 2.8). Patch quality measures the availability of food and other resources and is often strongly correlated with primary productivity of the patch. For a given patch quality, the number of trophic levels possible will increase with patch size. Changing quality, however, will affect the minimum size needed for support of an additional level. Added trophic levels can include parasites as well as predators. Patch isolation could also be a factor in the absence of higher trophic levels (Ripple et al. 2014). A strongly isolated patch such as an oceanic island or isolated mountaintop may not be detected by dispersing predators or parasites. In the epidemiological literature, the minimum size of a host patch that can support a disease parasite is called the *critical community size* (Grenfell and Harwood 1997). Below that size, the disease is present only intermittently or not at all. In the well-studied case of the measles virus infections in humans, host populations of 250,000 or fewer escape persistent outbreaks of the disease (Grenfell and Harwood 1997; Keeling and Grenfell 1997).

Communities, like populations, often exhibit seasonal or longer-term fluctuations. Such changes need to be accommodated by fragments so that the species they harbor can persist. Otherwise, even species that are often abundant can be lost during seasonal or multiannual periods of low numbers.

Another seasonal phenomenon that may be important in community dynamics is the periodic arrival and departure of migrants or overwintering

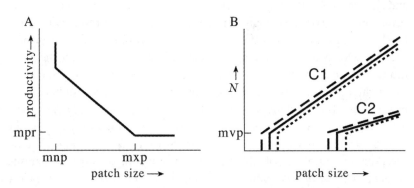

FIGURE 2.8 Graph A shows the interaction of habitat patch size and quality (productivity) on its suitability for a target species. *Mpr is* the minimum level of patch productivity that can support the target species; *mnp* is the minimum patch size that can sustain the target species regardless of the productivity level; *mxp* is the smallest size of patch that can support the target species at *mpr*. Graph B shows the effect of patch size and quality on the number of trophic levels that can be supported. C1 and C2 are successive consumer trophic levels, N is size of the population, and *mvp* is the minimum viable population size. Each curve is shown with dashed, solid, and dotted lines representing good, medium, and poor quality, respectively, for the focal patch (Lidicker 2000).

species. Less easy to accommodate are stochastically occurring catastrophes. Ideally, a patch needs to be large enough that portions of the patch will survive such devastating events, and the effects will be reversible.

One path to patch extinction is the gradual change in community structure through the loss of species and perhaps the gain of new ones invading from the surrounding matrix. Eventually, the community may become so modified that it can no longer be considered representative of the original community type. Such altered communities can sometimes be restored to their original type by succession, by gradual recolonization of characteristic species that were lost, or by human intervention. Whether that is possible will depend not only on the size and quality of the patch, but on the degree of connectedness to other patches of the same sort. Species losses from habitat patches are not random. Large species, those on higher trophic levels, and less vagile species will tend to be lost first. Moreover, corridors connecting patches will filter out those species for which the corridors are inaccessible or unsuitable. The role of corridors in metacommunity maintenance will be explored further in chapters 5 and 6.

Managers and conservationists should also keep in mind that natural or nearly natural community types provide so-called ecosystem services that benefit all life, including humans (see chap. 1). Processes such as carbon dioxide sequestration, production of oxygen, cleaning and storage of water, soil formation, decomposition of wastes, breakdown of toxins, harboring of pollinator species, control of agricultural pests, maintenance of biodiversity, stabilization of climate, and many more (Daily 1997) are increasingly critical as humans continue to degrade their home planet. These functions may also be provided by anthropogenically generated community types but often less efficiently and with concomitant loss of biodiversity.

If community patches are severely isolated, maintenance of their original composition may require that they be quite large (fig. 2.8). Rolstad (1991) estimates that isolated spruce forests in northern Finland need to be at least 20,000 hectares (49,400 acres) to retain their original passerine bird assemblages. Retention of non-passerines, such as grouse, may require even larger protected areas, perhaps 100,000 hectares (247,000 acres). If a protected area targets species of special conservation concern, its design can be tailored to favor the particular requirements of those species. One design will not likely be optimal for all the species in the community. In all cases, the large patches or, alternatively, the array of fragments, need to be large enough to withstand the occasional but unpredictable catastrophes that are unquestionably part of life on this planet. A new challenge is to arrange reserves so as to allow distributional adjustments that can accommodate climate change requirements (Pearson and Dawson 2005).

Beyond Metacommunities: Landscape and Ecoscape Concepts

While the metacommunity concept has moved us to new levels of complexity and corresponding new insights, ecological science has now progressed to encompass even more complexity and is thereby improving our understanding of reality with new and exciting developments (Lidicker 2008; Levchenko and Kotolupov 2010). These new developments are also providing us with a perspective that will significantly enhance our abilities to cope with the ongoing decline of humanity's life support system on this planet.

The landscape ecology perspective can be traced back to the nineteenth century when large landowners and some biologists began to see the value of thinking about land management for agriculture, game management, and forestry on large spatial scales. It wasn't until the mid-1980s that landscape concepts were rediscovered (Forman and Gordon 1986; Turner 1989; Turner 2005; Lidicker 2007; Wiens 1992, 2007). At first, it seemed that the landscape concept with its emphasis on large spatial scales, and with study areas containing multiple community types or habitats, was precisely the perspective that was needed. Indeed, this development led to much intellectual excitement and new research approaches. For more traditional ecologists, however, this approach led to some uneasiness. Because of its relevance to conservation and human welfare generally, there was incentive to somehow connect the new landscape thinking to more traditional and widely accepted concepts familiar to scientists, and that is the idea of viewing reality in terms of systems of increasing complexity. This worked for both physical and biological scientists. For ecologists the levels of interest begin with individual organisms. Gradually, the population level of complexity was recognized and then the community level. These were the biological systems of interest to ecologists, and they were therefore referred to as "ecosystems." Now the landscape concept with its large spatial scale and generally encompassing multiple community types seemed like a good candidate to add to the generally acceptable levels of system complexity (Urban et al. 1987; Lidicker 1995a, b, 2008; Turner 2005; Wu 2013).

However, levels of biological complexity are based on major changes in system composition such that emergent properties are necessarily generated (table 2.2). Inconveniently, landscapes were increasingly defined on

the basis of spatial scale. In some cases, minimum sizes were actually specified. No mention was typically made of emergent properties. Landscapes
of this type clearly do not qualify as a higher level of biological complexity.
Size is not a feature that justifies a new higher level of organic complexity.

This dilemma was solved by the suggestion that a new level could be
justified by defining it as a biological level featuring multiple community
types. This arrangement generates many new emergent properties (table
2.2). It was suggested that this new level be called the "ecoscape" level
(Lidicker 2008), thus combining "landscape," "seascape," and "riverscape"
into one term. These entities of any size could then be judged as to whether
or not they fit the ecoscape criteria. Most important, ecologists can now
explore the properties and insights inherent in this recently defined level
of complexity knowing that they are operating within the boundaries of
ecological science. A particularly vivid example of the potential for thinking
at this ecoscape level is provided by freshwater streams. Recent evidence
emphasizes the importance for fish to have access to multiple community
types needed for dealing successfully with different seasons and life-history

TABLE 2.2. Emergent properties characterizing two levels of ecological
complexity (Lidicker 2008)

Community	Ecoscape
Kinds (species)	Community types present
Diversity (numbers and proportions of Species)	Diversity (numbers and proportions of Community types)
Biomass	Biomass
Spatial distribution of populations	Spatial configuration of patches (dispersion, shapes, etc.)
Vertical stratification (trophic levels)	Ecotonal features (edge effects)
Dominance relations among species	Connectedness among patches
Coactions among species: types and intensities	Interpatch fluxes of energy, nutrients, information, organisms
Stability (variability, resilience, predictability, seasonal changes)	Stability (resilience, constancy, predictability, seasonal changes)
Trends (succession, degradation, extinctions, stability)	Trends (succession, stability, degradation, extinctions, invasions)
Energetic properties: food webs, primary production, production/ respiration ratios, trophic efficiency, nutrient transit times	Energetic properties: productivities and so forth
Historical context	Historical context

stages. Moreover it is increasingly clear that streams have important interactions with surrounding vegetation. Both communities receive important
imports from the other (Fausch 2015). Finally, with respect to the multi-
plicity of ways that the word ecosystem is currently being used, one of us
has argued that this term should be used as originally defined to reference
any living component of the biosphere that is at least as complex as a single
individual organism, and that we seek to understand with a holistic perspective (Lidicker 2008).

Chapter 3

Understanding Fragmentation

Habitat loss and consequential fragmentation are considered by many scientists to be the largest threats to preserving the world's biodiversity and a major cause of extinction today (Henle et al. 2004; Fahrig 2003). Chapter 1 discusses causes of habitat loss. Much of conservation biology is focused on examining the consequences of fragmentation for species' persistence and exploring options for mitigation. *Fragmentation* is defined as the transformation of a continuous habitat into habitat patches that usually vary in size and configuration. The effects of habitat loss and fragmentation have not always been carefully analyzed or realistically appreciated. Cumulative research, however, indicates that habitat fragmentation can lead to changes in species composition and abundance and generally results in negative impacts for biodiversity. Both habitat loss and fragmentation can be due to natural or anthropogenic disturbance. Although we emphasize human-induced fragmentation, we will discuss and analyze both kinds.

In this chapter we examine the effects of habitat fragmentation on biodiversity and explore options such as improved connectivity to enhance survival of native species. As we explore the implications of concepts like fragmentation, corridors, and matrix, we need to remind ourselves that temporal and spatial scales critically affect how we think about these and related phenomena.

Natural versus Human-Induced Fragmentation

One of the greatest challenges in our understanding of fragmentation concerns the differences between naturally heterogeneous and human-fragmented landscapes. Whereas some species are adapted to natural fragmentation, it appears that many others suffer adverse consequences from human-induced fragmentation. It is because of this difference that maintaining and recreating connectivity has become a central issue on many conservation agendas. We need to understand how we humans affect biodiversity in order to mitigate the impacts. As examples, do natural boundaries between intact and burned forests or between avalanche zones and montane forests function similarly to those boundaries created by humans? What are the similarities and differences?

There are commonalities between naturally fragmented and human-fragmented systems. Regardless of the mechanisms involved, smaller patches of habitat contain fewer species as well as fewer specialists (i.e., species that depend on specific habitats, foods, or other limiting factors to survive). For example, herb communities that occupy the naturally patchy serpentine soils of California are likely to have a lower diversity of endemic species (species found nowhere else) in small patches than in larger patches (Harrison 1999). Similarly, research in a rain forest fragmented by human activities in New South Wales, Australia, illustrated that smaller patch size was correlated with lower overall species richness, as well as lower diversity of native species of ground-dwelling small mammals (Dunstan and Fox 1996). Two of the most important factors influencing species survival in retained patches include the size of the patch and the potential or realized connectivity among the patches. Here we will discuss all types of fragmented landscapes, bearing in mind that human-induced and natural fragmentation often differ in three important ways: (1) speed and pattern of change, (2) scale of change, and (3) ability of the resulting fragments to recover from perturbations (resiliency).

Speed and Pattern of Change

Natural fragmentation involving large areas tends to occur slowly, in contrast to the rapid human-induced fragmentation. Glaciation, for instance, causes slow fragmentation across large areas. Evidence suggests that many once continuously distributed populations of species became fragmented into multiple sites or habitat refugia during glacial advances. Species in refugia then evolved separately, contributing to species differentiation over time (e.g., DeChaine and Martin 2004). During interglacial periods, refugia often become reconnected. On the other hand, the rapid human-induced habitat loss and fragmentation affecting much of the globe has largely occurred in recent history (fig. 1.1).

FIGURE 3.1 Aerial photo showing the western edge of Yellowstone National Park (left), which is protected from logging, and Targhee National Forest (right), where the forest was logged in the 1960s and 1970s. (Photo by Tim Crawford.)

Many natural processes that cause fragmentation, ranging from fire and avalanches to volcanoes and windthrow, occur with different frequency and result in an altered landscape mosaic as compared to human-induced fragmentation. The impacts of these natural perturbations are often significantly different from those of human-induced changes in the same area. Tinker et al. (2003) compared landscape dynamics in Yellowstone National Park, where natural fires have been allowed to burn, to those in the adjacent Targhee National Forest, which experienced fire suppression along with intense clear-cutting over the past thirty years (fig. 3.1). The researchers found the post-harvest spatial characteristics in Targhee to be very different from the spatial characteristics created by the natural fires (the relatively small pre-1988 fires and the large 1988 fire) in Yellowstone. For example, before logging, the number and size of forest patches were similar between Yellowstone and Targhee, but after logging, the Targhee contained significantly more and smaller patches than Yellowstone.

Understanding such differences is important in directing future management and restoration efforts, including those of wildlife corridors. In general, the landscape mosaic that results from human-caused fragmentation, from clear-cutting of forests, plowing of grasslands, construction of reservoirs, and development of agricultural and urban artifacts, is different from the mosaic that results from naturally occurring types of fragmentation. The differences in fragment patterns have consequences for biodiversity conservation and implications concerning the need for connectivity.

In addition to the differences in the speed and pattern of change between natural and human-induced fragmentation, humans have contributed to substantially more habitat loss and fragmentation, particularly in the last century, than would naturally occur. That difference can best be seen by comparing the quantities of human-caused fragmentation and natural fragmentation in different biomes. Fragmentation due to human-induced causes impacts 49 percent of the original tropical and subtropical dry broadleaf forest range, whereas only 6 percent of that same forest type is naturally fragmented. The amount of human versus natural habitat fragmentation is 53 to 4 percent for temperate broadleaf and mixed forest and 55 to 17 percent for Mediterranean systems. With the single exception of boreal forests, where the ratio is 4 to 13 percent, human-induced fragmentation is higher than natural levels of fragmentation in every major biome in the world (Wade et al. 2003). Looking toward the future, we can anticipate that the human caused changes that have been mentioned will continue, probably at faster rates. In addition, we are faced with climate change as well as the associated increased frequency and intensity of fires. In chapter 8, climate change issues will be discussed more fully.

Consequences of Human-Induced Fragmentation

Scientists have been working to understand the impacts of human-induced fragmentation on the world's biodiversity. Cumulative work that has been achieved so far indicates that consequences of human-induced fragmentation for native flora and fauna can be extensive. Impacts range from a decline in the numbers of species, population sizes, and species' ranges to increases in exotic species and predation on native flora and fauna (Haddad et al. 2015). Here we describe many of the factors that influence biodiversity survival in fragmented landscapes.

The configuration of the landscape after habitat loss determines the impact of resulting fragmentation on the area's original biodiversity. Habitat loss from human activities may result in one remnant patch or multiple habitat patches. The number of patches tends to decline as a result of the original community being progressively destroyed (fig. 3.2). Two of the most important factors influencing species survival in retained patches

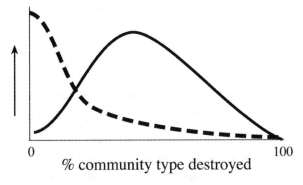

% community type destroyed

FIGURE 3.2 Schematic representation of how average fragment size and number of fragments change as the percent of a particular community type in a landscape progresses from complete cover (none destroyed) to 100 percent destroyed. The y axis is numbers or size. The dashed line is the mean size of fragments, and the solid line is the number of fragments.

include the size of the retained patch and the potential or realized connectivity among patches. Other related factors that affect species conservation include edge effects, impact of edge species, nature of the matrix, species loss, and genetic consequences. Good connectivity may ameliorate some of these impacts and allow a species to utilize more than one habitat patch, thus helping to maintain regional species' persistence.

Species richness and diversity are strongly influenced by patch area, patch characteristics, ecological continuity, surrounding habitat, degree of isolation, and the matrix (Humphrey et al. 2015). Larger patches generally harbor more species than smaller patches (Laurance et al. 2002), which translates to a higher probability of conserving more native species. The Biological Dynamics of Forest Fragments Project (BDFFP) was initiated in 1979 to assess the effects of fragmentation on biota in Amazonia. Researchers' findings, not surprisingly, often showed a positive correlation between fragment size and species richness such that intact forests contained more rain forest species per unit area than fragments. Extinction rates were negatively correlated with fragment area, such that more species went extinct within smaller fragments (Laurance et al. 2002). Other research results from BDFFP were that species also were lost from the larger fragments studied (e.g., 100-hectare, or 247-acre, fragments). These extinctions happened over relatively short time spans, indicating a need for immediate conservation action in fragmenting landscapes (Ferraz et al. 2003). Cumulative research from the BDFFP suggests that rain forest species, ranging from

---

BOX 3.1

THE SLOSS DEBATE

The desired number of patches and their size for biodiversity conservation are encompassed in the SLOSS (single large or several small) debate (Rebelo and Siegfried 1992; Tjørve 2010). Given the limited amount of conservation resources, should we be focusing on one large reserve or several smaller reserves in an area? The debate has raged, with proponents of the single large reserve emphasizing survival of the area-demanding species, survival following catastrophic events, and the like. Proponents of multiple reserves contend that each reserve is likely to have a unique set of species and that more species can therefore be conserved by setting aside multiple reserves (Diamond 1976; Simberloff and Abele 1976). Moreover, it is preferable not to tempt extinction by having all individuals of a given species in a single patch, in case an event such as a disease were to eliminate all the individuals in that patch. There is no one answer to the SLOSS debate, as each species will respond differently to the size, number, and location of patches. The debate has simmered down in recent years in recognition that the best plan is to set aside as many reserves as possible, and that they should be as large and connected as possible (Rösch et al. 2015).

---

mammals and birds to beetles and butterflies, are sensitive to fragment area, and that responses to fragmentation are individualistic by species, with some species responding negatively and others responding positively (Laurance et al. 2002).

While retaining larger fragments may help some species survive, retaining only one large residual fragment of a formerly continuous natural community may be risky (box 3.1). Disease, a catastrophic storm, or some other stochastic event could potentially wipe out all remaining individuals in a single habitat patch. Unfortunately, there are numerous species in which all the individuals currently exist in only a single patch. These include, for example, the Gir lions of India discussed in chapter 1 (Oza 1983). This dilemma is illustrated by a long-standing topic of discussion in the conservation literature, namely, is it preferable to save one large patch of habitat or many small ones (SLOSS debate). See box 3.1 for further discussion.

Extensive literature demonstrates that larger fragments that are closer or more connected to source populations have higher species richness than smaller habitat fragments that are more isolated (e.g., Hanski et al. 2000). In part this is because the relative size and isolation of habitat patches can influence species' persistence (Åberg et al. 1995; Dunn 2000). Ultimately, many species that require large areas to maintain functional populations will need to move among remaining habitat patches to survive, whether many small patches or several large patches remain. The location of patches relative to one another and the connectivity among patches will play a critical role in their survival. Isolation of habitat fragments from one another can ultimately lead to population declines. Some species may be able to survive as metapopulations in multiple habitat fragments but only if sufficient interchange among the patches occurs (chap. 2). Researchers have documented local extinctions of species in small isolated habitat patches where access to large core habitat areas (large continuous habitat patch) or other habitat fragments is restricted (fig. 3.3). For example, mountain lions (*Puma concolor*) in the Santa Ana Mountains of California have already become extinct in a 75-square-kilometer (29-square-mile) habitat fragment and were expected to become extinct in another 150 square kilometers (58 square miles) of habitat if a housing project were to sever possible connections with other fragments (Beier 1993). A similar situation also occurs in the nearby Santa Monica Mountains as well as farther north in the Santa Cruz Mountains. In the Santa Monica case, serious inbreeding is combined with unusually aggressive behavior that was observed among family members (Riley et al. 2014). Research in the Northern Territory of Australia showed that rose-crowned fruit doves (*Ptilinopus regina*) and pied imperial pigeons (*Ducula bicolor*) were particularly affected by isolation from other rain forest habitat (Price et al. 1999). Fewer habitat patches within a 50-square-kilometer (20-square-mile) area resulted in fewer species of doves and pigeons within any particular area. This result indicates that the presence of those birds is likely influenced by the presence of other habitat patches nearby. Likewise, red-backed voles (*Clethrionomys californicus*) living in remnant forest patches in the US Pacific Northwest were found to be essentially isolated, in that almost no voles were detected outside of patches, suggesting that dispersal among patches is rare (Mills 1996). DNA work suggests that voles remaining in isolated patches have less genetic variation than those in large continuous populations, which may make them more susceptible to extinction from inbreeding depression and lack of immigrants to help boost populations in fragments. Similarly, California voles (*Microtus californicus*) have low genetic variation in small populations that persist after population crashes as compared to large panmictic populations that prevail at high densities (Lidicker 2015).

Also, distances among remaining habitat fragments may influence species' use of adjacent human-modified habitat (Perault and Lomolino 2000). Species may have different spatial scales of movements such that some species may venture farther from remaining habitat patches than others (e.g.,

FIGURE 3.3 The grizzly bear (*Ursus arctos*) is an example of a large, wide-ranging species that exists at low densities and is especially susceptible to human-induced fragmentation. (Photo by Jeff Burrell.)

Ricketts et al. 2001). Diversity of nearby habitats and structure or cover remaining in the human-impacted areas also may affect species' presence at a given place in the landscape (Hilty and Merenlender 2004). For example, wolves in northern Italy were most active in agricultural land adjacent to forested areas (Massolo and Meriggi 1998), and the same was found to be true for carnivores in northern California's oak woodland and vineyard landscape (Hilty and Merenlender 2004). Species unable to use heavily modified landscapes at all may be more prone to extinction (e.g., Bentley et al. 2000).

Species with higher mobility should theoretically survive better in a fragmented landscape and among more isolated patches of habitat. Such species can move between patches and thereby minimize the negative demographic and genetic consequences associated with small populations (see chap. 2). However, this is not always the case. Many species possessing the physical capability do not move across human-impacted landscapes, particularly when natural habitat patches are far apart. Those species with limited mobility or other behavioral impediments to moving are often restricted to single patches in fragmented habitats, and the associated risks of single-patch occupancy can lead to local extinction (Laurance 1991; Gascon et al.

FIGURE 3.4 An Idaho license plate reflecting anti-grizzly bear sentiment. (Photo by Joel Berger.)

1999). Species such as grizzly bears (*Ursus arctos*) come into conflict with humans by eating garbage, harassing pets, even attacking people. Such conflicts often lead to their removal or at least reduced survival and increased risks of interpatch movements in the human-occupied landscape (e.g., Haroldson and Frey 2002; fig. 3.4). Therefore, in spite of their ability to travel long distances, bear populations can become isolated in habitat fragments.

Changes in Species Composition of Patches

The species composition of habitat patches should not be assumed to be stable. There may be seasonal, successional, or serendipitous changes that can influence extirpations and invasions of the patch. Documenting these compositional changes is critically important as they are likely to significantly influence species persistence and the meta-community as a whole (Naeem et al. 1994). For example, weedy species and generalist predators often become more prevalent, while rare, marginally adapted, and specialist species tend to decline (Kemper et al. 1999; Ye et al. 2013). Research in the Atlantic forest of southeastern Brazil suggests that there is a shift in plant structure and species composition as tropical forest fragments are reduced in size, with small fragments containing more weedy plant species (Tabarelli et al. 1999). In situations in which a habitat patch is progressively degraded or destroyed, the number of species at first increases as new species move into newly created habitats and others invade the original patches.

Eventually, some of the species that made up the original community will disappear, and only species that can survive in the degraded habitats will persist (fig. 3.2).

While weedy species may invade or increase in abundance in habitat fragments, species that were once widespread may disappear. The loss of species may then lead to multiple additional impacts. A good example of this was documented in western Australian eucalyptus (*Eucalyptus salmonophloia*) woodlands where mistletoes (*Amyema miquelii*) were more likely to be found in large fragments of habitat than in small fragments (Norton et al. 1995). This is likely due to the sensitivity of both mistletoe and avian dispersers to fragmentation. Loss of mistletoe in fragments ultimately may contribute to a decline in two butterfly species (*Ogyris* sp. Lycaenidae and *Delis aganippe* Pieridae) that feed on mistletoe, thereby contributing to a further cascade of species loss in these habitat fragments.

Like plants, native fauna often disappears from small patches, which can also lead to cascading effects (box 3.2). Large predators are often the first species to go extinct as fragment size diminishes, which can cause a cascading loss of trophic interactions (Beier 1993; Ripple et al. 2014; fig. 3.3). Terborgh et al. (2001) showed that the loss of large predators began an ecological meltdown on islands created by dams in Venezuela. Consequent to the loss of predators, the researchers documented that herbivores increased greatly, leading to a cascading loss of much plant and other animal diversity. Parallel to the Venezuelan study, the unraveling of an ecosystem was documented on the Pacific Coast of North America, where sea otters (*Enhydra lutris*) were extirpated, then herbivores became overpopulated, and kelp forests began to disappear (Estes and Duggins 1995). In other places, such as in southern California, loss of large carnivores has led to the release of pressure on middle-sized predators, with consequent shifts in the dynamics at herbivore and plant trophic levels (Crooks and Soulé 1999).

After isolated islands or fragments of habitat are created, species loss can occur gradually over time, which is sometimes called *species relaxation*. A good example of this phenomenon comes from lizard fauna that experienced cumulative loss in species diversity over time on islands in Baja California that were once linked to the mainland and are now isolated (Wilcox 1978). Even national parks may lose species as parks become more isolated due to surrounding human activities. Newmark (1995) documented that species loss was greater than species recruitment in western North American national parks. Older chaparral and coastal sage scrub habitat fragments in urban San Diego County, California, contained fewer chaparral and scrub specialist bird species than did younger stands (Bolger et al. 1997; Crooks

BOX 3.2

A SAMPLE OF THE IMPACTS OF HABITAT FRAGMENTATION ON BIRDS

Because birds are comparatively easy to study, much research effort has been directed toward examining different impacts of habitat fragmentation on birds. Impacts of fragmentation on birds, as with most species, are often confounded by concomitant habitat loss. Few studies have successfully teased apart the effects of fragmentation and reductions in habitat availability (Fahrig 2003; Schmiegelow and Mönkkönen 2002; Fahrig 2017). Despite that caveat, cumulative research indicates that there are a variety of impacts but that they are often community or species specific. Some examples of changes in bird communities include the following:

- Less overall available habitat generally means lower species richness and fewer specialists (e.g., Schmiegelow and Mönkkönen 2002).
- Larger habitat patch sizes will have more individuals as well as species of birds, with some species being absent from smaller patches altogether (e.g., Beier et al. 2002). Some reasons for absence from smaller patches may be food shortages and limited nest sites (Burke and Nol 1998).
- Habitat fragmentation can impact bird density and fecundity, which may be influenced by patch size, distribution of patches across the landscape, and landscape composition (Donovan and Flather 2002).
- Bird species' distribution in fragmented habitat is species specific; some birds occur across all patch sizes and throughout patches, while other birds may be more likely to use only certain patch sizes or occur only in the interior or in the edge of patches (e.g., Schmiegelow and Mönkkönen 2002).
- Edge effects can affect avian nesting success (Paton 1994). The impact of edges, including increased nest parasitism and nest predation, is community, scale, and species specific (Stephens et al. 2004).
- The type of vegetation and the location in which birds nest can affect predation rates (e.g., Aquilani and Brewer 2004). Overall landscape composition also influences nest predation (Zanette and Jenkins 2000).
- Matrix quality is positively associated with higher levels of patch biodiversity (Boesing et al. 2018).
- Species living in patches may be influenced by a complexity of interacting factors (Freeman et al. 2018).

et al. 2001). Likewise, bird species occurring at lower densities were more likely to go extinct than species with higher densities. Studies documenting species attrition are important, because they indicate that species currently found in habitat fragments, especially relatively new fragments, may not be able to survive in those fragments over the long term.

Genetic Considerations Affecting Species Extinction

Small isolated populations may experience genetic drift that can cause genetic erosion, inbreeding, and reduced lifespan and fitness that can ultimately lead toward population extinction (Lohr et al. 2014). A long-term study of an isolated population of greater prairie chickens (*Tympanuchus cupido pinnatus*) in Illinois documented both decreased fertility and decreased egg hatching rate, with a decline in genetic diversity (Westemeier 1998). When birds from other populations were introduced into the isolated study group, increased fertility and egg hatching rates resulted, indicating that low genetic diversity was impairing population recruitment. Similarly, song sparrow (*Melospiza melodia*) reproduction was studied on islands of various size and isolation on the North American continent, where researchers documented that natural selection favored noninbred birds after the populations experienced bottlenecks or population crashes (Keller et al. 1994). These studies are examples illustrating that low genetic diversity can be a major factor contributing to a decline in species. Because of such evidence, researchers are showing that conservation efforts may be most successful when focusing on larger populations. For example, research indicated genetic erosion in small populations of an endangered tetraploid pea (*Swainsona recta*) in Australia, such that conservation efforts should focus primarily on populations of fifty or more reproducing plants rather than the smaller inbred populations (Buza et al. 2000). One way to diminish genetic inbreeding impacts and boost effective population size is through reconnecting disjunct populations, such as through corridors.

While genetic issues contribute to population declines, genetic erosion is not always the ultimate factor causing extinction for many mammal species. Extirpation can also be the result of change in the environment that may reduce population size or distribution and may lead to extinction due to demographic collapse. Also, remaining residual populations may be eliminated when a small population lacks adults of each sex or suffers some stochastic environmental event such as a severe drought. For example, only small populations of black-footed ferrets (*Mustela nigripes*) remained by the 1980s. An outbreak of canine distemper nearly wiped out the species, and ultimately the population plummeted to eighteen in captivity, which was probably all that remained of the species (Seal et al. 1989). In that case, the population escaped total extinction because of human intervention, but it illustrates how small populations are vulnerable to disease and other natural catastrophes.

In summary, human-induced fragmentation can happen quickly across large areas and can permanently affect biodiversity. Fewer, smaller, and more isolated habitat patches with increased negative edge effects can lead to species loss and changes in community composition.

Role of the Matrix

We will now expand our perspective to explicitly consider the context in which we find the fragmented metapopulations or metacommunites previously discussed. That context is the *matrix*, the various kinds of communities in all of their physical and biotic dimensions that surround habitat patches, and as we will see can profoundly influence them. Note that, so far, we have largely ignored the matrix considering it mainly as simply marking the boundaries of the habitat patches (fragments) of interest. Tacitly, we assumed that the matrix for a given metapopulation or metacommunity was of uniform quality. Now we will examine matrices in detail and discover that they can be very complex and exert significant influences on the habitat fragments of concern. Conservation biologists and agroecologists refer to the matrix as the human-influenced or human modified land use that surrounds native communities and habitats. While this view remains the main focus of our concept of matrix, we will also generalize the definition to a broader ecological context, and thereby consider matrix to be any community that either borders a habitat patch of interest or is located between the focal patch and another patch of the same habitat.

We will examine the role of the matrix in influencing connectivity among fragments or patches, the importance of edge effects generated by specific juxtapositions of community types, effects of the matrix on the population dynamics of target species, and the influence of the matrix on the invasions of exotic organisms. Lindenmayer and Franklin (2002) give us a thorough treatment of the importance of the matrix to the conservation of biodiversity in forests and Prevedello and Vieira (2010) conducted meta-analyses to assess the effect of different types of matrix on connectivity. Prugh et al. (2008) also emphasize the central importance of matrix size and quality in affecting the fate of habitat fragments.

Since any investigation of the role of the matrix necessarily involves at least two communities or habitats, we will be operating at the multi-community or ecoscape level of complexity (see chap. 2). We will examine the role of the matrix in influencing connectivity among fragments, the importance of edges or boundaries between community types, effects of the matrix on the population dynamics of the target species of interest, and the role of the matrix in influencing invasions of exotic organisms.

As it often comprises many different kinds of land use and cover types, the matrix can be complex, indeed. Some of the matrix or all of it may be human-modified communities such as agricultural fields of various kinds, clear-cut forests, grazed grasslands, and towns. Various barriers to movements of organisms may be present as, for example, roads, fences, irrigation ditches, airports, and urban developments. Finally, the matrix can be dynamic, with seasonal changes of many kinds, as well as changes caused by successional processes, fire regimes, and human activities. As we will see, organisms living within the community fragments of interest will be profoundly influenced by the size and characteristics of the matrix to which they are exposed (Prugh et al. 2008; Humphrey et al. 2015). For example, Drapeau et al. (2000), in a study of bird communities in northwestern Quebec, illustrate this point. They demonstrated that the nature of avian assemblages was about equally influenced by habitat features and by the landscape context in which they occurred. Similarly, Boesing et al. (2018) studied bird species diversity in patches of forest in the Brazilian Atlantic Forest, and report that the quality of the matrix was positively correlated with bird biodiversity in the patches. Reunanen et al. (2000) conclude that for flying squirrels (*Pteromys volans*) to persist in the boreal forests of Finland, it is not sufficient to provide large patches of mature mixed deciduous and conifer forests (optimal habitat); the patches must also be connected by dispersal corridors of young secondary forest. Freeman et al. (2018) studied bird communities inhabiting fifty-nine forest fragments in South Africa and Mozambique and report that the patch/matrix relations can be quite complex with interactions occurring among patch size, matrix type, life-history features such as forest specialist or generalist, and species' unique species–area relationships.

The matrix poses special challenges for those wishing to mathematically model the behavior of organisms in complex spatial arrays such as metapopulations or metacommunities. While accepting that actual matrices are likely to be quite complex, modelers will necessarily have to simplify them in order to limit the number of parameters incorporated in particular models. We have already seen how the original metapopulation models by Levins (chap. 2) assumed a dimensionless and uniform matrix that could be characterized by a single parameter, namely, the probability of an emigrant being successful in founding a new colony in an empty patch. Other models take distances between patches into account but assume that only one uniform community type is involved. Still others may assume that matrix influences extend only a limited distance from fragment edges. Such simplifications, while necessary on practical grounds, will influence the model outcomes to varying degrees. Modeling more realistic patterns of biodiversity within the matrix show that degree of matrix suitability impacts species richness in patches (Prevedello et al. 2016). Another study was performed by Cantrell et al. (1998) in which they treated the matrix not as a constant but as declining in quality. Their model considered the fate of two competing species living in habitat patches and found that the competitive coefficients between the species shifted as the matrix deteriorated and could even reverse in sign, so that the previously subordinate species became competitively dominant. This dynamic resulted from differing negative influences on dispersers of the two species in the matrix as it gradually degraded.

A special kind of matrix is known as "working landscapes" by the conservation community. These are terrestrial or aquatic places that have a dual function. First, they are places that are used for economic gain such as agricultural production of food or fiber; forests for harvesting wood, mushrooms, or game animals such as deer; and fish ponds or oyster beds. The second function is to significantly benefit humanity in ways that are not primarily economic. This requires active management in order to achieve such benefits. These might include significant contributions to preservation of biodiversity, carbon storage, oxygen production, serving as corridors for organisms to disperse through, decomposition of organic wastes and other so-called ecosystem services. It could also include recreational benefits such as hunting, fishing, hiking, camping, as well as various types of playgrounds. Large parks would also fit into this category. The role of working landscapes will be discussed in more detail in chapter 10.

Traveling the Matrix

To begin with, it is critical to acknowledge that real matrixes can be very complex, composed of a variety of different types of communities, variously arranged spatially, and of differing dimensions. If landscape function concerns connectedness among the communities that make up the landscape, movements of organisms among the habitat patches, that is, across the matrix, are of fundamental importance. These movements can be dispersal, within-home-range movements, or exploratory excursions. Longer-range movements will primarily be dispersal (chap. 2), and those are the movements that will be of most concern to conservation planners. Avenues where travel through the matrix is relatively easy, that is, corridors, may or may not be present and available to those with wanderlust. Moreover, each species in a community will perceive the matrix in its own way and respond uniquely. The assessment of matrix permeability to movements will therefore not generally be easy to discern and will be especially difficult to predict in a management context.

It may be useful to think of the matrix surrounding a particular community patch as being on a gradient with respect to its permeability for a given species living in the patch. We can devise six levels or grades of traversability:

- No impediment exists except for distance to the next available patch.
- No impediment exists at certain times; various levels of impermeability exist at other times. Favorable times might be seasonally organized, multiannual events, or connected to major unpredictable perturbations. For example, a rare flooding event may connect the headwaters of one stream system to another.
- Minor inhibitions exist for travelers to overcome, thus magnifying distance effects. These, of course, can also be intermittent.
- Moderate impediments to movements exist.
- Strong barriers to movements exist, but some movements do occur. Successful crossings may be limited to certain age, sex, or health groups, or to seasons.
- Complete barrier to movements prevails.

A particularly nefarious complication with this gradient would occur when a matrix provides moderate or good permeability but is bisected by a nonpermeable barrier such as a major highway or river that prevents travelers from reaching otherwise available patches. Laurance et al. (2004) have shown that even narrow, low-traffic roads in Amazonia can greatly inhibit movements of many rain forest species of birds. This is similar to results reported by Diamond (1973) for rain forest birds in New Guinea and southwest Pacific islands.

Permeability or traversability of the matrix will depend on a species' access to the matrix (more about this later), the quality of the matrix with respect to survival and facility of movements, and the distance to neighboring patches. The variables of distance and quality interact, as shown in figure 3.5. If distances are short and quality is good, individuals may even be able to combine several patches into a single home range (fig. 3.6; Andreassen et al. 1998; Noss and Csuti 1997). Therefore, a potential traveler will need to assess both the qualities of the matrix and its dimensions. A poor-quality matrix may be crossed if the distance to the next patch is not

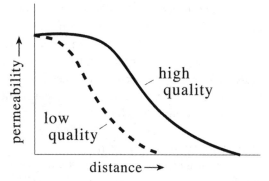

FIGURE 3.5 Permeability of the matrix as experienced by dispersing organisms as a function of distance to neighboring habitat and matrix quality.

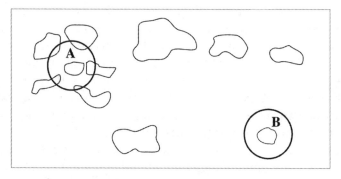

FIGURE 3.6 Several patches of habitat assembled into a single home range (A), and a home range composed of a single isolated patch (B).

too great. Longer travels may be ventured if the matrix is more favorable. Several studies with voles (*Microtus*) in experimental landscapes illustrate this discrimination. Distances of 1 meter (39.37 inches) of matrix are readily crossed, whereas 4 meters is considered risky but still crossed frequently, and 9 meters is rarely crossed (Andreassen et al. 1996; Lidicker and Peterson 1999; Wolff et al. 1997). An important variable here is the ability of an organism to perceive distance across the matrix.

Of paramount importance is the recognition that matrix permeability is particular to the various species in the community. What is easily crossed by some may be impenetrable to others. Thus, the matrix may serve as a filter, allowing good connectivity for some components of the focal community but not for the rest. When this situation applies, and it is likely to be common, progressive local extinctions of disconnected species will allow community structure and function to drift away from their initial conditions. Eventually, extinctions occur, and the original community type will transform into something else. These delayed extinctions can be a slow process and hence difficult for managers to detect. Communities in some landscape contexts can therefore appear healthy but actually be moribund. This condition has been called the *extinction debt*. One grassland study found that the current number of specialists matched the area and connectivity from seventy years earlier, reflecting an estimated extinction debt of 40 percent (Helm et al. 2006). Ultimately, it must be paid.

Differential use of the matrix is illustrated by Gascon et al. (1999), who studied birds, small mammals, frogs, and ants in forest fragments in central Amazonia. They found that populations of species using the matrix were stable or even increased following fragmentation, but those that did not use the matrix declined or became extinct. Similarly, Ricketts et al. (2001) documented the moth biota in a 227-hectare (561-acre) forest fragment surrounded by four types of agricultural development in Costa Rica. The various species of moths moved different distances into the agricultural matrix depending on the type of matrix crossed, and overall the moth species richness was greater within 1 kilometer (0.62 mile) of the forest than at more distant sites. A meta-analysis on many taxa, from butterflies to mammals, found that movement is greater through a matrix of more similar structure to a species' habitat (Eycott et al. 2012).

Because the matrix may filter out those species less able to travel to other patches, it is appropriate to consider what features of organisms favor movement abilities. The ability of organisms to travel is termed their vagility. Generally, large, nonsedentary species are good travelers. It helps if they are habitat generalists and thus comfortable in a variety of community types. Species with superior movement capabilities include runners and fliers. Individuals who are trophic generalists or who can go for long periods without eating can relatively easily cross non-habitats. Some small species, and especially seeds, can attach themselves to larger species and get transported by them (phoresy). An interesting case involving blind snakes (*Leptotyphlops dulcis*) illustrates how bizarre phoresy can be (Gehlbach and Baldridge 1987). These snakes are picked up by screech owls (*Otus asio*) and placed in their nests. Here they feed on nest arthropods and ectoparasites of the nestling owls. Young owls grow better and suffer less mortality when snakes are resident in their nests. When the owls fledge, the snakes leave the nest and discover that they have dispersed to a new environment. We humans are, of course, world champion phoretic agents. We intentionally and unintentionally transport living creatures all over the world,

even across major ocean basins. Sometimes these transfers benefit humanity, such as when the potato was brought from South America to Europe.
Many, however, are detrimental to agriculture, forestry, human health, and
native biotas, and this has become a huge conservation problem, second in
importance only to loss of habitat for species.

Even very small creatures may be good dispersers. Many microorganisms and even very young spiders can use air or water currents to passively
drift to new patches. Other things being equal, asexual species will have
better odds of successfully establishing in an empty patch than those that require more than one sex or mating type to reproduce successfully. Those organisms with high reproductive rates will also improve their rate of success
by sending out large numbers of dispersers. One of the largest litter sizes
found among voles of the genus *Microtus* occurs in the taiga vole (*M. xanthognathus*) of northern Canada and Alaska. Its preferred habitat is recently
burned spruce (*Picea*) forests. As succession proceeds following fires, the
forest gradually becomes unsuitable for the voles. They succeed by having
offspring that find new burned patches in unpredictable locations (Wolff
and Lidicker 1980). Species involved in complex, especially obligate, social
relationships are unlikely to be able to penetrate matrix successfully. And
if they do successfully reach a new patch, they will likely be fended off as
unwelcome immigrants and not be able to integrate into existing social
groups. More subtle but potentially critical attributes are a species' propensity to enter non-habitat and its capacity for finding and using movement
corridors. We will discuss these two issues later.

Conservation planners can, with caution, as many exceptions are
known (Lidicker and Koenig 1996), use morphological, behavioral, and
life-history features like those discussed to effectively design landscapes that
will connect community patches most advantageously. Conversely, they can
use them to predict where problems are likely to occur.

Edges and Edge Effects

When a patch of some habitat or community type is juxtaposed to a different type of patch or community, an edge is generated between them.
Such edges can naturally occur or can be the result of human activities, and
species responses may differ. Human-created edges and edges created by
natural processes in forests function differently; natural edges have a higher
species richness than the adjacent interior forest and function to keep matrix species from invading the forest in contrast to human-created edges

(Magura et al. 2017). Inevitably there are fluxes across such boundaries, and these movements of organisms, or their gametes, and abiotic materials such as water, sources of energy, or even information can have profound influences on the functioning of the two neighboring communities. When a rabbit peers out of its brushy hiding place and observes a predator walking by, information has been transferred across the border. One abiotic flux generally ignored is that of artificial light. Bird et al. (2004) demonstrate that artificial lighting adjacent to the habitat of beach mice (*Peromyscus polionotus*) in Florida significantly inhibited their foraging activities.

Movements of nutrients across community boundaries can be extremely important. It has long been recognized that streams, rivers, and estuaries often receive large inputs of nutrients from adjacent terrestrial communities (Wallace 1997). Less appreciated are the sometimes significant movements of nutrients out of the aquatic environment to the terrestrial. A particularly spectacular example is given by black bears (*Ursus americanus*), who remove salmon (*Onchorhynchus* spp.) from streams in the Pacific Northwest and carry them inland up to 150 meters (almost 500 feet). In one study, T. E. Reimchen (pers. comm. 1997), bears removed 63 percent of a salmon run and left half of that in the forest to benefit a myriad of scavengers, decomposers, and plants (Ben-David et al. 1998).

Thus, when patch meets matrix, new dynamics are generated, and these are called *edge effects*. The border area, which manifests these effects, is often called an *ecotone*. Animal ecologists generally attribute this idea to Aldo Leopold, which he wrote about in1933 (Lidicker and Peterson 1999). He considered edge effects to be beneficial features of landscapes. Edges were often sites of high productivity and enhanced biodiversity and so were much desired by wildlife managers. Among plant ecologists, the notion of ecotone dates to the end of the nineteenth century (Clements 1897). Boundaries between plant communities were thought of as "tension zones," often featuring enhanced productivities. Thus it was that ecotones became associated with new or emergent properties, not necessarily predictable from knowledge of the two adjacent communities separately. Another feature of ecotones is the possibility that their biodiversity will be enhanced by the occurrence of edge species, which are organisms whose primary habitat is the ecotone itself. In spite of their importance ecotones are not necessarily obvious to the human eye and can be difficult to measure in length and width.

The realization has grown that edges can have negative effects on the participating communities as well as positive. Particularly vulnerable are species that live in the interior of patches and avoid or do poorly on edges,

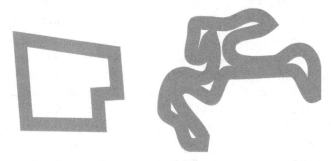

FIGURE 3.7 Two differently shaped habitat patches of the same size showing different edge encroachment. The edge zone is indicated by the gray borders. The patch on the right has a vastly greater edge-to-area ratio than the one on the left and therefore has a much smaller effective patch size for edge-intolerant species.

and there are increasing numbers of examples reported of this behavior. Since edge zones may or may not be used by species living in the patch, the area encompassed by edge may need to be subtracted from the total patch area to give the effective patch size for those species. This results in reduced patch areas and therefore smaller population sizes for the inhabitants (fig. 3.7), thereby increasing the risks of all the negative effects of small population sizes (chap. 2). If patch residents actually avoid going into the ecotone, it will reduce their chances of dispersing out of the patch. In fact, we can grade a species' behavioral response to edges ranging from refusal to even enter the ecotone, and hence being unable to find the actual patch edge and its adjacent matrix, all the way to entering the matrix without inhibitions (fig. 3.7). And, of course, a given edge will provoke various responses in different species (Garmendia et al. 2013).

Because both physical and biological changes are associated with edges, the amount of edge for a given patch or corridor can profoundly alter ecosystem function and structure. Microclimate alteration is a physical effect of edge that can negatively affect forest interior species while positively affecting other species. At the edge of a forest, direct sun increases, wind exposure is higher, and snow loads can increase (Laurance et al. 1997), although the strength of these effects can vary depending on the surrounding matrix and edge orientation (Aragón et al. 2015). In forested fragments, soil temperature also increases to resemble that in nonforested areas (Mills 1996), and overall temperature and vapor pressure can vary more along edges. Humidity and depth of humus also often differ on edges compared to interior habitat (Mills 1996; Carvalho and Vasconcelos 1999). Such influences can extend as much as 50 meters (164 feet) into forest patches (Laurance 1997; Sizer and Tanner 1999). These altered conditions can inhibit regeneration of vegetation where seeds are particularly sensitive to desiccation and can increase mortality due to windthrow (Laurance 1991, 1997). As a result of these microclimatic alterations, vegetation composition is often different at the edge compared with the interior of a forest. Altered vegetation can be detected up to 500 meters (1,640 feet) into a tropical rain forest (Laurance 1995). In Amazonia, tree species along the edge of habitat fragments suffered high mortality (Mesquita et al. 1999). For very small fragments of natural habitat or narrow corridors, microclimatic changes associated with the edge may permeate throughout.

Such changes in microhabitat and consequently to natural vegetation can be one of the contributing reasons for corresponding faunal changes in composition and density. Strongly edge-intolerant species may have higher densities in interiors compared to edge areas. Such negative edge effects may ultimately result in smaller functional habitats within retained remnants. If a patch is too small, the entire patch could be considered edge habitat, thus eliminating favored habitat altogether. Likewise, narrow corridors may not be functional for such species.

In contrast, some species benefit from or prefer edge and increase in abundance there. Edge offers access to resources within multiple habitats, helping a number of different species to thrive (Ries et al. 2004). The altered abundance or distribution of species near the edge can further affect species within habitat patches through negative interactions (Murcia 1995). Generalist predators and exotic species are often edge species and sometimes outcompete specialists and native species. They can also contribute to increased predation, competition, and parasitism on native interior species (Beier 1993; Murcia 1995). Smaller habitat fragments with higher edge-to-area ratios provide increased access of weedy species into fragments and can enhance movement of edge-loving exotic species and pests (Panetta and Hopkins 1991). Brown-headed cowbirds (*Molothrus ater*), raccoons (*Procyon lotor*), opossums (*Didelphis virginiana*), and crows (*Corvus* spp.) are examples of species that thrive in edge habitat and can have a large impact on forest interior species. Such edge species often act as nest predators, nest parasites, or cavity competitors of interior species, and they can contribute to decreased populations of ground nesting birds, forest songbirds, reptiles, and amphibians in remaining habitat fragments (e.g., Hartley and Hunter 1998; Dijak and Thompson 2000; Paton 1994). In the Purcell Mountains of British Columbia, increasing fragmentation and logging contributed to edge-loving deer (*Odocoileus virginianus*) moving into a region not previously occupied by deer, followed by mountain lions. The mountain lions now have found mountain caribou (*Rangifer tarandus caribou*), a naive prey and easy target, and are significantly contributing to the declining numbers of an already tenuous caribou population (Kinley and Apps 2001).

Species associated with humans can also reduce native biodiversity in retained fragments by invading the edges and smaller fragments. Domestic and feral animals, such as cats and dogs, damage native species populations in remaining habitat by chasing and preying upon them (Crooks and Soulé 1999). Similarly, pioneer plant species that did not previously occur cause problems for the original community of species. For example, such plants invaded up to 10 meters (33 feet) into tropical rain forest patches in Brazil (Sizer and Tanner 1999). Likewise, livestock can encroach upon remnant habitat and either directly compete with other herbivores or alter the habitat characteristics, leading to decreased use by wildlife (Kemper et al. 1999). Fragments can also experience direct increased human impacts, especially when adjacent to high human density. Increased edge allows humans greater access for recreation, including legal and illegal hunting of animals (Simberloff and Cox 1987). The habitat fragments may then become devoid of species that are sensitive to human activity or heavily hunted, which can have cascading impacts on the remaining ecological community.

Adjacent human activities can also affect the overall integrity of remaining habitat patches. In agricultural zones, pesticides and fertilizers can drift from the fields into habitat and affect flora and fauna alike. In addition, research on small fragments in South African renosterveld shrublands suggests that grazing, trampling, and fires in the human-occupied lands affect the remnant habitat (Kemper et al. 1999). Adjacent roads also can pollute retained habitat. For example, nitrogen deposition from air pollution threatens native grasses and associated biodiversity on serpentine soils in northern California. Because nitrogen is a limiting nutrient in serpentine soils, deposition of nitrogen changes the composition of grasses, facilitating generalists and contributing to a decline of serpentine-associated grasses and forbs (Weiss 1999).

In addition to fragmenting habitat, roads can also be a source of light, noise, and mortality for mobile species that occupy adjacent habitat patches. Studies indicate that light pollution can disorient animals, such as turtles, often imperiling species (e.g., Tuxbury and Salmon 2005). The combination of light, noise, and high human activity can deter the presence of some species in adjacent habitats. For example, female grizzly bears in Alberta, Canada, showed a negative relationship to areas with more vehicles, traffic noise, and human settlements. Male grizzly bears in the same study were more likely to use high-quality habitat near roads at night, especially where cover existed (Gibeau et al. 2002). While many species show a tendency to avoid roads with high vehicle usage, roadkill remains a large source of mortality for wildlife. Vehicle collisions with large mammals are increasing in developed countries, and several million collisions per year occur worldwide (Malo et al. 2004). Researchers in Saguaro National Park on the United States–Mexico border estimate that fifty thousand animals, including reptiles, amphibians, birds, and mammals, are killed by vehicles annually (Gerow et al. 2010). Dispersing and young individuals of many species may be particularly susceptible to becoming roadkill due to their inexperience, a factor that may be important to consider for rare or endangered mobile species. For example, automobiles killed four of nine radio-collared mountain lions observed dispersing in Southern California (Beier 1995). Species occupying patches next to or bisected by roads also have an elevated risk of mortality. Florida scrub jays (*Aphelocoma coerulescens*) with home ranges adjacent to roads had a significantly higher mortality rate than jays far from roads (Mumme et al. 2000).

Examples of reduced use of edge by nesting birds are increasingly being reported for a variety of habitats (see examples in Lidicker and Koenig 1996). The ovenbird (*Seiurus aurocapillus*), a deciduous forest interior species, nests successfully but at 40 percent lower densities within an edge zone that extends 150 meters into the forest (Ortega and Capen 1999). King et al. (1996) report that nest survival was higher inside of a 200-meter (656-foot) edge zone. In tropical forests of Uganda, some interior species suffered reduced densities up to 450 meters (1,476 feet) from edges (Dale et al. 2000). In boreal forests, also, four species of birds did not respond to territorial calls across a forest gap, and in fact responses declined starting 40 meters (131 feet) from the gap (Rail et al. 1997). Edge effects are not limited to forest birds, as demonstrated by a particularly informative study on bobolinks (*Dolichonyx oryzivorus*), a grassland species. Nesting density and success rates were reduced within 100 meters (328 feet) of forest or hedgerow edges, and birds with failed nests on edges moved farther from the edge when renesting; roads elicited decreased nesting densities but without reducing success rates. Finally, edges with old fields or pastures were not avoided, and nests on those edges enjoyed success as good as or better than interior grassland nests (Bollinger and Gavin 2004). Albrecht (2004) reported on edge effects in shrubby wet meadows bordering on crop fields in the Czech Republic. There, scarlet rose finches (*Carpodacus erythrinus*) exhibited 41 percent nest success within 100 to 200 meters of the edge compared to 83 percent survival for interior nests. Productivity of nestlings per capita was also 63 percent higher in interior nests. Driscoll and Donovan (2004) report an interesting complexity in which the reduced nesting success that characterizes wood thrushes (*Hylocichla mustelina*) nesting in forest edges bordering agricultural fields (in central New York) disappears if the fields are merely patches embedded in otherwise continuous forest.

A second kind of edge affects the ability of species to move through the edge. This effect is sometimes called edge permeability or edge hardness (Stamps et al. 1987). Permeability is a feature central to models that involve the relative amount of edge and its effects on dispersal dynamics. It is important, therefore, that this parameter be accurately measured. A report by Sieving et al. (2004) describes a fascinating situation in north-central Florida in which forest interior species of birds readily crossed an edge into open habitat when foraging in mixed-species flocks. Specifically, when the tufted titmouse (*Baeolophus bicolor*) was present, other species were more inclined to move into open habitat to mob a stuffed screech owl (*Otus asio*) equipped with recorded calls. The authors speculated that this facilitation resulted from a perception of reduced predator threat in the presence of the socially dominant and highly alert titmouse. This raises the interesting possibility that land managers in this region could actually increase the connectivity of a landscape for forest interior birds by encouraging the presence of tufted titmice.

The modeling of matrix permeability should logically incorporate edge permeability, matrix quality, and distance that must be traveled. This is not an easy assignment. If edge enhances dispersal, it is likely that this effect is caused by the edge itself. On the other hand, if crossing movements are inhibited, the effect may be a product of both edge characteristics and matrix characteristics. Relevant here for modelers as well would be knowledge of the ability of focal organisms to assess matrix quality and its dimensions. Poor judgment about the matrix that needs to be traversed will negatively affect the travelers' chances for success.

Both edge use and edge permeability can be expressed in a summarizing metric of performance measured across the edge (fig. 3.8). Performance can be measured in terms of behavior, physiology, or numbers, as seems appropriate. Two major categories of edge effects are immediately apparent. One of these comprises cases characterized by the absence of emergent properties for a given target species. In such cases, the response of organisms across the edge can be explained strictly by the organism's response pattern in the two adjacent communities separately (fig. 3.8A) and is referred to as a *matrix edge effect* (Lidicker and Peterson 1999). Diagnostic of this pattern, the organism's behavior on the edge changes abruptly from that associated with one community to that found in the other community, or the change is more gradual but symmetrical across the edge, or the change is accurately reflective of the intermediate character of conditions in the edge. In the last case, the response pattern corresponds to the degree of mixing of the two communities. The second type of edge effect is associated with emergent properties; that is, the edge will elicit a response that is not predictable from knowledge of the organism's behavior in the two community types when they are not adjacent to each other (fig. 3.8B). This category of behavior is called the *ecotonal edge effect*. The new response to edge may be an enhancement of function, a diminution of response, or an asymmetrical pattern not attributable directly to the mixing of two community types.

In practice it may not always be easy to distinguish these two categories of effects. For example, pattern b in figure 3.8A may appear quite similar to patterns in figure 3.8B. Operationally, the null hypothesis will be the matrix effect. It can, in principle, be predicted from known performance of a species in the two habitats separately, plus the measured blending of the communities on the edge. Any significant deviation from this prediction will suggest an ecotonal effect and indicate directions for further investigation of the phenomenon.

Matrix effects would be anticipated when two adjacent communities are quite different, such as with aquatic-terrestrial borders or with abrupt changes in soil type. In a study of forest-farmland edges in Illinois, Heske (1995) found only matrix effects for five species of carnivores and four species of rodents. On the other hand, Mills (1996), investigating small mammal responses to edges in Douglas fir (*Pseudotsuga menziesii*) forest with clear-cuts in Oregon, found that among four species, three different edge responses occurred. One species (Townsend's chipmunk, *Tamias townsendii*) exhibited a matrix effect; two species (red-backed vole, *Clethrionomys californicus*, and deer mouse, *Peromyscus maniculatus*) showed ecotonal behavior, with effects extending 45 meters (148 feet) into the forest; and one (Trowbridge's shrew, *Sorex trowbridgei*) showed no recognition of an edge at all. In the absence of fire, taiga voles (*Microtus xanthognathus*) in Alaska persist mainly on the edges of taiga forest and swales of horsetail (*Equisetum*). This combination of communities supplies both good burrowing conditions (above the permafrost) and a supply of rhizomes essential for successful overwintering (Wolff and Lidicker 1980). Similarly, adult male cotton rats (*Sigmodon hispidus*) in South Carolina preferentially overwinter where patches of good cover (*Rubus* sp.) adjoin grasslands, which presumably have better food resources (Lidicker et al. 1992). Sex-biased use of

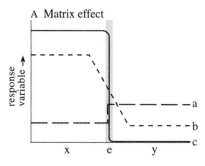
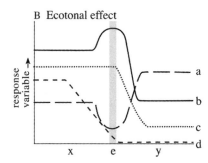

FIGURE 3.8 Idealized graphs representing matrix and ecotonal edge effects. Response variables (y axis) such as numbers and intensity of use are plotted against space traversing two habitat patches, x and y, and the edge (e) between them. The matrix effect (A) is depicted without emergent properties, and the ecotonal effect (B) shows the presence of emergent properties (Lidicker 1995).

matrix is also shown by the Eurasian flying squirrel (*Pteromys volans*) in Finland, where males readily crossed a matrix of low-quality forest unsuitable for breeding, whereas females foraged in the matrix but did not cross it (Selonen and Hanski 2003).

With all of the profound effects that edges can have, it is important to assess the amount of edge relative to the area of the habitat fragment. More compact patch shapes will have relatively less edge (fig. 3.7). For a given patch area, those with more edge will experience stronger edge effects overall and edge intolerant species will experience reduced patch size compared to those with proportionally less edge. Moreover, abrupt or "hard" edges such as those caused by human alterations of the landscape or aquatic/land boundaries will usually have stronger ecotonal effects. Although this ratio is clearly important, measuring it is often not so easy. Area can be relatively easily measured, but edge length is definitely tricky. The measured length of any boundary that is not absolutely straight will depend on the length of the measuring unit used. The more complex a boundary is, the shorter the measuring segment must be to capture that complexity. Fractal geometry is a method for analyzing the shape and hence the perimeter of complex objects, which is exactly what ecologists typically contend with. For complex edges, the measured length increases as the measuring segment decreases. The larger the measured perimeter relative to the size of the estimator segment, the more complex the boundary and the greater its fractal dimension. This relationship is illustrated for the country of Norway in figure 3.9; the same principle applies to small habitat patches. For the biologist and land manager, what is important is that the measuring segment be relevant to

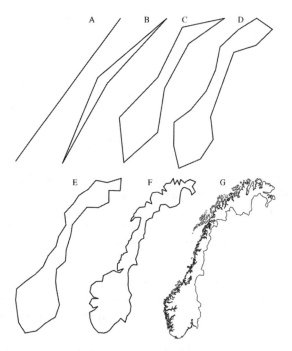

FIGURE 3.9 Representations of Norway's perimeter (from A to G) become increasingly complex, longer, and more accurate as the measuring segment used becomes shorter. Starting with the longest axis of the country (1,784 kilometers, or 1,115 miles), which is the longest possible measuring unit, A gives such a poor representation of the country that 62 percent of the line actually runs through Sweden and Finland. The line touches the Norwegian border in only four places. In the five representations of B–F, the measurement unit length is progressively halved. By the fourth halving (E), Norway is clearly recognizable and the measured perimeter is 2.3 times greater than in A. With further halving (to 1/32 of the starting length), we begin to see a suggestion of the numerous fjords and islands that characterize the western and northern coasts of the country (F). In G, the actual perimeter is shown in as much detail as possible given the scale of data resolution on the original map.

the organisms and processes of interest. For example, if one is interested in landscape boundaries of relevance to a wide-ranging carnivore such as a gray wolf (*Canis lupus*), an edge measure of tens of kilometers might be appropriate. On the other hand, for a beetle, one would have to use a length of a meter or less. When measuring perimeters from maps, the minimum length of measurement will be determined by the scale of data resolution. Boundary complexity not resolved by the available data will not be revealed and hence cannot contribute to boundary calculations.

Matrix as a Resource

Whereas a species' habitat may be largely confined to a particular community type, there may be situations in which matrix communities are used for something other than for traveling to another habitat patch. One possibility is that there may be occasional opportunities for abundant food resources in the matrix that can be exploited accordingly. An example is provided by the American marten (*Martes americana*), which, in the Sierra Nevada of California, requires old-growth forests for its existence. Nevertheless, it prefers forests within 60 meters (197 feet) of meadows, where it regularly exploits dense populations of *Microtus*. It rarely will penetrate more than 10 meters into open meadows but readily uses lodgepole pine (*Pinus contorta*) riparian areas with abundant herbaceous ground cover (Spencer et al. 1983). Meadow voles (*Microtus pennsylvanicus*) do something similar. When artificial patches of grassland are created by mowing, these voles prefer to live on the edges of patches, presumably because that allows them to forage into the mowed areas to take advantage of tender new grass sprouts stimulated by the mowing (Bowers and Dooley 1999). The case of bears making heavy use of salmon on a limited seasonal basis has already been mentioned.

The matrix may also offer access to some resource that is rarely needed by a species, perhaps seasonally, such as a forage plant rich in a scarce mineral, a favorable hibernation site, or access to a pollinator. Ricketts (2004) has shown that tropical forest fragments in proximity to coffee plantations (in Costa Rica) enhance pollinator activity in the coffee, especially within 100 meters of the edge. Edges between different communities may also be especially productive and consequently be attractive places for organisms to live in or near (see preceding discussion). Furthermore, we know that some species change habitats according to the season, making ready access to both habitats essential for survival. Rodents may move into the floodplain of a river or lake as it dries out, and then retract to higher ground when high-water conditions prevail (Sheppe 1972). The endangered salt marsh harvest mouse (*Reithrodontomys raviventris*) lives in the salt marshes of San Francisco Bay. But when the highest spring tides occur, it needs some adjacent upland to retreat to until the water levels decline. Norwegian lemmings (*Lemmus lemmus*) breed in alpine tundra but move into subalpine brush for winter (Kalela et al. 1971). There the snow is supported by the brush, providing more subnivean spaces for foraging. In temperate climates, commensal house mice (*Mus musculus*) live in agricultural fields in the summer and move to barns and houses for the winter. Rice rats (*Oryzomys palustris*) living in the Texas coastal salt marshes use upland areas as refuge from high tides, for limited foraging in the winter, and as a hideout for juveniles (Kruchek 2004). Many species perform altitudinal migrations between summer and winter ranges, as does the mule deer (*Odocoileus hemionus*) in California. Of course, there are long-distance migrants, especially among birds, bats, and ungulates, and including the monarch butterfly (*Danaus plexippus*). In those cases, however, the concept of patch-matrix interactions is not applicable, as the two or more required habitats are not adjacent.

Finally, there are cases in which organisms completely change habitats in the course of their ontogeny. Many amphibians, for example, need an aquatic habitat for reproduction and growth of larvae and then switch to a terrestrial habitat as metamorphosed juveniles. Bluegill sunfish (*Lepomis macrochirus*) live as juveniles in the vegetated littoral zone of ponds and then move into open water as they become adults. Juxtapositions of particular combinations of habitats are therefore often important for species, and failure to accommodate this in conservation planning can lead to extinctions or reduced population viabilities.

Matrix as Secondary Habitat

So far, we have considered situations in which the matrix was more or less favorable to dispersers but was not suitable for residency. It is possible, however, that the matrix may serve as secondary habitat. That is, it may be able to support resident individuals of a focal species, although at lower numbers or only intermittently. The reduced densities may be caused by higher mortality rates, reduced reproductive output, or both. Such a population may more easily become extinct and need to be replaced by immigrants from the good habitat patches. This situation has been called a *source-sink axis*. A source population is one in which reproduction is adequate to balance mortality and usually to export surplus individuals, as well. Such a population supplies the residents for a secondary habitat. Sink populations are those living in secondary habitat. Their reproductive output is generally insufficient to maintain the population in most years, and hence its continued existence depends on input of immigrants from source patches.

Occasionally, sink populations may produce enough offspring that a few will succeed in immigrating into a source population. This back dispersal may have significant genetic consequences for the recipient populations but is rarely demographically meaningful. There is one circumstance, however, when dispersal from sink to source is critically important. If the source population should unexpectedly become extinct, such as after a severe density crash following a peak in numbers, if a predator or disease should discover the patch and decimate the population, or if it is wiped out by a weather-based catastrophe, the patch can be recolonized from survivors in the sink habitat. A documented case of this happening in the bank vole (*Clethrionomys glareolus*) has been reported by Evans (1942). Another example is the Alabama beach mouse (*Peromyscus polionotus ammobates*), whose optimal habitat is on the sand dunes immediately inland from the coastal beaches of the Fort Morgan Peninsula. A few individuals live in more inland scrub (marginal) habitat. Following a hurricane, the optimal habitat can be completely destroyed by storm waves, leaving the residents of the interior scrub the only survivors available to recolonize the optimal dunes when that habitat is restored (B. J. Danielson, pers. comm., 2005).

The existence of source-sink dynamics raises a practical caution for land managers. Because population densities in the sink habitat may be as high, or almost as high, as in the source patches (Van Horne 1983), tragic mistakes could be made if density alone is used as a measure of habitat suitability. A manager might conclude that the source patch is not needed because of the large amount of secondary habitat containing good numbers of the target species. Destruction of the source patch would then lead to loss of all of the sink populations, because their long-term persistence was dependent on dispersers from the source population. It is essential, therefore, to ascertain whether populations in secondary habitat are in fact self-sustaining, that is, not a sink, before management decisions of that sort are made. Murphy (2001) describes an interesting case involving the eastern kingbird (*Tyrannus tyrannus*) in central New York. There the best habitat (riparian) serves as a source only intermittently. The bulk of the population nests in sink habitat (uplands and floodplains), which is only marginally poorer than the source. Reproduction in the sinks contributes importantly both by providing immigrants to the high-quality patches and by helping to sustain the overall population size in the face of slow long-term decline.

Still another way that matrix as secondary habitat can influence population dynamics has to do with its spatial extent or amount relative to the extent of high-quality habitat patches (Andrén 1994). This ratio of optimal to marginal (secondary) habitat patch area (ROMPA) has been postulated to be a major factor in determining the pattern of population dynamics exhibited by vole (*Microtus*) populations (e.g., Lidicker 1988, 2000; Ostfeld 1992; Delattre et al. 1999; fig. 3.10). The idea is that at low ratios (little optimal habitat) numbers of voles remain low, as reproduction in the favorable sites is just sufficient to maintain modest numbers within the patches and to provide enough dispersers to keep the widely spaced good patches connected. At high ratios, where optimal habitat predominates in

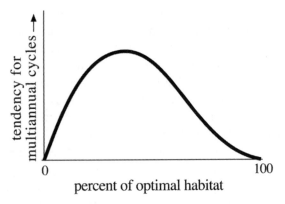

FIGURE 3.10 Hypothesized relationship between the ratio of optimal to marginal patch areas (ROMPA), here expressed as a percent of optimal habitat in the landscape (x axis), and the tendency of voles to show multiannual cycles in abundance (y axis) (Lidicker et al. 1992).

the landscape, overall densities will be high, since each patch is very productive and connectivity among patches is high. Such populations will likely show annual changes in numbers corresponding to increases during the breeding season and decreases during nonbreeding periods. Overall, they will average high densities. At intermediate values of ROMPA, multiannual cycles in abundance will likely be produced. How this works is that at low-density periods, voles will be restricted to the good habitat patches. As reproduction commences, dispersers will be produced, as voles generally are presaturation dispersers (chap. 2). These dispersers will find any empty good patches and begin to colonize the matrix of secondary habitat, as well. By the end of the breeding season, and perhaps even the next one, densities will remain low in all of the landscape, as successful reproduction in the optimal patches will have been funneled largely into dispersal. As the matrix and good patches begin to fill up in subsequent breeding seasons, densities will start to increase, producing moderate numbers, a pre-peak high. The landscape would then be poised so that in the following year densities would reach peak numbers. Such peaks are typically followed by a crash to low numbers, with survival restricted to the optimal areas. In this way multiannual cycles in abundance are generated. Of course, it is more complicated than this (Lidicker 2000), but this scenario illustrates how the matrix can play a key role in the demographic behavior of these rodents.

Evidence for the ROMPA effect comes mainly from studies on the California vole (*Microtus californicus*) and European common vole (*M. arvalis*; Lidicker 1992; DeLattre et al 1999). It is possible that ROMPA-like behavior is limited to voles, but that needs to be investigated further. In any case, the phenomenon is of some general interest because voles tend to be keystone species in many temperate, boreal, and arctic landscapes, where they form the basis of the predator food chain.

Matrix as a Sink or Stopper

If a matrix is readily permeable to dispersers, much of the reproductive output of a habitat patch may be funneled into emigration. When this output is balanced by immigration, there is no net loss in numbers and many benefits associated with this patchy type of metapopulation will be manifest (chap. 2). On the other hand, if the matrix is large relative to the size of patches, if it does not support a resident population of the target species, and especially if few other habitat patches are available to dispersers, such a matrix will function as a dispersal sink. Dispersers will be continuously enticed to leave, and there will be little or no immigration back into the patch. Especially in species with presaturation dispersal, the density of the patch population may be chronically depressed below what it could be if dispersal was more limited (fig. 3.11A).

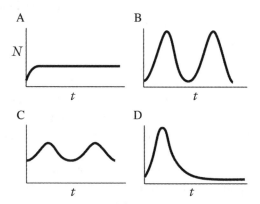

FIGURE 3.11 Hypothetical population density dynamics for a focal species living in an optimal habitat patch surrounded by four different kinds of matrix habitat. For all four graphs, N is the size of the focal population (y axis), and t is time (x axis). (A) Matrix is a large sink habitat; (B) matrix is a complete barrier to dispersal, effectively isolating the habitat patch; (C) matrix is inhabited by a generalist predator that feeds opportunistically in the focal patch; (D) focal patch includes a specialist predator on the focal species, but the matrix allows some dispersal of the predator in and out of the patch. Immigration would occur when prey densities are increasing rapidly, and emigration would happen when prey are extremely low in numbers.

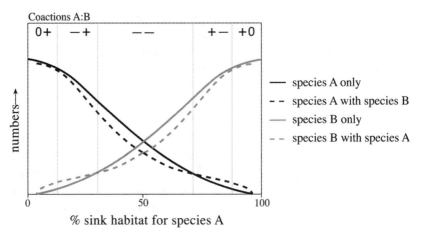

FIGURE 3.12 Schematic representation (Danielson 1991, and pers. comm.) of the interactions between two competitor species that share the same habitats. Each species is able to exclude the other from its source patches. However, the sink habitat for each species is the source habitat for the other. The x axis represents the proportion of a landscape occupied in the sink habitat for species A and is the reciprocal of the proportion of its source habitat; source habitat for species A is sink habitat for species B, and vice versa. The y axis is population numbers. For simplicity, the responses of both species are assumed to be symmetrically reciprocal. The resulting coactions (species A:B) are shown along the top bar with 0 for no interaction effect, + for a positive effect, and – for a negative effect.

The matrix as sink is a dangerous trap for dispersers. In an important contribution, Danielson (1991) argues that changes in the proportion of source to sink patches in a landscape (another ROMPA example) can significantly affect the nature of the interaction between two species. For example, if two species are potential competitors, but each does better in the sink habitat of the other, their interactions may change qualitatively as the proportion of sink to source varies. This occurs when species A excludes species B from significant portions of B's sink habitat, thereby facilitating B's search for its own source habitat and overall reducing its losses to the sink. Figure 3.12 illustrates how the interaction coefficient between two such species changes as the landscape shifts from 0 to 100 percent sink for one species, and concomitantly from 100 to 0 percent for the other. Their coaction goes from commensalism to exploitation in favor of the rare species, to competition, and finally to exploitation and commensalism favoring the other species, which has now become the rare one in the landscape.

If, on the other hand, the matrix surrounding a habitat patch is a complete barrier to dispersal (a "stopper"), the patch is effectively isolated from other patches and will suffer accordingly. Dispersal will become frustrated (Lidicker 1975), and numbers will build up within the patch, perhaps reaching high densities (fig. 3.11B). This could produce a chronically abundant population. However, if population growth leads to densities that exceed the patch's carrying capacity for that species, resource damage as well as opportunistic predators will lead to physiological and behavioral pathologies that will generate a likely crash to very low numbers. Such a population will show strongly fluctuating population numbers. The risk is that at the low points of these cycles in numbers, the population will be subject to the negative effects of small populations (Allee effects; chap. 2). Unless the patch is quite large, extinction of the deme is a likely outcome.

In an overall perspective, we can conclude that the matrix really does matter. In particular, we can appreciate that habitat fragments, the matrix, and the edges between them generate fascinating and important scenarios that can challenge ecologists, conservationists, and land managers worldwide. Moreover, in the future, we need to anticipate the environmental challenges that are inherent in large-scale and extremely complex conservation initiatives, even involving multinational cooperation (Ascensão et al. 2018; Chester et al. 2012).

Chapter 4

Approaches to Achieving
Habitat Connectivity

Habitat fragmentation resulting from increasing human activities in natural areas poses a great threat to the long-term conservation of biodiversity worldwide, as discussed in earlier chapters. Corridors are important because they can be a tool for maintaining viable populations of biota in fragmented landscapes by enhancing connectivity between larger core areas of habitat (Forman 1995; Kubeš 1996; Bennett 2003; Perault and Lomolino 2000). In recent decades, research on habitat connectivity and corridors has increased substantially. Likewise, as we discussed in chapter 1, policy and conservation implementation has increasingly incorporated corridors. Because corridors are being promoted and implemented worldwide, it is important that we use our knowledge from the research to date to direct our conservation efforts.

The purpose of this chapter is to examine methods for achieving connectivity among habitat fragments by synthesizing corridor research and real-world conservation examples. To begin, we review the broad use of the terms *connectivity* and *corridor* and introduce landscape elements that may function as corridors (also see table 1.1). We then review the different types of corridors that may improve connectivity and discuss the applicability of the corridor concept at various spatial scales. Finally, we introduce some of the known and theoretical benefits of corridors, both to conservation of biodiversity and to human quality of life.

What Is a Corridor?

A number of definitions of corridors and connectivity have been proposed over time. In earlier references, corridors were defined as routes that

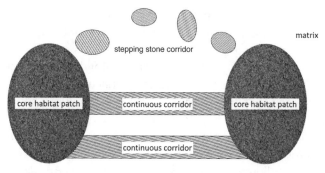

FIGURE 4.1   A graphical representation of two core habitat patches connected by two continuous corridors and a stepping-stone corridor, all surrounded by matrix (see chap. 3 for a definition of matrix).

enhanced speedy and unselective spread of biota between regions (Perault and Lomolino 2000). The Ninth US Circuit Court of Appeals defined corridors as "avenues along which wide-ranging animals can travel, plants can propagate, genetic interchange can occur, populations can move in response to environmental changes and natural disasters, and threatened species can be replenished from other areas" (Walker and Craighead 1997). Others have described corridors as linear landscape elements that connect two or more patches of natural habitat and function to facilitate movement (Soulé and Gilpin 1991). Connectivity has been said to describe the extent to which flora and fauna can move among patches, rather than requiring the linear landscape element described as a corridor (Hansson 1995; fig. 4.1). Some countries have developed their own legal definitions of corridors, emphasizing different objectives and approaches to biodiversity conservation (Jongman 2004; Tanzania 2018). Sometimes corridors are referred to as habitat corridors, wildlife corridors, linkages, or ecological structures. They can be part of ecological or habitat networks, which encompass core areas, corridors, and connecting nodes; or they can be synonymous with greenways, greenbelts, or open space, depending on the context.

For this book, we define a corridor as any space that facilitates the movement of populations, individuals, gametes or propagules, and plant parts capable of vegetative reproduction in a matter of minutes, hours, or over multiple generations of a species (box 4.1). Corridors may encompass altered or natural areas of vegetation and provide connectivity that allows biota to spread or move among habitat fragments through areas otherwise devoid of preferred habitat. Landscape elements that function as corridors may also serve other purposes, providing aesthetic amenities, ecosystem service values, cultural heritage protection, and recreational opportunities. Some landscape elements provide connectivity for biota without being designated for that purpose.

Corridors can be viewed over broad spatial and temporal scales. At one extreme, we have corridors that connect continents, including the Isthmus of Panama, and land bridges such as the Bering Strait that appear and disappear with sea level changes over time scales of millions of years. On a subcontinental scale there are efforts to connect forest communities from southern Mexico into Panama (Kaiser 2001) and a grand plan called Two Countries One Forest that seeks to keep the northern Appalachian Mountains connected in eastern North America. In reality, these large landscape efforts are intended to create protected area networks that are composed of a series of protected areas and corridors at a more regional scale. A successful corridor on a regional scale is the Braulio Carrillo National Park in Costa Rica, which was established to connect Atlantic lowland rain forest (La Selva Biological Station) with high-altitude protected areas. Today, Costa Rica represents a country that has planned and worked to conserve a network of corridors and protected areas throughout the country (Fagan et al. 2016).

Fragmented communities take time to equilibrate to current conditions, generating an extinction debt or lag effects. Lindborg and Eriksson (2004) found that species richness in fragments of seminatural grasslands in Sweden reflected patterns of connectivity one hundred years ago rather than the current or recent landscape configuration. Considering this, proactive planning to maintain connectivity over time can be important.

Many land planning efforts involve a particular site making connections and can range in size from less than one to a few hundred kilometers. The Donaghy's Corridor project in Australia involves a 1.2-kilometer (.74-mile) long corridor designed to link the World Heritage reserves of the Lake Barrine section of Crater Lakes National Park with the Gadgarra State Forest in Queensland's tropical forest (Tucker 2000; https://site.emr projectsummaries.org/2016/03/05/donaghys-corridor-restoring-tropical -forest-connectivity/). Since inception of the project, over one hundred species of rain forest plants have been established from local seed sources along Toohey Creek, a waterway connecting the two protected areas that run across Donaghy's property. Local volunteers, private landowners, and government agencies worked together to maximize planting success. The planted corridor is now protected with a conservation agreement that establishes the area as a nature refuge and excludes livestock. The corridor is regularly monitored and is being maintained. The rapid restoration of vegetation has clearly contributed to movement and occupation of the riparian corridor by smaller mammals, and long-distance mammal movements have been detected using genetic analyses.

At even smaller scales, a corridor may be a structure only tens of meters or yards wide that crosses a road or canal, or goes through the matrix. For example, in British Columbia, a wooden passageway under Highway 31A is helping western toads (*Anaxyrus boreas*, aka *Bufo boreas*) move safely from breeding lakes to upland habitat (Valhalla Wilderness Society 2017). Small-scale corridors might be trails or paths that guide organisms through thick vegetation or over difficult topographies. Even odor trails, pathways marked or scented by wildlife, such as those established by mammals and ants, could qualify as corridors (Kozakiewicz and Szacki 1995; Liro and Szacki 1994). They are linear, enhance the movement of organisms, and may connect habitat patches. Although conservation planners generally deal with medium-scale corridors, functionally meaningful corridors may transcend a broad range of spatial and temporal scales.

Various countries, collaborative conservation efforts, NGOs and others have created their own definitions of corridors. Global standards for landscape connectivity and corridors are needed to direct policy and practice just as the world has a standard definition of protected areas (Dudley 2008). The IUCN Connectivity Specialist Group initiated this effort with the goal to support increasing policy and practice around corridors and to ensure that the best science is accessible and incorporated in connectivity conservation efforts around the world. More information on this group and this effort is available at https://www.iucn.org/theme/protected-areas/wcpa/what-we-do/connectivity-conservation.

Types of Corridors

Different kinds of landscape elements can enhance connectivity (fig. 4.2). Many elements serve as corridors that are not explicitly designed for the purpose, such as roadside vegetation, hedgerows, and greenways. In other cases, corridors are purposely retained, maintained, and restored to facilitate landscape connectivity for individual species, groups of species, or entire ecological communities. In later chapters, especially chapters 6 and 7, we will address some factors that make these landscape elements more or less effective and sometimes even detrimental to native species.

Before we dive into the different types of corridors we need to clarify two concepts: structural and functional connectivity. Structural connectivity is a measure of habitat permeability based on the physical features and arrangements of habitat patches, disturbances, and other landscape elements presumed to be important for organisms to move through their environment (box 1.1). It can be recognized on aerial photos or satellite images of Earth's surface, and it is pretty straightforward to understand, measure, and communicate. Functional connectivity, on the other hand, is the degree to which evidence indicates that landscapes or seascapes facilitate or impede the movement of organisms (box 1.1). It is connectivity from a species' perspective with the consequence that a landscape that is connected for one species may be full of barriers for another. A stark example would be a cliff: it is perfectly manageable for a lizard to move up the rocks, but impassable to (most) humans. Functional connectivity can be difficult to measure because it is species-specific and depends not only on the species' movement ability in relation to the landscape features, but also on other factors such as the individual's internal motivation to move and the level of risk encountered when traveling (Bélisle 2005; Elliot et al. 2014). There now are several approaches to studying functional connectivity, including behavioral experiments such as translocation studies and resource selection function models that describe habitat suitability (see chap. 7).

In some landscapes, such as in the agricultural regions of Europe in which remnant forest patches in a matrix of fields and pastures are connected by hedgerows or riparian corridors, structural connectivity equates functional connectivity for forest-dependent species. In other cases, structural connectivity may exist without functional connectivity. For plant species with poor dispersal abilities, a corridor that is visible on a map may be too long or too narrow to allow dispersal between the structurally linked core areas. Similarly, habitat patches may be functionally connected for some species without any visible landscape linkages. This is often the case for mobile species such as many birds or habitat generalists that are able to move through the matrix, provided that the expanse of matrix between habitat patches is not too vast.

Early on in connectivity conservation, corridors were mostly designed based on structural connectivity. Only in the early 2000s, the argument was made that taking species' needs into account will result in better corridors, and methods were developed and employed to measure functional connectivity and use it for corridor design (e.g., Chetkiewicz et al. 2006). A paradigm shift occurred again when ecologists realized that connected landscapes are essential for facilitating range shifts of many species as an adaptation to a changing climate. Designing structural connectivity as a coarse-filter approach that accommodates the need of many species is now often recommended as the first step; the connectivity needs of specialist species that slip through the coarse filter should be considered in the second step (e.g., Beier 2012; chap. 8).

Landscape elements that enhance connectivity but exist for other reasons are de facto corridors. These are often locations where optimal habitat or even marginal habitat is left undisturbed, providing a different vegetative structure from the surrounding matrix. In highly modified environments, such remnant habitat may be disturbed, invaded by exotics, or sparsely vegetated, but some plants and animals may still be able to disperse through it or survive within it. Fencerows, windbreaks, roadside vegetation, and ditches may serve to enhance connectivity (Bennett 1990; Kubeš 1996; Kasten et al. 2016).

Roadside corridors (vegetation strips along roads) are an example of de facto habitat that can have both positive and negative effects on connectivity for native biota. They can offer habitat to both plants and animals and can act as a conduit for movement among habitat patches. The presence of native vegetation should enhance the ability of roadside corridors to act both as a conduit and as supplementary habitat. One example where roadsides provide connectivity is in Southern California, where revegetated highway edges enhance connectivity for native rodents and urban-adapted birds between habitat patches, although more sensitive bird species do not use them (Bolger et al. 2001). However, downsides to roadside corridors may include not serving specialist species and drawing species to roads, where they may ultimately be killed by automobiles.

In agricultural landscapes, fencerows, unmanaged ditches, streams, and shelterbelts can all serve as de facto corridors. Often native or nonnative vegetation along fences is not managed, offering vertical vegetative structure that some species of plants and animals use to live in or travel through. One experimental study, for example, showed that mice (*Peromyscus leucopus*) preferred structurally complex fencerows over other landscape elements (Merriam and Lanoue 1990). Similarly, vegetation along ditches and streams is often left and can serve as both habitat and a conduit for species traveling among larger habitat patches. In northern California, detection of mammalian predators was elevenfold higher along streams than in vineyards (Hilty and Merenlender 2004), and bird and small mammal diversity was higher along structurally complex and wider remnant vegetation adjacent to streams than in denuded riparian areas or vineyards (Hilty 2001). Shelterbelts, tree rows planted to prevent soil drift and delay snowmelt on fields, are another element in agricultural landscapes used by some species of wildlife. For example, most movements of studied migratory bird

FIGURE 4.2 Different landscape elements can function as corridors. (A) A tunnel designed for stopping amphibians getting hit on the road; (B) A hedgerow in an agricultural landscape; (C) Aerial view of field windbreaks in North Dakota; (D) A section of the European Green Belt, a corridor connecting the Baltic Sea to

the Adriatic Sea, located where the Iron Curtain used to be. ([A] Photo by John Cleckler, USFWS (tunnel). [B] Photo by Brian Keeley. [C] Photo by Erwin Cole, USDA Natural Resources Conservation Service. [D] Photo by Klaus Leidorf.)

species that breed in agricultural shelterbelts in North Dakota were found to occur in shelterbelts and in connected habitat patches rather than unconnected sites (Haas 1995). Restoration of corridors in agricultural areas can boost pollinator services (Adams 2016). These studies all suggest that vegetation structures within agricultural landscapes can function as movement corridors and even provide habitat for some species of wildlife. These linear elements can be problematic for native wildlife, however, by inhibiting movement of some species and harboring or boosting the presence in the landscape of exotic species and predators that might not otherwise be able to persist (see chap. 6 for more discussion on pitfalls of corridors).

Identifying landscape elements that may already, or could with some restoration, serve as corridors, is important in planning species' conservation. Variables such as dimension, vegetative structure, and overall landscape context will affect their utility (chap. 5).

For the most part, the landscape elements described here are most likely to facilitate generalist species and may not serve specialists; they could also cause a net loss for some species and result in mortalities (chap. 6). Still, in heavily human impacted landscapes where setting aside or restoring larger corridors is not feasible, enhancement of de facto corridors may help retain what species do remain in the landscape. Because most evidence of de facto corridor use is based on short-term observational studies, future evaluation and monitoring will be important to refine our understanding of their utility.

Even when corridors are planned, connectivity for biodiversity may be only one of the purposes. For example, greenways, also referred to as open-space systems or greenbelts, can potentially provide connectivity. These are areas that are set aside for recreation, cultural events, and ecosystem services, usually within a densely developed landscape and often in cities, suburbs, and the adjacent countryside. They can include everything from natural habitat to farmland to areas unfit for development because of susceptibility to floods, topography, or other factors (Zube 1995; Arendt 2004).

Frederick Law Olmsted is credited as the founder of the greenway concept. In the early 1900s when he planned the Boston Emerald Necklace, Boston's park system, he used greenways to link the city's parks (Fábos 2004). The 1987 report of the US President's Commission on Americans Outdoors launched today's greenway movement by describing a future vision of greenways that offer people access to open spaces near where they live, as well as linkages to rural and urban open spaces (Fábos 2004). Current greenway planning around the world generally concentrates on multiple goals including water management, cultural resources, and recreation.

Attainment of connectivity for biodiversity in the context of these other goals depends on site-specific variables and the species of focus. In some cases, human activities may inhibit the corridor's effectiveness for wildlife (Haight et al. 2005). Trampling of vegetation, purposeful or inadvertent introduction of nonnative species, and wildlife harassment by pets are some examples of factors that may impair a greenway's connectivity value for biodiversity. Even so, greenways not explicitly focusing on conservation should be evaluated for their potential as corridors by providing habitat, acting as conduits, and even harboring source populations. For more on the history, functions, design, and further examples of greenways we refer the reader to the book *Designing Greenways: Sustainable Landscapes for Nature and People* (Hellmund and Smith 2013). This book emphasizes the need for greenway planning to include landscape-level considerations, promoting conservation by using design guidelines developed with sound ecological principles such as those discussed in chapters 2 and 5. In general, the smaller, more heavily used and less biologically intact greenways, will likely have less biodiversity value compared to larger, more intact greenways with less human activity. Most greenway planning is occurring in areas where species sensitive to human disturbance have already disappeared, so that the greenways are likely serving more human-adapted and generalist species.

Increasingly, research and engineering efforts are developing new methods to mitigate the risk of wildlife deaths caused by roads and to develop mechanisms to retain connectivity despite roads. Roads are a ubiquitous landscape feature throughout the world and have the potential to sever once-connected wildlife populations. Increasingly, studies are showing that roads limit genetic interchange and contribute to wildlife mortalities for many species. Figure 4.3 illustrates the density of roads in Europe. Such infrastructure can decrease the abundance of species (Torres et al. 2016). Even if species are willing to try to cross roads, they may end up as roadkill. For example, within one month, roadkills tallied in five US states included 15,000 reptiles and amphibians, 48,000 mammals, and 77,000 birds; and those are likely underestimates (Havlick 2004). Improving connectivity across roads can maintain genetic interchange, decrease wildlife deaths, and diminish vehicular damage. To those ends, crossing structures often are designed to serve a variety of species.

Researchers in Europe found that one relatively cost-effective way to reduce bird deaths was to build an incline on the side of the road at least

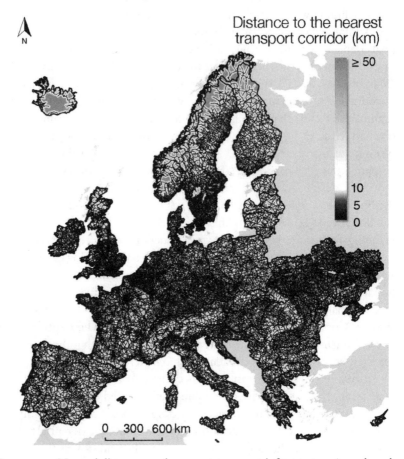

FIGURE 4.3 Mapped distances to the nearest transport infrastructure (paved roads and railways) in Europe based on the small-scale pan-European topographic dataset EGM v7.0 (2014) using a Lambert azimuthal equal area projection. Distances were quantified at a resolution of 50 m for inland Europe and islands larger than 3,000 km^2 and ranged from 0 to 83.5 km (Torres et al. 2016).

1.5 meters (5 feet) in height to provide birds lift above oncoming cars (Van Bohemen 2002). Other species might require underpasses or overpasses. When highways are being upgraded or new roads are being installed, it is a good opportunity to assess potential impacts to wildlife and plants and mitigate for them. Extensive research, strategy, and design of road-crossing structures have been conducted in Banff National Park in Canada (e.g., Clevenger and Waltho 2005; fig. 5.3), in the Everglades of Florida (e.g., Foster and Humphrey 1995), in the Netherlands (e.g., Van Bohemen 1996), and elsewhere. These efforts provide increasingly specific design specifications for a range of species, from tunnels for amphibians to overpasses for ungulates and bears (Forman et al. 2003).

Some corridors focus solely and explicitly on ecological needs. They may buffer linear landscape elements of particular importance to biodiversity such as riparian zones, conserve priority areas for individual species conservation, or promote community integrity across broad regions. In contrast to de facto corridors, these are often designed using scientific principles, biological surveys, and models to help determine landscape location. The major assumption in designating such corridors is that they will enhance conservation by promoting one or more connectivity goals (box 4.1; see chap. 5 on corridor design).

Riparian Areas

Arguably one of the most important landscape elements for biodiversity and connectivity is the riparian corridor. Riparian corridors are made up of vegetation growing adjacent to streams and rivers that are sometimes retained in human-dominated landscapes (fig. 4.4). Riparian areas support a disproportionately large amount of biodiversity and ecological processes compared to other landscape elements, and conserving these sites can provide multiple natural resource benefits (Hauer et al. 2016). Increased focus on riparian connectivity may help lead to a conservation network (Fremier et al. 2015). Maintaining functional river systems including adequate vegetation along the rivers also protects in-stream biota by controlling erosion and providing shade to keep water temperatures cool. Retaining buffers along streams can benefit terrestrial biota as well. Bird species richness and abundance appear to be greater where adequate riparian buffers are retained, according to studies of forests ranging from boreal forests in Sweden to riparian forests in California and Georgia (Hodges and Krementz 1996; Hilty 2001; Hylander et al. 2004). Buffer zones around wetlands and riparian habitats also have been found to be important for amphibian and reptile populations (Semlitsch and Bodie 2003), and are often selected as travel corridors by carnivores. Riparian buffers may be explicitly retained for conserving species, or they may be de facto, the result of policies such as those oriented toward water quality enhancement.

Establishing and preserving vibrant riparian corridors is a good approach to conserving species, but conservation of riparian corridors alone

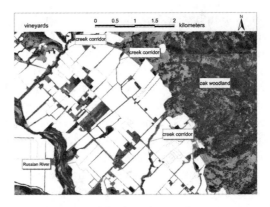

FIGURE 4.4 Stream corridors composed of remnant riparian vegetation in Alexander Valley's oak woodland and vineyard landscape (Sonoma County, California). To emphasize the vineyard matrix, a map compiled in 1997 of vineyard blocks (white) has been superimposed onto the ortho photo.

is inadequate. First, the landscape context is important: corridors within a less intact landscape will be less effective. Second, stream and river corridors can lead wildlife into areas of human activity instead of to other habitat patches. When this occurs, corridors are essentially dead ends. In Bozeman, Montana, black bears (*Ursus americanus*) follow streams into town each year, only to discover that streams are eventually funneled underground or that their riparian vegetation has been stripped away. The bears suddenly find themselves in human neighborhoods and often become disoriented or begin eating trash (Haines 2004; McMillion 2004).

Corridors for Individual Species Conservation

In addition to protecting specific landscape elements such as riparian areas, corridors may be mandated in individual species' management plans. For example, dispersal corridors were proposed and identified through logging areas for spotted owls (*Strix occidentalis caurina*) in the US Northwest (USDA/USDI 1994). Similarly, an important part of the recovery of panthers (*Puma concolor coryi*) in Florida has been to identify and create safe corridors for them to move among remaining habitat fragments (Cramer and Portier 2001; Frakes et al. 2015). This has involved both selecting specific locations such as road underpasses and conducting broad-scale landscape

BOX 4.1

PLANNING A CORRIDOR: BIODIVERSITY, SCALE, AND GOALS

The following lists contain the hierarchical levels of biodiversity commonly considered when planning a corridor, the scales at which corridors are implemented, and the potential goals that can result from corridor implementation.

Levels of Biodiversity
  Gametes (mature male or female haploid germ cell that can unite with
    another of the opposite sex)
  Propagules (pollen and seeds)
  Individual (of a species)
  Deme (of a species)
  Species
  Community
  Ecoscape (landscape or seascape)
Spatial Scale (of corridor)
  Local (e.g., underpass)
  Regional (e.g., river corridor)
  Continental or cross-continental (e.g., mountain range)
Potential Goals
  Daily movement (e.g., access to daily resources)
  Seasonal movement (e.g., migration and access to seasonally available
    resources)
  Dispersal (e.g., genetic exchange, mate finding)
  Habitat (e.g., wide corridor)
  Long-term species persistence (e.g., adaptation to global warming)

connectivity analyses. Road underpasses along Florida's Interstate 75, also known as Alligator Alley, enhance connectivity and reduce panther deaths on roads because fencing inhibits road crossing and guides the animals to the underpasses. Farther north in Florida, the Pinhook Swamp was identified as a regional link connecting the Osceola National Forest to the Okefenokee National Wildlife Refuge for panthers as well as black bears (Harris et al. 1996).

Whether designating a planned or an unplanned corridor to help conserve a species or a community, the purpose is to enhance functional connectivity (box 4.1). What can provide connectivity for one species may act as a barrier to movement for other species. To avoid unintended negative consequences, it is important to evaluate how a corridor may affect species with different life histories in the landscape. Later in this chapter, we discuss benefits of corridors to conservation of biological diversity, and in chapter 6 we discuss potential negative consequences.

Corridor Complexities

As background to discussing goals that corridor projects seek to achieve, we briefly review a few points about corridors that often cause confusion. First, corridors can target some or all levels of biodiversity (box 4.1). Second, they occur at many different spatial scales. For example, some corridors may be a few meters or yards long to facilitate movement of smaller species, while others may span one or several countries to provide a conduit for biotic movements over a long time period (Norton and Nix 1991). As will be discussed in chapter 7, researchers and land-use planners must be explicit about the scale of a proposed corridor and the species that it is designed to benefit. Third, corridors may provide connectivity for one species and not another due to species' different operational scales and habitat requirements. Finally, because the integrity of a community may affect individual species' survival, connectivity planning for entire communities should be considered where possible, rather than focusing on individual species. With the above issues addressed, corridors may achieve any of five connectivity objectives: daily movements, dispersal, seasonal movements, habitat connectivity, and long-term species persistence (box 4.1).

Individuals often move from one patch to another each day. One common reason for daily trips is to access resources such as water. Corridors can ensure that travel routes to necessary resources are not severed to avert imperiling a population. Corridors may also protect their key habitats or help animals avoid predation they might suffer in crossing modified habitat or human-dominated landscapes (Noss 1987). Riparian buffer zones, for example, provide habitat for otters (*Lontra canadensis*) and allow them to move along stream courses. Corridors can allow individual bears to move around a large home range (Jordán 2000).

Enhancing survivorship of dispersers is another common goal. Dispersal among relatively isolated populations in a metapopulation is important for preserving genetic diversity and reestablishing populations in habitat fragments where a species has become locally extinct (chap. 3; McCullough et al. 1996). Understanding how organisms move across the landscape, including the distances traveled will help in conserving key landscape elements and potential travel pathways. A scientific review illuminated that

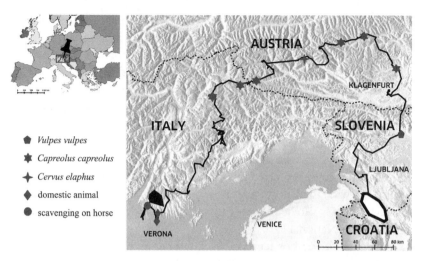

Vulpes vulpes

Capreolus capreolus

Cervus elaphus

domestic animal

scavenging on horse

FIGURE 4.5 Long-distance dispersal of male grey wolf Slavc from a transboundary pack in the Dinaric Mountains through Slovenia and Austria to the Eastern Alps in Italy between December 20, 2011, and March 26, 2012. The natal territory in the Dinaric Mountains is marked in white (442 km^2 from July 17 to December 19, 2011) and the new territory in the Italian Alps in black (150 km^2 from March 27 to August 27, 2012). The cumulative distance of the dispersal route is 1,176 kilometers, and the straight-line distance between the natal and new territory is 233 kilometers. See legend for detected predation/scavenging event locations along the dispersal route (Ražen et al. 2016).

animal movements in human-dominated landscapes are shorter than in undisturbed areas (Tucker et al. 2018). Unfortunately, few studies have adequately documented dispersal events, although global positioning systems (GPS) are making such studies more feasible for larger animals (Kays et al. 2015). A Greater Yellowstone Ecosystem study using GPS collars for wolverines (*Gulo gulo*), a species that exists naturally at extremely low densities, found that dispersing individuals may move hundreds of kilometers or miles when looking for mates or establishing a new territory. One individual moved from the Greater Yellowstone Ecosystem in Wyoming to Rocky Mountain National Park in Colorado, and years later was killed in North Dakota (Inman et al. 2004; https://www.huffingtonpost.ca/entry/wolverine-killed-north-dakota_us_573545ffe4b060aa7819ef8a). Likewise, a wolf fitted with a GPS collar in Slovenia undertook a 2,000-kilometer (1,243-mile) journey over the Alps to Italy, where he found a mate and successfully raised a litter of pups (Ražen et al. 2016; fig. 4.5).

In addition to corridors being important for dispersal, some individuals, demes, and species require corridor conservation to facilitate annual or seasonal migrations. For example, pronghorn (*Antilocapra americana*) migrate up to 270 kilometers (167 miles) between Grand Teton National Park and the Red Desert in Wyoming, a route used for over six thousand years. This route has several natural topographic bottlenecks (fig. 4.6; Berger and Cain 2014). Human developments such as roads, rural residences, and energy development could accidentally and permanently cut off this migration route. Researchers characterized the exact migration route, and this led to the first federally protected migration corridor 70 kilometers long and 2 kilometers wide to protect the longest terrestrial migration in the lower 48 United States. It also helps ensure that this species does not disappear from Grand Teton National Park.

Beyond the local or regional level, efforts are being made to identify and maintain connectivity on a much coarser level, such as across continents. Connectivity on this scale focuses less on individual species survival and more on keeping healthy ecological communities and their landscapes connected. Generally speaking, such connectivity is achieved through functional ecological networks that are a mix of protected areas and corridors, a configuration that increases the total amount of habitat available for wildlife. In addition to providing habitat for wildlife and connectivity among core habitat areas, these webs of cross-regional ecological networks may allow unidirectional range shifts in the event of massive global change, such as climatic warming, thereby increasing the chances of long-term species persistence (Groves et al. 2012). Such large landscape networks may also serve to facilitate processes such as the movement of chemicals (e.g., nutrients and pollutants), energy (in the form of organisms), and materials (e.g., sediments and debris).

Examples of such connectivity plans that amalgamate into ecological networks are the Paseo Pantera Project in Central America and the Great Eastern Ranges Corridor in Australia, which addresses connectivity issues across Eastern Australia. When introduced in 1990, the Paseo Pantera Project (www.afn.org/~wcsfl/pp.htm) represented a new approach to conservation in Central America with the vision of conserving a Central American Biotic Corridor more than 900 kilometers (558 miles) long. The Wildlife Conservation Society and the Caribbean Conservation Corporation initially spearheaded the effort, with seven countries agreeing to cooperate as signatories. Now many other groups have joined the effort, which is also referred to as the Mesoamerican Biological Corridor. Researchers modeled the optimal biological connections to assess the best potential

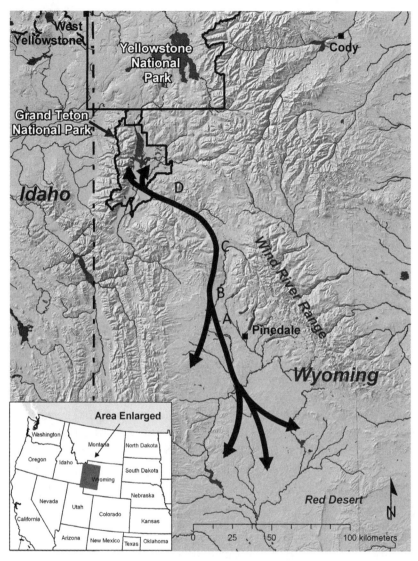

FIGURE 4.6 The migration route of pronghorn (*Antilocapra americana*) of up to 270 kilometers (167 miles) between Grand Teton National Park and the Red Desert in Wyoming. Natural topographic bottlenecks along the route are noted: (a) Trapper's Point, a 0.8-kilometer (0.5-mile) natural constriction that has been used by pronghorn for 5,800–6,800 years; (b) Sagebrush Gap, a corridor 100–400 meters (328–1,312 feet) wide between a river floodplain and forest; (c) a high-elevation divide; and (d) a 100–200-meter constriction between sandstone cliffs, a road, and the Gros Ventre River (Berger 2004; permission to use by Joel Berger).

configurations of this network of connected protected areas. The strategy was to buy land that would result in a series of linked protected areas. Over time, sustainable development priorities have hampered progress on the biological targets, such that progress for connectivity has occurred in only a few countries (Holland 2012).

Similar to Paseo Pantera, the Great Eastern Ranges connectivity conservation corridor is intended to ensure that adaptation in response to climate change can occur (http://www.greateasternranges.org.au/). The endeavor is an evolving partnership of many different entities from governmental and nongovernmental organizations, businesses, and private landowners. The landscape is currently mostly unfragmented and key connectivity areas have been mapped out. A major strength is the vision of a connected landscape, empowering the stakeholders to engage in conservation and plan for climate change with the goal of helping humans and their natural support system. It has also served as an inspiration to drive Australia toward legislation that would institutionalize continental-scale connectivity conservation. The latter is still aspirational but continues to be promoted (Pulsford et al. 2012).

In North America, the Wildlands Network spearheaded initial scientific and design work on the Yellowstone to Yukon Conservation Initiative (see the case study in chap. 10); the Maine Wildlands Network; the Sierra Madre Occidental Biological Corridor in Mexico; and the Sky Islands Wildland Network in the US Southwest—all ambitious conservation efforts on a large spatial scale. In these efforts, the vision is to identify, retain, and restore wildland networks in various regions across North America using science-based tools to select critical sites (Noss 2003). Many new large landscape efforts continue to form across North America (http://largeland scapes.org/projects/projects/large-landscape-practitioners-network/) and the world (e.g., http://www.globescapes.org/).

Biological Benefits

Ecological connectivity may benefit biota in a number of ways. Maintaining and restoring connectivity often means maintenance or enhancement of natural habitat, so one obvious benefit can be more available habitat. Additional habitat should permit greater species richness, as predicted by the theory of island biogeography (see discussion in chap. 3). In support of that theory, a study in French Guiana revealed increased species richness of bat communities within forest patches connected by forested corridors compared to isolated patches (Brosset et al. 1996). Additional habitat also means the potential for more individuals within a species by providing more home range sites. In Alberta, Canada, a study of forest birds revealed that forest corridors contained many resident adult birds and a large number of juveniles, suggesting that the corridors serve both as habitat and as conduits for dispersal following logging (Machtans et al. 1996). Increased numbers of individuals within a population can be very important for conservation of small populations that are constrained by human activities. In some cases, corridors may support actively reproducing populations that can disperse to other populations (Perault and Lomolino 2000). However, quality of habitat can range within corridors from intact to impacted, such as by edge effects including human activities. As we discuss in chapters 3 and 6, these factors can affect the ability of populations to survive or even pass through corridors.

Beyond the benefits of additional habitat, corridors can increase overall species' persistence compared to equivalent patches of habitat that lack connectivity. They do this by assisting in the movement of species among otherwise separate populations. By facilitating movements of species, corridors may serve to buffer groups of small populations from extinction by increasing persistence of species within a given patch and enabling recolonization of a patch after local extinction (chap. 3; Laurance 1991; Beier and Loe 1992; Newmark et al. 2017). For example, a study in Queensland, Australia, found higher survival of tropical rain forest mammals in forest fragments connected by corridors, indicating that corridors enhanced persistence (Laurance 1995). Similarly, in South Carolina, Bachman's sparrow (*Aimophila aestivalis*) was less likely to colonize isolated patches of habitat than patches that had a higher degree of connectivity (Dunning et al. 1995).

Increased connectivity can facilitate dispersal and thus increase genetic interchange among both plant and animal populations, reducing the risks of inbreeding depression (Beier and Loe 1992; Bennett 1999). Dispersal may increase levels of genetic variability within populations and reduce fixed differences between populations (chap. 3; Christie and Knowles 2015). Even a low level of gene flow will avoid the chance fixation of deleterious genetic traits (Hedrick 1996). Genetic variability can increase species resilience to environmental change (Christie and Knowles 2015), although some hypothetical concerns about dilution of locally adapted genes have been posed (chap. 6). Spatial analyses of the impacts of fragmentation on a marsupial carnivore, *Antechinus agilis*, in Australia examined gene flow in continuous habitats, fragmented habitats, and fragmented habitats with corridors (Banks et al. 2005). The results offered evidence that the surrounding matrix was a barrier and that corridors provided increased gene flow among connected fragments.

Some of the most well-studied barrier crossing structures are along the Trans-Canada Highway in Banff National Park. Concerns about the highway dividing the park led to the construction of thirty-eight underpasses and six wildlife overpasses that have helped more than 152,000 animals cross the road safely (http://calgaryherald.com/news/local-news/how-do -the-animals-cross-the-road-in-banff-national-park; fig. 4.7). Long-term research has demonstrated that the structures are used by many different species such as but not limited to grizzly bear, black bear, wolf (*Canis lupus*), lynx (*Lynx canadensis*), elk (*Cervus elaphus*), and moose (*Alces alces*) (Clevenger and Waltho 2005). Detailed research demonstrates differential use by sex of species, such as grizzly bears, which is important information for ensuring that structures designed to facilitate population level connectivity are most robust (Ford et al. 2017). Given increasing evidence that crossing structures are used by wildlife, many other countries are considering the provision of wildlife crossing structures in road construction and reconstruction projects. The Netherlands has developed *ecoducts*, passageways that facilitate wildlife movement over roads, which are successfully used by red deer (*Cervus elaphus*), as well as other species (Friedman 1997). The Montana Department of Transportation has completed over forty fish and wildlife crossing structures along a 90-kilometer (56-mile) stretch of highway on the Flathead Indian Reservation, which has facilitated continued movement of wildlife across a major highway and a decrease in human–wildlife vehicular collisions (Huijser 2004).

Corridors may also help dispersers avoid predation or human-caused death in attempting to cross matrix or human-occupied lands. For cheetahs (*Acinonyx jubatus*), a major cause of death is humans, so identifying corridors is critical for maintaining populations. Targeting community tolerance of cheetahs within the corridors could be a critical step forward (Acton et al. 2018). Because so few studies document dispersal of wide-ranging species, evidence of this is indirect. A study showing poor dispersal success of mountain lions across a heavily humanized landscape in Southern California indicates that human-caused deaths may limit potential dispersal (Beier 1995). Of nine dispersers, three were killed due to vehicle collisions, one was killed in an urban area by a police officer, and three died from disease and other natural causes. The overpasses and underpasses in Banff decrease animal–vehicle collisions by 80 percent and the mortality rate for elk on roads is almost zero as compared to 100 such collisions per year prior to crossing structure installation (Jarvie 2017). There also is some evidence

FIGURE 4.7   This culvert undercrossing is one of thirty-eight wildlife underpasses and six wildlife overpasses constructed because of concerns about the Trans-Canada Highway bisecting Banff National Park. (Photo by Anthony Clevenger.)

that corridors could direct the movement of species, the implication being that corridors could decrease human-related deaths and wildlife–human conflicts. For example, a study examining butterfly movements in South Carolina found that corridors between larger patches of habitat helped direct movements of specialist species among patches (Haddad 1999). This behavioral study demonstrated that focal butterfly species were deflected off of corridor edges and were more likely to move between patches through corridors than across matrix habitat.

Continental corridors may be critical to species survival as climate fluctuates through the millennia. It is clear that flora and fauna alike have shifted historically, for example, between glacial and interglacial periods (DeChaine and Martin 2004). In some cases, corridors may diminish the risks of extinction by facilitating range shifts among biota in response to catastrophic events or long-term environmental change (Bennett 1990; Robillard et al. 2015). A number of studies have attempted to predict how global climate change will affect species survival. They document how precipitation and temperature currently influence where species can occur and model where populations may occur in the future given potential changes (e.g., Dilts et al. 2016). Models can show which populations would be most immediately affected by the changes. For example, a study of bighorn sheep (*Ovis canadensis*) that live on isolated mountain ranges in Southern California indicated that climate change could affect which populations persist in the long term (Epps et al. 2004). Simulation and empirical studies suggest that adding corridors between protected areas can be an effective strategy to facilitate range expansion, but the effectiveness depends on the size and the elevational gradient in the corridor (Imbach et al. 2013); the degree of landscape fragmentation (Renton et al. 2012; Mokany et al. 2013; Gimona et al. 2015); the amount of available habitat (Collingham and Huntley 2000; Synes et al. 2015; Hodgson et al. 2011); climate velocity (Renton et al. 2012); a species' dispersal ability (Meier et al. 2012; Kubisch et al. 2013; Mokany et al. 2013; Gimona et al. 2015); and habitat preferences (Hodgson et al. 2011).

Finally, corridors can help retain healthy functioning ecosystems. For example, where riparian zones are buffered to create natural corridors, the broad strip of natural habitat can retain the overall functioning of river systems. Corridors also can keep low-density predators in habitat fragments, the loss of which can result in a cascade of ecosystem impacts (Power et al. 1996). Further, they can help maintain species and essential services such as pollination (e.g., Kremen and Ricketts 2000; Kormann et al. 2016), and corridors everywhere can be seed sources for revegetation and recruitment of the diversity of plant species.

Benefits to Humans

Designing, maintaining, and restoring a network of connectivity across a landscape can directly benefit humans, as well as biodiversity. If such zones of connectivity are open to public access, open spaces can be important places for recreational hiking, biking, and relaxing. For example, the city of Boulder, Colorado, offers recreational and outdoor opportunities in the surrounding open spaces and mountain parks (www.ci.boulder.co.us /openspace/openspace). Also, its urban greenway trails serve as an alternate commuting route for those not driving cars. Countryside corridors offer human amenities as well. Naturally vegetated field margins not only increase the aesthetic appeal of the countryside, they can also incorporate horseback riding, hiking, and biking trails for recreation (Fry 1994).

Depending on the species, suite of species, or natural community that a corridor is supposed to support, combining connectivity for biodiversity with human recreation may or may not be appropriate. Research examining the influence of recreation on wildlife indicates that many species of wildlife are sensitive to the presence of humans and may not use an area heavily used by humans (e.g., Larson et al. 2016, but see Reilly et al. 2017). Also, foot, horse, and bike travel can facilitate transport of nonnative species. When they are brought in by human activities they can out-compete native plants, including food sources for wildlife, compromising the integrity of the corridor (discussed further in chap. 6).

Beyond recreation, rural and urban greenways play an important role in limiting urban expansion and sprawl and retaining distinct boundaries around different urban areas (Ahern 1995; Kubeš 1996). For example, one of the objectives of Sonoma County's Agricultural Preservation and Open Space District (SCAPOSD; California) is to identify and retain community separators and keep various communities from growing together (SCAPOSD 2000). The scenery can be an important part of the attraction of people to a community or region and can increase property values. The effectiveness of greenways can contain development within, but development often just leapfrogs beyond greenbelts. Whether greenways are effective at containing development likely relates to the strength of other tools that can direct or limit growth, including zoning and other regulations, as well as private land conservation tools (see chap. 10).

Corridors also provide free ecosystem services. Retaining buffer corridors on steep hillsides, for example, can limit hillside slumping, landslides, and erosion. Greenbelt corridors can limit pollution, such as from busy highways into adjacent neighborhoods. Likewise, retaining buffers along streams and around wetlands can help sustain the natural water-filtering process and limit damage to humans and structures due to flooding, which is predicted to increase in frequency and extreme in many places due to climate change. Because human disturbances change the flow of materials, riparian corridors can serve to moderate flows of such materials as sediment, fertilizer, toxic residue, and pesticide into river systems. Where anthropogenic inputs to streams are regulated because of the presence of endangered species, or to protect human water supplies, natural vegetation filters are far less costly than installing high-technology water filtration systems that are often used to ensure that water quality standards are met.

Corridors or strips of natural habitat also can be beneficial in agricultural systems. Hedgerows and other linear habitats can help limit soil loss due to wind and water erosion. In addition, corridors can help retain snowpack in windy areas, increasing total water accumulation and storage. Especially noteworthy is a long-term research program extending more than thirty years on agricultural landscapes in western Poland. This program found that belts of vegetation such as shelterbelts, strips of meadow, and hedgerows in agroecosystems provided multiple benefits. Rainfall increased in crop areas adjacent to shelterbelts, and winds were ameliorated. Lateral movements of pesticides were reduced; soil and air temperatures were cooler in crops during hot weather, reducing evapotranspiration. Adjacent belts of vegetation helped reduce soil erosion by water and wind, helped economize overall water needs, and decreased leaching of nutrients such as nitrate ions, which reduced the need for added fertilizers. The reduced export of pesticides and nutrients had the secondary effects of reducing pollution in groundwater and adjacent watercourses and preventing eutrophication of nearby lakes and rivers. In addition, landscapes with these corridors had much higher densities and diversity of animals, including game animals, pollinators, and predators of crop insect pests. Shelterbelts offered a supply of wood, shade, and aesthetic amenities; hedgerows served as effective fences for livestock (Ryszkowski et al. 1996). Comprehensive investigations such as this are leading the way toward management of agroecosystems for sustainability, maximum production, control of nonpoint sources of pollution, and conservation of native biota.

Corridors within agricultural areas can provide other direct services to farmers. Where agricultural lands and residences are adjacent to one another, natural vegetation buffers at field margins can reduce the drift of pesticides into residential communities, into waterways, and between fields (Fry 1994; Ryszkowski et al. 1996). Because monoculture farming is most susceptible to pest damage, retention of natural vegetation along field margins can reduce the effect of pest populations in two ways. First, it can intercept searching behavior of pests, especially those with poor dispersal ability, and thereby reduce pest damage (Fry 1994). Second, field margins can serve as reservoirs of natural enemies of crop pests, potentially reducing the need for pesticides although field margins can also harbor pests (Altieri 2010). Many predators that are helpful in reducing pests in agricultural areas have higher densities close to the edge of fields (Fry 1994; Nicholls et al. 2001). One study, for example, showed that the presence of natural vegetation in corridors suppressed populations of leaf- and stem-sucking pests in soybean monocultures, although those corridors did not diminish defoliator pests (Rodenhouse et al. 1992).

Species harbored within more natural vegetated corridors may also play important pollinating roles. A study in California demonstrated that maintaining functional migratory corridors for pollinators that are negatively affected by habitat loss and fragmentation helped to sustain native pollinator ecosystem services in agricultural and natural systems (Kremen and Ricketts 2000). Similarly, a study in Costa Rica found that coffee plantations within 1 kilometer (0.62 mile) of a forest had 20 percent higher coffee yields because of improved pollination (Ricketts et al. 2004). Further, the results of a large-scale experiment in Florida demonstrated that thin strips of habitat among patches facilitate two types of plant–animal interactions, namely pollination and seed dispersal (Tewksbury et al. 2002).

In summary, corridors can take many forms and can be beneficial for natural and human systems on a variety of different scales. From road-crossing structures to watershed corridors to continent-wide connections, they can facilitate dispersal and migration and increase the overall quantity of habitat available, helping individuals move and populations retain connectedness.

Chapter 5

Corridor Design Objectives

Careful corridor design can help avoid the pitfalls discussed in the next chapter and increase the chance that connectivity goals will be achieved. Though general recommendations about design are always difficult because of the immense variety and even idiosyncratic nature of situations where corridors are implemented, recent research reveals some important considerations of general applicability for establishing and maintaining connectivity. In general, habitat quality, continuity, and dimensions of corridors, as well as landscape context, will affect overall corridor utility.

If facilitating movement of a specific species or community is the focus then their habitat preferences, dispersal behavior, and other life-history features should be considered in the design process. In particular, the spatial and temporal scale associated with these life-history features should inform corridor design specifications. Landscape context, target species, social context, and institutional missions will all influence the project scale. Connectivity science has a longer history in terrestrial ecology and conservation biology than in freshwater ecology, and thus most of the focus of this book is on terrestrial corridors. However, it is imperative to mention the connection between freshwater environments and their importance for terrestrial and of course freshwater taxa. At the terrestrial–freshwater interface are riparian communities, semiterrestrial areas adjacent to water bodies and influenced by freshwater, which often serve as effective wildlife corridors. Connectivity science has been increasingly applied to stream ecology and watershed science (Freeman et al. 2007). This is in part because hydrologic connectivity is often impeded by stream alterations such as dams and water removal for agricultural and other uses.

Temporal scale influences corridor design considerations, and in this case refers to the time of year the corridor needs to function or the time period over which it should facilitate species movements. Corridors may be needed for seasonal movements such as migrations or for range shifts over decades in response to climate change. The time over which corridors need to be functional may also affect design. In the case of a seasonal corridor, there may be human activities within the corridor that might be incompatible with the species during migration but are otherwise acceptable.

The time period in which a corridor should function and its conservation objectives will affect the dimensions required. Some corridors are important for metapopulation persistence. Harris and Scheck (1991) suggested that these corridors should be at least 100 to 1,000 meters (328 feet to .62 mile) wide. Other corridors are supposed to function for decades or even centuries and serve entire ecological communities. The researchers speculated that these corridors should be more than a kilometer (.62 mile) wide. However, if a corridor is necessary only during a temporary human disturbance of less than one year, then a width of 1 to 10 meters (about 3 to 33 feet) might be sufficient. The inference from these examples are that the spatial and temporal scales need to be specified because they will influence the planning objectives.

Another factor that will influence corridor conservation plans is the designation of focal species. The premise is that corridors should incorporate, where possible, attributes that might enhance use by the focal species. We define *focal species* as those that warrant special protection; they may be keystone, umbrella, flagship, indicator, specialist, or vulnerable species. *Keystone* species are those whose impact on the landscape is disproportionately large relative to their abundance. *Umbrella* species are those whose needs overlap with other species such that their conservation should result in conservation of the other species, while *flagship* species have public appeal for social or economic reasons. *Indicator* species are those whose status is used as a proxy measure of ecosystem conditions. *Specialist* species may be limited by available habitat or other resources, and *vulnerable* species are those listed as endangered or threatened by governments or other groups.

Focal Species Considerations

If corridors are designed around a particular suite of focal species, then these should be carefully selected. Beier et al. (2008) suggested selecting a suite of species that together will represent the movement needs of most species. The suite should include species requiring dispersal for meta-population persistence, species with limited dispersal capabilities, habitat specialists, species important for ecological processes such as predation or pollination, and species sensitive to barriers, and keystone species. Suites of species selected based on these criteria often consist of ten to twenty species, including plants, invertebrates, fishes, amphibians, reptiles, and mammals (fig. 5.1). Alternatively, Lechner et al. (2017) suggest an approach to select a suite of species that represent different dispersal guilds, where species are grouped based on similar dispersal behavior and habitat requirements.

Many corridor designs have been based solely on large, charismatic species. This may be appropriate when and if these species are suffering from loss of connectivity and the objective of the corridor is to recover the species. These charismatic species can also be powerful flagship species, able to rally the support of the public and other stakeholders. Several studies have tested whether corridors designed for these species will also provide connectivity for other species and thereby serve as umbrella species for connectivity planning purposes; however, the studies produced varying results. In one study, overlap in dispersal habitat of a bird, a butterfly, and a frog in a fragmented landscape was high for places identified as important for dispersal, suggesting a corridor designed for one of these taxa will serve the other two as well (Breckheimer et al. 2014). A regional conservation network designed for a jaguar (*Panthera onca*) serves as a good umbrella for other mammal species according to calculations of the overlap between the network and species richness, habitat quality, and fragmentation indices of about 1,500 co-occurring mammal species (Thornton et al. 2016). Similarly, Epps et al. (2011) determined that conserving movement corridors for African elephants will likely preserve habitat and potential landscape linkages for other large mammal species inhabiting Tanzanian protected areas.

Another corridor network plan designed based on the life history of mountain lions (*Puma concolor*) in California was found to support several other important biodiversity elements including serpentine rock as a surrogate for rare plants, old-growth forest, different types of oak woodlands, and watersheds containing an endangered fish (Thorne et al. 2006). However, in this same study the researchers found that endemic amphibian, reptile, and mammal populations were not well represented by the path of the puma. Likewise Cushman and Landguth (2012) confirmed that habitat specialists and species with limited dispersal abilities are only weakly represented in corridors designed for other species. They also tested how well

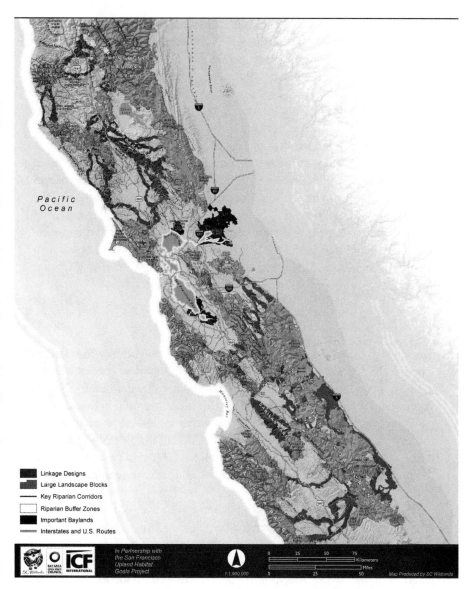

Legend:
- Linkage Designs
- Large Landscape Blocks
- Key Riparian Corridors
- Riparian Buffer Zones
- Important Baylands
- Interstates and U.S. Routes

Pacific Ocean

FIGURE 5.1 The San Francisco Bay Area Critical Linkages is an example of using expert knowledge to map the habitat requirements and movement needs of 66 focal species including 21 mammals, 13 birds, 6 amphibians, 5 reptiles, 4 fish, 2 invertebrates, and 15 plants. (Permission to use by SC Wildlands.)

three carnivore species can function as umbrella species, and concluded that they performed poorly as connectivity umbrellas, because they were associated with high-elevation forests, while low-elevation, nonforest species were most at risk to lose habitat and suffer from fragmentation. The poor umbrella function of carnivores for corridor designs was also noted in a study specifically comparing location of corridors between protected areas (Beier et al. 2009). The corridors used by carnivores did not overlap with the corridors favored by most other focal species, and even posed relatively high resistance for noncarnivores. In conclusion, using umbrella species in connectivity conservation needs to be carefully considered. Safer approaches are to either select a suite of focal species as discussed above, including the flagship species, or focus on conserving structural connectivity based on natural land cover types and supplement the corridor designs when necessary with species designs for habitat specialists (see chap. 8). This will help ensure that the selected linkages are useful for a larger assortment of species.

Habitat Requirements

The quality of habitat within an ecological corridor often will relate to whether the species of interest will use the corridor. Some species will not use a supposed ecological corridor because of low habitat quality. Native vegetation can enhance the probability that a corridor will be used, while exotic vegetation could deter biota. Nonnative riparian vegetation can sometimes dominate a community and exclude native flora and fauna, as in the case of *Arundo donax* (Boose and Holt 1999) and *Tamarix chinensis* (Stromberg 1997). In other cases, however, nonnative vegetation can serve a structural role or provide food.

Species that evolved in relatively continuous as opposed to naturally fragmented patches are less likely to be able to adapt to human-induced fragmentation and may require more natural corridors. This is especially true where vegetative structure changes dramatically because of humans, whether by the loss or the addition of structural components. A study in South Carolina sprayed fluorescent powder on berries eaten by birds to learn how birds moved and dispersed seeds in an experimentally designed corridor-patch forest and found that birds followed the forested corridors more often than they flew across clearings (Levey 2005). Arboreal species are extremely sensitive to the level of fragmentation, corridor quality, and spatial isolation. In one study in Australia, arboreal marsupials declined by 97 percent in both tropical forest fragments and corridors (Laurance 1995).

In such a scenario, maintaining or restoring corridors may suffice only if they encompass intact forest characteristics.

Whereas some species require access to large areas of relatively homogenous habitat, other species need connectivity among different types of habitats at various life-history stages. Many ungulate species move back and forth each year from montane summer habitat to valley winter habitats, sometimes passing through naturally constricted areas. Blockage of such passageways could eliminate an entire population of ungulates from a region. Similarly, the survival of some wetland species is imperiled where efforts focus solely on wetland conservation with little or no attention to upland buffer zones or corridors. Though a species may spend most of its life within a wetland, retaining upland connectivity is important for species that require upland habitat for a portion of their life cycle or for dispersal (Semlitsch and Bodie 2003). One such species is the bog turtle (*Glyptemys muhlenbergii*), which has become endangered not only because of wetland loss but also by loss of surrounding upland habitat corridors needed for the turtles to move among wetlands (Klemens 2000).

In some cases, species may require specific natural elements for their survival, and it may be necessary to incorporate those elements into corridor design. They include features such as snags, denning and hibernation sites, and salt licks, as well as other features. For instance, to retain a continuous population of woodpeckers, snags or standing dead trees should be included in the corridor because they are an important resource for woodpeckers (Farris et al. 2004). Similarly, sugar gliders (*Petaurus norfolcensis*), a threatened arboreal marsupial, rely on large old trees for both food and breeding resources and are correspondingly found only where such trees persist, including in linear habitats along streams and roads (Van der Ree and Bennett 2003).

Given the potential importance of habitat type and quality, corridor design should incorporate intact natural habitat when possible. Specific requirements of focal species should be reviewed to ensure that those species will continue to have access to needed habitat types over time. If known, special elements required by focal species for survival should be retained in the core habitat areas and corridors likely needed by those species. If the corridor is large relative to species movements, and the species can become resident within the corridor, it will be particularly important to incorporate those habitat needs within the corridor and not just in the habitat patches being connected. In some cases, active management of both the corridor and the surrounding matrix may be required to retain appropriate habitat within a corridor in the long term.

Dispersal Considerations

Understanding the periodic dispersal of a species may also inform the temporal scale that should be addressed so that a population does not become isolated and suffer from associated impacts. Little is known about dispersal patterns for most species, including the range of distances traveled, the success of dispersers, and the likelihood of dispersers using different types of human-occupied landscapes (chap. 3). Northern spotted owls (*Strix occidentalis caurina*), for example, disperse in random directions, so discrete corridors might not be beneficial for them (Simberloff et al. 1992). It is for these reasons that planning for minimum dispersal should not be the one and only design criterion.

Like that of animals, plant dispersal strategy can influence movement through corridors. A computer simulation model examining seed flow demonstrated that plant dispersal ability influenced the number of species that may occur in riparian corridors across a fragmented landscape (Hanson et al. 1990). In addition, species with higher dispersal capabilities are often aggressive weeds that can displace native species (Kubeš 1996), making it more challenging for native species in the long term.

In some cases, successful plant dispersal may depend on mutualistic relationships. For example, in Western Australia, frugivorous birds, such as the mistletoe bird (*Dicaeum hirundinaceum*) disperse mistletoe seeds (*Amyema miquelii*). As mistletoe birds prefer large habitat fragments, seeds are not dispersed into small remaining habitat fragments (Norton et al. 1995). Similarly, flying foxes (*Pteropus* spp.) in Samoa play an important role in dispersing and pollinating plant species, and hence their survival and behavior in fragmented systems will affect the overall forest structure (Cox et al. 1991; Cox and Elmqvist 2000). In cases where mutualistic relationships influence dispersal, it will be important to identify and consider both the focal species and the need of the species it depends on for dispersal when planning for connected landscapes.

The ability of species to disperse is an important determinant of how likely they are to effectively move in response to climate change. While good dispersers will be more successful, species that disperse slowly face greater challenges. Assisted migration and assisted evolution (helping species adapt to hotter conditions through heat tolerant genes) are being discussed as necessary strategies for species that will not be able to shift their ranges by themselves. This book discusses landscape conservation strategies to allow species to shift their ranges in response to climate change and we will not go into further detail on these in situ management topics.

Generalist versus Specialist

Corridors may act as filters (also see chap. 6) in that generalists, species that use multiple habitat types or have relatively broad diets, will pass through, while specialists will be impeded. One reason for this is that human-induced fragmentation tends to lead to an increase in weedy species and habitat generalists, whereas rare and sensitive forest-interior species decrease in numbers (Dijak and Thompson 2000). It is also more likely that generalist species will dominate narrow corridors, and invasive exotics may also be present. It is specialists and those species prone to human–wildlife conflicts that most need well-designed and functional corridors to ensure species survival in a fragmenting landscape (Kozakiewicz and Szacki 1995).

One reason it is more difficult to conserve corridors for specialist species is that corridors need to be wide enough to retain the specific habitat needs of specialists and prevent a host of edge effects (chap. 3). Edge effects may preclude the availability of specific habitat elements or food items, especially if those resources are dependent on interior, undisturbed conditions. For instance, red-backed voles (*Clethrionomys californicus*) may inhabit primarily interior habitat rather than edge or clear-cut areas because the interior forest provides cool, moist conditions that enhance the growth of fungi, a preferred food item (Mills 1996). In some cases, generalist and exotic species may limit use of corridors by specialist species, so that regular control of such species may be necessary. An option requiring less intensive land management would be to retain wide enough corridors that generalist and exotic species, which generally are edge species, do not penetrate the entire corridor.

What we find is that specialists are likely to need wider corridors or larger stepping-stone patches with more intact and native habitat in order to maintain the habitat conditions that favor their survival (e.g., Perault and Lomolino 2000). For specialists with specific habitat or food limitations, the design of corridors may need to ensure that those elements will be incorporated and retained.

Behavioral Factors

Intraspecific and interspecific interactions are important factors to consider in connectivity planning. Social species might not use corridors unless they can move in groups (Laurance 1990; chap. 6). Likewise, mutualistic relationships, antagonistic relationships, and density-dependent survivorship may influence corridor use (Lidicker and Koenig 1996). Earlier, we mentioned an example of a mutualistic relationship: the spread of mistletoe being dependent on frugivorous bird behavior. Antagonistic relationships are those in which one species interferes with another species, which may happen when species from the matrix invade a corridor. Brown-headed cowbirds (*Molothrus ater*), a species often associated with edges, parasitize nests of songbirds, decreasing their reproductive success in small fragments and narrow corridors (Hansen et al. 2002). In fact, surprisingly high abundances of cowbirds were associated with narrow forest-dividing corridors in southern New Jersey, USA (Rich et al. 1994). Density dependence is illustrated by root voles (*Microtus oeconomus*), whose dispersal was determined to involve movement toward patches with lower densities, indicating the importance of spatiotemporal demographic variability (Andreassen and Ims 2001).

Another type of variable to consider is behavioral factors that may limit a species' movement. A species may be physically capable of living in and moving across human-impacted regions through corridors or even the matrix but may avoid edges and fail to travel across any sort of disturbed habitat because of behavioral constraints (e.g., St. Clair et al. 1998). Surprising though it may seem, some birds will not fly across even narrow water gaps or forest clearings (e.g., Machtans et al. 1996), and some large and wide-ranging mammals like mountain lions avoid crossing deforested areas (e.g., Opdam et al. 1995). Similarly, American and European species of martens (*Martes*) generally avoid open areas and are considered to be forest dependent (Bissonette and Broekhuizen 1995). California red-backed voles also appear to avoid cleared areas, perhaps because their primary source of food, truffles, is found generally within forests and rarely in cleared areas (Mills 1996). In an experimental study of butterflies in Natal, South Africa, stands of exotic trees and mowed grass had a highly negative influence on butterfly flight paths, causing butterflies to switch directions (Wood and Samways 1991). Efforts to understand behavioral constraints can help ensure that corridors will function well for focal species.

Sensitivity to Human Activity

Some species do not accept close proximity to humanized landscapes, either because of their intolerance of human activities or because of lack of human tolerance toward them. Air, ground, and water pollution, as well as noise and lighting can affect whether focal species use a corridor (Van Bohemen

2002). One species whose movements have been observed to change be-cause of lights is the mountain lion. A study in Southern California showed that the species avoided corridor sections illuminated by artificial lights and instead used densely vegetated corridors (Beier 1995).

Many top carnivore species that are generalists could theoretically survive and fulfill their energetic needs in heavily human-impacted areas, but such adaptation requires human tolerance of the species. Wolves in the Great Lakes region of North America seem to survive relatively well in a re-gion dominated by agriculture and ranching on private land, despite some conflict with humans (Treves et al. 2004); whereas wolves as well as grizzly bears are less tolerated on private lands and multiple-use public lands in the Rocky Mountains (Clarke et al. 2006; Gude et al. 2012).

A review of focal species' tolerance to human activities will be important in designing corridors, especially if human activities occur in the corridor (box 6.1). In cases where little information exists about species' sensitiv-ity to humans, it is best to err on the side of caution. The most cautious approach is to limit human activities within corridors and place corridors as far away from high-density human areas as possible. If human activities must within or near corridors, they should at least be directed farther away from elements of higher biodiversity value, such as riparian zones. Other steps to limit overall human disturbance include limiting noise, light, pet and livestock activity, feeding of wildlife, and degradation of existing vegetation. For example, earth berms can reduce noise and lighting from sources such as roads, and thick vegetation can filter air particles and water runoff from developments. In circumstances where there are human-car-nivore (or other wildlife) interference problems, sometimes conflicts can be minimized through such mechanisms as buffer zones, physical barriers such as fences, active protection, or compensation schemes, all tools to con-sider in designing linkages. Corridor design may also require encouraging compatible human behavior adjacent to corridors, such as storing trash and bird feed properly, discouraging feeding of wild animals, and limiting the planting of vegetation such as fruit trees that may encourage some species to wander out of corridors into human-occupied parts of the landscape.

Physical Limitations

Some species have clear physical limitations that are important to be aware of when designing corridors. For instance, fencing that is too high and has strands close together near the ground can impede pronghorn antelope's

FIGURE 5.2 The carcass of a mule deer (*Odocoileus hemionius*) caught in a wire fence along Highway 87 in Utah. (Photo by David House.)

(*Antilocapra americana*) and other species' movements (Hailey and DeArment 1972; fig. 5.2). Similarly, researchers speculate that turtles may suffer difficulties in round culverts because of their propensity to try to climb the sides and end up flipped upside down. Moreover, corrugated culverts could snare juveniles or high-center adults. Consequently, square culverts without corrugation may be the most prudent type of underpass to facilitate turtle movement under roads (K. Griffin, pers. comm. 2005). Other types of barriers to consider range from roads, railroads, and canals to cliffs, high mountain passes, and large waterways.

All such potential barriers should be identified in the planning process so as to avoid corridors that look good on paper but in fact do not adequately serve species of interest (see chap. 7). In some cases, it may be necessary to replace human-created barriers such as fences or poorly designed culverts with wildlife-friendly designs. It is hard to generalize what makes a wildlife-friendly fence, because that can depend on the species. While a plethora of guidelines exist (e.g., Hanophy 2009), most generally accept the following standards: three strands of smooth wire, with the bottom strand at least 16 inches above the ground, the second strand at 24 inches, the third wire at 32 inches, and a pole on top at 40 inches. The top pole provides an important visual barrier that wildlife can detect and prevents them from getting tangled in a top wire. Generally, the more visible the materials, the better, so wood, recycled plastic, and PVC are better than wire. Living fences or hedges present an alternative approach as exemplified by farmers in Mexico who set organ-pipe cacti (*Stenocereus thurberi*) in the ground as closely spaced fence posts. Consideration of guidelines, regulations, and incentives for managing the corridors in the long term is important, as these variables influence the long-term success of corridors.

Topography and Microclimate for Climate-Wise Connectivity

Forecasted changes to global climate predict that many species need to shift their ranges from a few kilometers to hundreds of kilometers over the course of this century to track suitable climates (Lawler et al. 2013; Loarie et al. 2009). Numerous studies have documented that terrestrial species have already shifted their ranges with climate change (e.g., Parmesan 2006; Moritz et al. 2008). However, landscape fragmentation inhibits dispersal movement and therefore constrains range shifts. When species are lagging behind, it results in a climate–biota mismatch. This can, for example, lead to massive tree die-offs during extreme weather events or prolonged droughts, because the area is no longer climatically suitable for the species; climate change surpassed the species' physiological tolerance. Below we discuss several corridor design concepts that may facilitate species' range shifts over decades.

In efforts to connect the landscape to increase population viability of focal species, the objective is to link core habitat areas to facilitate more continuous populations, or where that is not possible, to enhance movement among distinct populations. The objective changes when the goal is to facilitate range shifts, because now core habitat areas need to be connected to areas that will become suitable habitat in the future. This objective can be framed in two ways. We can consider it from a species perspective, asking where today's species are going to track suitable climate conditions. Alternatively, we can look at it from a land perspective and ask where tomorrow's species are coming from (Hamann et al. 2015). These two perspectives can inform the selection of corridor endpoints when prioritizing conservation action. We will go into more detail for selecting corridor endpoints to facilitate range shifts in chapter 8.

This shift in objectives with respect to corridor endpoints highlights another design principle important for climate-wise corridors: directionality. While not all species are moving upslope or poleward, there frequently is a predominant direction in which species will shift their ranges. Maps of climate velocity, or current and predicted species distribution models for large suites of species can inform planners about predominant range shift directions in their areas (Loarie et al. 2009; Lawler et al. 2013).

In addition to corridors connecting protected or natural areas, Olson et al. (2009) suggest managing what they call adaptation corridors. These are corridors following natural environmental gradients such as elevation, precipitation, or soil gradients, which are embedded within natural landscapes, and contain multiple habitat types. A good example of systematic prioritization of corridors that span environmental gradients comes from South Africa where selected corridors represent biological gradients (north-south upland-lowland and east-west macroclimatic gradients) within each biogeographically distinct water catchment to ensure biodiversity persistence (Rouget et al. 2006). Corridors that include climatic gradients are thought to enable species to adapt to climate change by providing conditions for persisting through variable climate conditions (natural vegetation retains moisture and provides thermal refugia), short-distance distribution shifts, and local adaptation. Adaptation corridors should be wide enough to reduce edge effects and be managed for increased resilience, which may include invasive species control, maintaining a healthy mix of natural vegetation, and allowing natural disturbance events (e.g., fire, flooding) to occur.

Variation in microclimates over short distances offers plants, invertebrates, and other taxa opportunities to respond to changing climatic conditions with short-distance dispersals (Opedal et al. 2015). Newly suitable habitat may only be a short way away, making long-distance dispersal toward the poles unnecessary. Spatial heterogeneity also results in greater genetic and species diversity, which are prerequisites for successfully adapting to new conditions. For example, Guarnizo and Cannatella (2013) studied eight families of tropical frogs in Central and northern South America and found that there is higher genetic diversity within frog species occurring in topographically more complex regions. Topographically complex sites in the alpine zone also had a higher richness of plant species than flatter sites (Opedal et al. 2015).

Topographic and elevational diversity results in local differences in solar radiation, air circulation, water runoff patterns, and, in cold climates, snow distribution, and can bring about wide ranges of temperature over short distances. For example, in one small area differences of up to 8°C existed between sites in close proximity, brought about by differences in slope and aspect (Ackerly et al. 2010). Because topographic diversity allows for short-distance dispersal to newly suitable habitats and increases adaptation capacity through genetic and species diversity it makes sites with high spatial heterogeneity more resilient to climate change than topographically simple sites (Anderson et al. 2014). Therefore, topographically and climatically diverse sites should be prioritized when planning climate-wise connectivity.

Corridor Quality: Continuity, Composition, and Dimension

Beyond assessing potential corridors according to focal species' needs, general site characteristics should be assessed. Site characteristics can influence the kinds of species that use corridors, as well as the long-term viability of corridors and the communities that depend on them. Corridor continuity, including habitat quality and composition of plant species, should be considered. Dimensions of proposed corridors are also important.

Continuous Corridors

A good deal of research supports the importance of continuous corridors as opposed to corridors that are bisected by roads or other activities, especially for more sensitive species (e.g., Tilman et al. 1997; Brooker et al. 1999). Species with limited dispersal capabilities will need corridors that provide conditions suitable for the species to live in and reproduce over generations to facilitate climate change adaptation, allowing the following generations to move farther on the climate trajectory. Because range shifts take time, live-in habitat in the corridors is important. For designing climate-wise corridors, this translates into the need to create wide corridors that contain sufficient habitat to accommodate home ranges or populations of slow-dispersing species. We expect that pinch points in the corridor will not stop dispersal and therefore can be part of a corridor, whereas barriers cutting through a corridor, such as highways, need to be mitigated. Unfortunately, few studies indicate threshold levels of connectivity below which various species or groups of species would be unable to use the corridor.

Generally, if the habitat within a corridor is too fragmented, passage of species through the corridor may effectively be stopped. Gaps and barriers, such as roads, should be avoided where possible. What is perceived as a gap varies from species to species. Gaps of greater than 5 meters (16 feet) in structural vegetation may deter the movement of transient chipmunks (*Tamias striatus*) (Bennett et al. 1994). For mountain lions, major highways that lack appropriate underpasses may prevent connectivity, but highway bridges that accommodate watercourses and some natural vegetation may help maintain connectivity (Beier 1995).

While maintaining a high level of connectivity is desirable, it is not always an option, especially in more developed landscapes. Fortunately, some species appear to be less affected by discontinuities in corridors than others, but again, tolerance of gap size is species specific. For tundra voles (*Microtus oeconomus*) and gray-tailed voles (*M. canicaudus*), gaps in corridors of less than 4 meters (13 feet) did not greatly change or increase transit time (Andreassen et al. 1996; Lidicker 1999; Wolff et al. 1997). Likewise, research on grizzly bears and mountain lions in the Crowsnest Pass on the border of Alberta and British Columbia indicates that at least some individual bears managed to cross a two-lane highway (Chetkiewicz and Boyce 2009). In that case, maintaining adequate corridors for the animals to approach the highway may be part of the long-term strategy, and as traffic levels increase and/or the road expands crossing structures will increase in importance. It is important to maintain and secure adequate road-crossing structures to ensure that populations on both sides of the highway do not become completely severed from one another.

An additional consideration is the nature of the habitat to be connected. This will depend in part on the focal species and their needs. It is important not to assume that connecting like habitat with like habitat is always the primary goal. As mentioned earlier, some biota require only one specific habitat type; others use several different resources or habitat types in various years or times of the year. For instance, amphibians and large mammals need pathways to water (Kubeš 1996), so ensuring that they have free access to upland habitat and riparian areas is important, especially in the dry season. Similarly, marsh rabbits (*Sylvialagus palustris*) spend their lives in wetlands but need upland forest-corridor habitats for dispersal (Forys and Humphrey 1996). Failure to recognize that a focal species needs multiple habitat types throughout its life can lead to poor delineation of corridors and potentially the decline of the species. Also, mammals might focus on scent trails or other more subtle attributes of matrix habitat that are difficult to assess. Until more is known about the life history and behavior of species of interest, such sensitivities to connectivity may remain undetected.

For some species and groups of species, general guidelines do exist as to the threshold levels of continuity needed to retain connectivity among otherwise disjunct populations. One modeling effort that assessed corridor functionality for plants suggested that if more than 50 percent of the habitat was destroyed, the percolation threshold (the point at which individuals may no longer be able to move through it) would be reached and a corridor would no longer function to facilitate plant species movement (Tilman et al. 1997). Other investigators have focused on how individual species cross nonhabitat to move among remaining fragments of habitat. For example, a study that examined how gliding marsupials moved among woodland patches in an agricultural system found that most of the movement occurred within 75 meters (246 feet) of the larger patches containing the focal species (Van der Ree et al. 2004). Generally, evidence from cumulative research indicates that willingness to cross gaps is species specific (e.g., Ricketts et al. 2001).

Given the importance of continuity for species conservation, increasing attention is being directed toward designing structures to bypass bottlenecks, such as road and railway crossings. With approximately 36.5 million kilometers of roads in the world, roads can be a formidable barrier and a source of mortality for wildlife, which is often associated with vehicle damage. Roads are perilous for resident species, especially for dispersers and introduced species that are unfamiliar with them. In chapter 3, we presented figures on numbers of species killed by roads, including a rare study documenting high mortality for dispersing cougars caused by automobiles. The attempted reintroduction of Canada lynx (*Lynx canadensis*) from the Yukon to the Adirondack State Park in New York, which has many more roads, illustrates how roads can significantly impact naive populations—of the thirty-two confirmed deaths in the park, twenty were by cars (McKelvey et al. 2000). Undoubtedly, roads, which caused a majority of the known deaths, contributed to the ultimate failure of the reintroduction. To reduce wildlife–vehicle collisions, species-specific information about crossing and mortality rates and preferred and successful crossing structures, ranging from vegetated overpasses to bridges and culverts, are needed (e.g., Clevenger et al. 2001; Parks Canada 2017; fig. 5.3). Fencing may also reduce wildlife roadkill. According to research, road and pipeline underpasses have successfully provided travel routes in many locations and for many species including migrating caribou (*Rangifer tarandus*) and elk (*Cervus elaphus*), grizzly bears, wolves, badgers (*Meles meles*), and boar (*Sus scrofa*) (Putmam 1997; Rodriguez et al. 1996; Clevenger and Waltho 2005; Stewart 2017). In chapter 10, we provide a case study of the Yellowstone to Yukon region where a high density of wildlife crossing structures facilitate wildlife movement. Literature pertaining specifically to such topics as the impact of roads on wildlife, the principles of road-crossing structure design, and examples have been compiled in two books about road ecology (Beckmann et al. 2010; Van der Ree et al. 2015).

Unless contrary evidence is available, the conservative approach of

FIGURE 5.3 Wildlife overpass across the Trans-Canada Highway, one of six established to increase connectivity for populations of wildlife in Banff National Park, Alberta. (Photo by Jodi Hilty.)

maintaining or restoring maximal continuity may be the best strategy for ensuring focal species survival, given the preponderance of evidence that lack of continuity in a corridor impairs its connectivity value. In addition to addressing continuity within corridors, this approach may include providing two or more corridor connections between habitat patches. The existence of multiple pathways can enhance the likelihood of a species moving from one patch to another (fig. 4.1). Further, redundancy provides a sort of insurance in case a corridor is affected or destroyed by unforeseen events ranging from natural impacts such as fire to human destruction. In some cases, it may be possible to incorporate different types of pathways such as upland and riparian corridors, each of which may be used by different species. Finally, monitoring corridor passage is important to allow for adaptive management when problems arise and to inform other corridor conservation efforts.

Stepping-Stone Connectivity

Functional connectivity for some species may not require a continuous connection of relatively intact natural habitat but could involve stepping-stones of habitat or protected areas that are not physically connected but that can facilitate dispersal or migration movements (Forman 1995; Bennett 2003; fig. 4.1). Species that are adapted to a habitat mosaic can more easily maintain a metapopulation structure because they are already adapted to disperse in fragmented habitat (Szacki and Liro 1991; Forys and Humphrey 1996). If a species has some ability to cross matrix habitat, clusters of small patches may be an alternative to continuous corridors. Modeling exercises suggest that for species with some ability to disperse in a fragmented landscape, stepping-stone patches can greatly increase the ability of the species to disperse (Söndgerath and Schröder 2002).

Stepping-stone connectivity can be sufficient depending on the life history of some species. Volant species such as Fender's blue butterfly (*Icaricia icarioides fenderi*) and whooping cranes (*Grus americana*) are two species whose life histories make them well adapted to stepping-stone connectivity (Schultz 1995). Frugivorous pigeons in New South Wales eucalypt (*Eucalyptus salmonophloia*) forests are also adapted to stepping-stone connectivity because eucalypt forests were naturally interspersed within rain forests. Currently, remnant patches of rain forest are surrounded by agriculture, but these pigeons appear to be able to move throughout the patchwork using exotic weed patches as stepping-stones, possibly because the pigeons evolved in a habitat type that was patchily distributed (Date et al. 1991).

Some species will need stepping-stones in which new populations can establish that then provide propagules for jumping to the next stepping-stone (e.g., wind-dispersed trees, butterflies). An extensive, long-term study of butterflies in England showed that gatekeeper butterflies (*Pyronia tithonus*) and small skippers (*Thymelicus sylvestris*) in a time of range expansion between 1983 and 2004 used patches as stepping-stones around dense urban areas. These two species have limited dispersal capabilities but are able to quickly establish new populations in empty habitat patches, characteristics that make range expansion through stepping-stones possible. These stepping-stones need to be sufficiently large to allow a population to establish, and spaced to enable propagules to reach neighboring suitable conditions (Collingham and Huntley 2000; Hodgson et al. 2012).

Marine protected areas can serve as stepping-stones for larval recruitment from distant populations if, in a protected area network, single areas are spaced at distances comparable to larval dispersal ranges and connected by ocean currents (fig. 5.4). It is important to note that where a community type has evolved in a more continuous array, a stepping-stone mosaic of patches may serve some species of the community but fail to allow for the maintenance of a successful metacommunity. Therefore, planning for

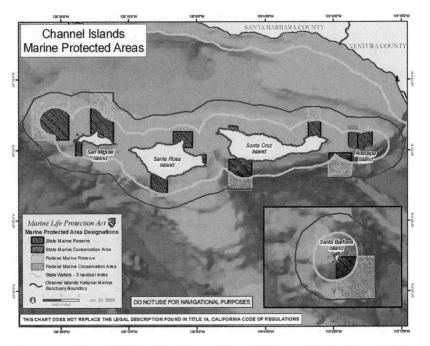

FIGURE 5.4 Networks of small marine protected areas are showing great promise for conserving marine biodiversity. Reserves in the middle can serve as stepping-stones to facilitate geneflow across the entire network. (From NPS: https://www.nps.gov/chis/learn/nature/marine-protected-areas.htm.)

maintaining connectivity through a stepping-stone approach should proceed carefully.

Habitat Quality

Earlier we discussed habitat quality with regard to focal species specifically, but ideally a corridor will serve more than just focal species. Corridors, be they continuous or stepping-stone, should function better for all native species if they retain a higher degree of community integrity. A computer simulation by Tilman et al. (1997) suggested that habitat in a corridor must be of equal or higher quality than that in larger core habitat patches to be effective. However, survival in marginal habitat might be feasible even given scarce resources. Some species may not use low-quality habitat at all, but for those that do the survival, reproduction levels, and ability to move through the corridor may be compromised. In one study, small mammals that used low-quality hedgerow corridors were found to be significantly less successful in reaching the other end than those that used high-quality corridors (Merriam and Lanoue 1990; Bennett et al. 1994). However, it still remains unclear whether corridors need to comprise high-quality habitat to function for dispersal. Some butterfly species are known to take advantage of corridors made up of marginal habitat as compared to traversing between habitat patches without any corridor. So especially for species that can move between patches within one generation, habitat quality may not be a key factor in corridor function (Haddad and Tewksbury 2005; Keeley et al. 2016, 2017).

Vegetative structure and composition both influence habitat quality. Vertical vegetative structure has been shown repeatedly to affect species richness and composition of biota in corridors (Forman 1995; Hilty 2001). Several studies found that for both native plants and animals adapted to structurally complex habitats, an adequate overstory cover influences species use (Fritz and Merriam 1993; Bennett et al. 1994). Studies of white-footed mice (*Peromyscus leucopus*) suggest that the mice prefer to move along fencerows that are more structurally complex (Merriam and Lanoue 1990), and chipmunks also were found to preferentially use fencerows with tall trees and woodland structure and to completely avoid fencerows associated only with grassy vegetation (Bennett et al. 1994). Likewise, in a northeastern North American farmland, ground-layer native woodland plant species needed overstory cover to survive within fencerow corridor habitat (Fritz and Merriam 1993).

Another habitat component that can increase corridor function is native vegetation, while exotic vegetation can deter use (Bennett 1991). Findings from a study in Queensland, Australia, indicated that corridors that were floristically diverse served arboreal species better than those that were primarily *Acacia*-dominated regrowth (Laurance and Laurance 1999). Nonnative vegetation can, in some cases, replace structures or functions on which the species of interest are dependent such that focal species can use corridors dominated by nonnative vegetation. In Queensland, rain forest birds used forested patches along a creek corridor within which lantana (*Lantana camara*), an exotic species, formed a dense shrub layer. Lantana provided another layer of structure, and the plant's flowers and fruit are potentially useful to birds. In that case, lantana appeared to contribute to desirable structure and serve as an alternative food source, so the corridor might function well for some birds despite the presence of common exotic species (Crome et al. 1994). Where plant restoration is required, however, native vegetation should be encouraged in order to maintain and increase biodiversity.

Low-quality or nonnative vegetation can sometimes be better than no corridor, depending on the focal species. Low-quality corridors enhanced populations of white-footed mice (Henein and Merriam 1990), and dispersing mountain lions in Southern California used a corridor with degraded vegetation (Beier 1995). However, fragmentation-sensitive birds in Southern California did not use revegetated highway corridors as much as remnant strips of native habitat (Bolger et al. 2001). Where connectivity of natural vegetation has been lost, planted corridors might serve at least some species.

Designated corridors ideally should be composed of relatively intact native vegetation. To retain native vegetation over the long term, wider corridors with less edge habitat are ideal. In some locations, control of nonnative species may be necessary to promote native species. Where habitat is too disturbed to retain native vegetation, simulating the characteristics of native vegetation such as its structure may be useful for some species, but that should be evaluated carefully. To increase resilience of corridors to climate change, some restoration efforts include planting trees and shrubs with greater diversity than had been there historically to make food and shelter available year-round under uncertain future climates. Plants were also selected based on their expected response to future climate conditions predicted for the site (Parodi et al. 2014).

Corridor Dimensions

In addition to continuity and habitat quality, studies indicate that the dimensions of corridors play an important role in determining what species occur within the corridor and the potential speed with which those species pass through the corridor. Here we examine what we know to date about the role of the overall dimensions of a corridor. Both length and width of corridors appear to play a role in the utility of landscape structures to facilitate movement. Vegetative structure could be considered a third dimension, but we discussed that earlier as a component of habitat quality, so we will not revisit it here.

Length is an important consideration. In general, shorter corridors are more likely to provide increased connectivity than longer corridors. Corridors that are too long might not contain some species due to increased distance from core habitat. A study of hedgerows in the Czech Republic concluded that three-quarters of forty-one forest plant species examined were not found beyond 200 meters (656 feet) away from the source woods, while the remainder of the species were detected as far as 250 to 475 meters (820 to 1,558 feet) away (Forman 1995). Configuration of forest corridors made up of linear strips of forest influenced species detection rates in forests in a thorough study done in southeastern Australia (Lindenmayer et al. 1994). Specifically, detections of a small carnivorous marsupial (*Antechinus stuartii*) were lower at the ends of the linear strips farther away from core forest.

Width is another determining factor in the number and composition of species found in corridors. There has been some debate in the literature about the benefits of narrow versus wide corridors. Scientists hypothesized that narrower corridors may increase the speed with which an animal moves through a corridor (Soulé and Gilpin 1991; Andreassen et al. 1996). However, the preponderance of data indicates that wider corridors are generally more effective for maintaining connectivity.

Wider corridors generally support more species and thus better uphold community integrity. For example, in the oak woodlands of northern California, fewer native carnivore and bird species were detected in narrower corridors, where activity levels of nonnative species increased (Hilty 2001; Hilty and Merenlender 2004). Wider corridors in Vermont (in the northeastern United States) also contained greater numbers of local bird species, and riparian corridors needed to be at least 150 meters (492 feet) beyond the river's edge to incorporate 90 percent of the bird species found in the community (Spackman and Hughes 1995). Similarly, bird richness in Western Australia and along the Altamaha River in Georgia (in the southeastern United States) was related to corridor width as well as vegetation degradation (Saunders and de Rebeira 1991; Hodges and Krementz 1996). Width also seems to be a good predictor of bird species richness, nest density, and breeding success in windbreaks (Forman 1995). In Washington State (in the northwestern United States), corridor width was among the variables that influenced faunal diversity (Perault and Lomolino 2000). In Alberta, Canada, forest-dependent birds declined in forest corridors through clearcuts less than 100 meters (328 feet) in width (Hannon et al. 2002), as did endemic understory birds in Chile (Sieving et al. 2000). In South Africa, even insect biodiversity as indicated by dung beetles and ants was higher in wider corridors (e.g., 280 meters) that encompassed high natural heterogeneity across the landscape (Van Schalkwyk et al. 2017).

As with fauna, use of corridors by native plants appears to be affected by width. The results of a spatial simulator used to explore the effect of dimensions suggest that width has a much greater effect than length on the probability that plant propagules will be successful, although length remains a significant variable (Tilman et al. 1997). If the corridor is too narrow, plants, which often move by sending undirected propagules, may lose them to nonhabitat or experience edge effects, further contributing to loss of propagules and limiting corridor effectiveness. Also, wider corridors are more likely to maintain microclimate variables desired by forest-interior plants (Forman 1991). Researchers in the Czech Republic concluded that species dependent on forest interior conditions such as many forest herbs, fungi, insects, and mollusks could not migrate through corridors 40 meters (131 feet) wide (Kubeš 1996). Likewise, surveys of river corridors in Vermont indicated that widths of at least 30 meters (98 feet) beyond the stream were necessary to encompass at least 90 percent of the riparian vascular plant species (Spackman and Hughes 1995). These studies clearly indicate that narrow strips of habitat may be insufficient to promote connectivity for many species due to the poor quality of habitat and edge effects.

Some researchers suggest that length and width interact such that longer corridors require greater widths. For instance, Beier (1995) suggested that dimensions of at least 100 meters (328 feet) wide and less than 800 meters (2,624 feet) long would be adequate for cougars. However, if the length were to exceed 1 kilometer (.62 mile), he recommended that the width should be greater than 400 meters (1,312 feet). One promising, more generalized approach bases minimum viable corridor dimension on the size of remaining patches of habitat (Kubeš 1996). Kubeš estimates that habitat patches measuring 0.5 to 5 hectares (1.2 to 12.4 acres) require corridors not exceeding 1,000 to 2,000 meters (.62 to 1.2 miles) in length and be at least 10 to 20 meters (33 to 66 feet) in width. Larger patches (5 to 50 hectares, or 12.4 to 124 acres) would need corridors of no longer than 400 to 1,000 meters (1,312 feet to .62 mile) long and at least 20 to 50 meters (66 to 164 feet) wide to maintain connectivity.

When designing corridors, planners should consider that corridors can provide habitat for species living in them. To facilitate this, corridor management should focus on reducing edge effects, eliminating pets, light, noise, nest predation and parasitism and other disturbances, and invasive species throughout the corridor to the extent possible. Because edge effects in terrestrial systems are significant up to 300 meters (984 feet) from the edge (See chapter 3), Beier et al. (2008) recommend first designing a corridor and then adding this distance to both sides as a buffer to minimize edge effects in the corridor, resulting in a corridor about 1,000 meters (0.62

mile) wide. In general, as Bentrup (2008) lays out, when the length of the corridor increases, so should the width; a corridor will generally need to be wider in landscapes that provide limited habitat or that are dominated by human use; and corridors that need to function for decades and are intended to facilitate range shifts should be wider. When designing a corridor, one question should always consider if the human footprint would increase on both sides of the corridor, would it be able to provide the ecological functions it was planned to fulfill?

Further research on corridor dimension is desirable to provide land managers and conservationists with definitive guidelines that may be widely applicable. Lack of clarity contributes to corridor width arguably being one of the most contentious issues, especially across private land, for example, where stream setbacks may be required and result in restrictions on land use (Dybas 2004). Available data do offer some generalities that can be useful in comparing one corridor design to another corridor design. Choosing sites that minimize length and maximize width is ideal for continuous corridors. For species that are sensitive to edge effects, maintaining corridors that are wide enough to retain some habitat not affected by edge is important. The same concept applies to stepping-stone corridors, where each patch should be large enough to maintain the integrity of the patch. Edge effects vary by habitat type, but, as discussed in chapter 3, the effects can often permeate several hundred meters beyond an edge. As mentioned previously, one of the inherent dangers managers and scientists face is that focal species might perceive corridors differently from our expectations and do not use them as anticipated. For that reason, perhaps we should err on the side of making corridors as large and encompassing as possible and incorporate as many favorable attributes as we can.

For illustrated guidelines for corridors and conservation buffers we refer the reader to Bentrup (2008). Of interest may also be the specific corridor design guidelines that were developed for the Bow Valley in the Rocky Mountains of Canada to accommodate movement by large mammals including grizzly bears and wolves (Bow Corridor Ecosystem Advisory Group 2012).

Landscape Configuration

Landscape elements such as rivers and rock outcroppings that may be barriers to focal species should be incorporated into planning. In addition, abiotic factors, which can deter use, should be considered. For example, wind direction can affect seed and pollen transfer. Similarly, transport of seeds by water should be considered, and migration routes of animals carrying seeds and pollen could be important in determining the location of successful corridors in the landscape.

In identifying corridors, it is also important to consider how the surrounding matrix may affect the utility of the corridor. It is clear from an increasing number of studies that species occurrence at any place in the landscape may not only depend on the immediate characteristics at that spot but also be affected by characteristics of the larger landscape. This means that in assessing and planning connectivity, it is important to consider current and forecasted landscape geometry, since the location with respect to other landscape elements can influence use. For instance, our research conducted in northern California's oak woodland and vineyard matrix examining the distribution of mammalian predators, demonstrates how the configuration of the landscape could influence the current and future probability of the occurrence of predators (Hilty and Merenlender 2004; Hilty et al. 2006). The results of our study indicated that the configuration of vineyard matrix appeared to influence the distribution of mammalian predators. In general, native species had lower probabilities of presence in large vineyard blocks than in isolated vineyards surrounded by natural habitat or within natural habitat. The highest probability of occurrence for nonnative species, in contrast, fell within large blocks of vineyard.

Other studies also indicate that species occurrence and richness in habitat patches and corridors may be influenced not just by local habitat variables in the patch but also by landscape variables farther from the patch. A study in the southern Appalachian Mountains of eastern North America indicated that relative abundance of songbirds was significantly correlated with landscape variables, including both composition and pattern variables up to 2 kilometers (1.2 miles) from the points surveyed. Local habitat variables were, however, more important than the landscape-level effects (Lichstein et al. 2002). Variation in the impact of landscape variables on songbird numbers could be related to the level of intactness of the surrounding landscape and is probably also species specific.

If the overall integrity of natural habitat is low in a given landscape, it may be more difficult to achieve connectivity with either continuous or stepping-stone corridors. If there are options, a good strategy is to place corridors in more intact landscapes. Because it is difficult to maintain and restore biodiversity in a heavily humanized landscape with core areas and corridors, Schmiegelow et al. (2006) propose a paradigm shift in which conservation lands are the supportive matrix and development activities are carefully managed so as not to erode other values. Thus, the islands are where more intensive human activities occur, set within a continuous conservation landscape. This concept has merit especially in biomes that are still relatively intact, such as the boreal forests. However, corridors will continue to be an important conservation tool, probably largely because they are intuitive to a broad audience and are especially relevant in landscapes that already experience a moderate-to-heavy human influence. Also, simple steps can be taken, in some cases, to make it possible for human-altered lands to enhance connectivity. Management of the timing and technique of mowing fields is an example of actions that can minimize disturbance and impact to nesting birds and other species. Mowing from the center of a field toward the edge, for instance, can improve wildlife survival (Natural Resources Conservation District 1999).

Riparian Corridors

Many corridors, both de facto and planned, encompass riparian zones, and maintenance of riparian corridors can protect hydrological and other processes. A disproportionate concentration of habitat diversity, nutrient cycling, productivity, and species interactions occur in the floodplains (Hauer et al. 2016), making them generally important for connectivity conservation. Riparian corridors are commonly used as movement corridors by many species of animals and plants (including terrestrial and aquatic species), support important ecological processes, provide cooler and moister microclimates than the immediate surrounding (especially important in summer or dry seasons), and tend to span climatic gradients as they are oriented along elevational gradients (Beier 2012; Krosby et al. 2014). In addition, riparian areas often enjoy popular support for water quality and recreation benefits, and do not require modeling, making them easy to convey for community conservation efforts (Townsend and Masters 2015). In many places, riparian zones often already have some legal protection (Fremier et al. 2015), though the legal requirements may not be wide enough to support a full suite of species that could potentially benefit from the corridors.

Various species from black bears to forest-dependent birds have been found to successfully use well-buffered stream zones. Riparian zones tend to have less exposure to wind disturbance, and vegetated streams facilitate the movement of many aquatic and semiaquatic animals, such as otters (*Lontra canadensis*) and mink (*Mustela vison*), that are sensitive to fragmentation (Harris et al. 1996; Laurance 1995). In addition, many species may require regular access to water, and stream gullies retain moisture and often contain deep alluvial soils that promote vegetative growth. Furthermore, riparian areas often contain diverse flora and fauna and are generally highly productive. In arid systems, the areas of highest biodiversity often fall within the riparian zones. The results of a study in Queensland, Australia, for instance, illustrate that riparian corridors harbor more bird species than human-planted windbreaks (Crome et al. 1994), and a study in California indicated that mammalian predators were eleven times more likely to be detected along riparian corridors than in the upland matrix (Hilty and Merenlender 2004). While the value of riparian corridors is widely appreciated there is no standard minimum riparian corridor width for ensuring maximum functionality. Minimum corridor widths range from 10 to 30 meters above the high water mark to include 90 percent of the streamside plant species and 75–175 meters are needed to include 90 percent of the bird species in Vermont, USA (Spackman and Hughes 1995).

Some species, however, do not use riparian zones and might need upland corridors. Voles such as *Arborimus longicaudus* and *Clethrionomys californicus* probably would not use riparian corridors, so upland corridors would be needed for them (Simberloff and Cox 1987). Also, riparian zone vegetative composition may vary dramatically from that of other terrestrial communities, so some species might not be represented within a riparian corridor. Disagreement remains as to the importance of maintaining upland corridors, with some researchers arguing that while upland corridors have an important function, wide riparian corridors that contain upland vegetation should function adequately.

Hydrologic Habitat Connectivity: Structural, Functional, and Ecological

It is widely recognized that maintenance of natural patterns of hydrologic connectivity is essential to the viability of populations of many riverine species (Bunn and Arthington 2002; Lytle and Poff 2004). Also, hydrologic connectivity is highly relevant, because hydrologic connectivity is the water-mediated transport of matter, energy, and organisms within or between elements of the hydrologic cycle (Pringle 2001; Freeman et al. 2007). Maintaining and restoring hydrologic connectivity requires that we address human activities such as roads, dams, and rerouting of water for agriculture and urban uses and their influence on freshwater ecosystems. The problem can be extensive, for example, dams have fragmented the tributary networks of six of eight major Andean Amazon river basins (Anderson et al. 2018). Hydrologic connectivity differs from terrestrial habitat connectivity in the fact that hydrologic connectivity is expressed in three dimensions: longitudinal (upstream–downstream), lateral (across the channel/floodplain), and vertical (surface–groundwater). The timing, duration, and magnitude of flow, along with other stream flow dynamics, determine levels of hydrologic connectivity. Therefore, restoring natural flow dynamics is the focus for recovering hydrologic connectivity. However, to recover functional freshwater connectivity hydrologic connectivity must be coupled with species habitat requirements. The integration of connectivity concepts across hydrology and ecology is well illustrated by Larsen et al. (2012), who take advantage of a graph theory construct used often by terrestrial landscape ecologists (Minor and Urban 2008) to measure connectivity for aquatic-dependent species.

At the landscape scale, watershed connectivity captures the importance of the upland processes such as land use and habitat fragmentation on stream processes, including water quality, quantity, and sediment dynamics (Pringle 2003; Tetzlaff et al. 2007). The importance of longitudinal connections from the headwaters down is also recognized in the river continuum concept (Vannote et al. 1980) and river corridor principles (Ward et al. 2002; Gangodagamage et al. 2007). A review of increasing numbers of landscape-scale watershed studies revealed that landscape context and heterogeneity can help to predict freshwater ecosystem composition, structure, and function (Johnson and Host 2010).

In hydrological connectivity, the physical characteristics of the catchment define structural connectivity while the more dynamic information such as rainfall and runoff patterns provides the basis for functional connectivity (Lexartza-Artza and Wainwright 2009). This is different than what these terms refer to in terrestrial connectivity, discussed earlier, where functional connectivity implicitly includes measurements of an organism's movement through the landscape or estimates of habitat suitability required for movement based on species preferences. However, Roy and Le Pichon (2017) define functional connectivity in riverscapes as species- and life-stage-specific capacity to travel from a habitat patch to another, which depends upon the organism's swimming capacities, dispersal behavior, energy costs, and mortality risks. As in terrestrial systems, both structural and functional connectivity and species habitat requirements at different life stages need to be considered to fully understand hydrologic connectivity

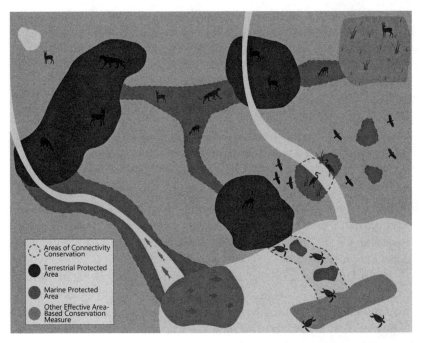

FIGURE 5.5 An ecological network for conservation is a system of connected natural and seminatural landscape elements that is managed with the objective of maintaining or restoring ecological functions to conserve biodiversity. It consists of protected areas and areas of other conservation value that are connected by ecological corridors. (Permission to use by the Center for Large Landscape Conservation.)

and to assess potential consequences of hydrologic alterations on species and ecosystem processes.

Ecological Networks for Conservation

We can create more resilient landscapes by increasing connectivity in all the different ways discussed above. At a landscape-scale level, a variety of conservation elements can function together to create ecological networks in which ecosystem processes are maintained, metapopulations stay connected through dispersal, seasonal migrations are uninterrupted, and species can shift their ranges in response to climate change. In general, networks consist of nodes or core areas that are connected by corridors or edges (fig. 5.5). In ecological networks the nodes are core areas that may or may not be protected but offer high-quality habitat. Ideally, these core areas should be surrounded by a buffer zone, which functions to decrease the negative effects of the matrix onto the core areas. To prevent isolation and all the resulting problems on ecosystems (see chap. 3), core areas need to be connected by corridors of appropriate dimension and configuration. Increasing the sustainability of use within the matrix will further improve ecosystem functioning.

While the goal of increasing habitat connectivity is clear, designing corridors is still a relatively new practice, and we don't have many examples where the influence of these design elements has been measured in the field to assess improvements on functional connectivity (see assessment, chap. 7). Nevertheless, a good deal of thought has been brought to bear on the information needed to design corridors that will improve landscape permeability and should help avoid some of the pitfalls we cover in the next chapter.

Chapter 6

Potential Pitfalls or Disadvantages of Linking Landscapes

Throughout this book we have emphasized the need, indeed the necessity, of maintaining or restoring connectivity in the increasingly fragmented landscapes on this planet. This imperative supports not only the conservation of biodiversity but nothing less than the sustainability of the human life-support system. In this chapter, we will explore the possibility that pursuing this essential endeavor can carry the risk of negative effects. Our stewardship must therefore be sufficiently sophisticated that we anticipate and avoid or mitigate unfortunate results that nullify or even worsen what we are trying to achieve in our conservation and planning programs.

We need to acknowledge that corridors may even include obstacles affecting movements. Moreover, not all linear landscape features are necessarily corridors. Human linear constructions such as windbreaks, roads, canals, riparian buffer zones, and greenbelts serve other functions, such as providing for the movement of people and goods, aesthetics, recreation, or filtering abiotic flows such as wind, pesticides, and fertilizer runoff. While such linear constructs may incidentally serve as wildlife corridors in addition to their primary functions, it is not appropriate to assume that they will inevitably function as corridors, as is sometimes suggested. Because these human-built linear landscape features are often quite different from natural habitats in structure, configuration, and size, it is especially likely that they may promote various negative effects (box 6.1). In evaluating these features as corridors, it is important to start with a clear understanding of their intended functions, keeping in mind that multiple functions may sometimes be in conflict.

Here we discuss what can go wrong with corridors and cause them to fail in their mission, possibly even making things worse. We hope careful attention to these potential pitfalls will allow planners to avoid them, or at

146

BOX 6.1

CAN GRIZZLY BEARS AND HUMANS SHARE CORRIDORS?

Canmore, a small town near the entrance to Banff National Park in Alberta, Canada, is home to about fourteen thousand people who generally care about the health of their environment. This gateway community lies in the Bow Valley, which links Banff and Jasper National Parks to the north with wilderness areas in the Rocky Mountains to the south.

The town is situated on both sides of the Bow Valley River, making the most likely wildlife travel corridor, the river, impassable. Under pressure of expanding residential development and golf courses two corridors were established in the 1990s, one on each side of the valley, higher on the mountainsides. Maintaining connectivity for wildlife in this area is significant both locally and continentally, at the Yellowstone-to-Yukon scale (see case study in chap. 10).

In recent years, increasing recreational use has been decreasing the effectiveness of the wildlife corridors, and human–wildlife conflict is becoming more frequent, especially during berry season (Katz 2017). In 2016, a biker was hurt when encountering a bear; the animal fled when the biker's friend used bear spray (Croteau 2016).

Given recent carnivore–human conflicts, including the story of Bear 148 in 2017 (see Introduction), Bow Valley scientists came together to develop recommendations for proactive measures to mitigate human–wildlife conflicts in the corridors and across the valley. Recognizing the importance of working in a coordinated fashion on these challenges across various jurisdictions they ensured that representatives from all levels of government, from town to province to national park, were involved (Town of Canmore, Town of Banff, and Alberta Environment and Parks 2018). Recommended actions include more education and outreach, keeping wildlife out of developed and high-use areas, consolidating trails, and having predictable trail closures during high-conflict seasons. Night closures were also discussed as an option in some corridors where conflict is likely. For land and wildlife managers and a town dedicated to coexistence, a growing human population and increasing visitor numbers will continue to be a challenge. The proactive recommendations will likely be tested and hopefully will enable wildlife to continue moving through these corridors in the future. For land managers, this grizzly bear case is at the very least a lesson in how the best of intentions to forge linkages through the designation of corridors may not turn out the way they were intended.

least to quickly recognize problems when they appear and rectify them. As the conservation community gains experience with corridors, there should be fewer and fewer unanticipated surprises. Early papers calling attention to possible negative influences of corridors include Panetta (1991), Hobbs (1992), Simberloff et al. (1992), and Bennett (2003). In evaluating negative impacts, it is important to consider (1) whether these impacts compromise the intended purpose of the corridor; (2) whether they overwhelm the beneficial impacts of the corridor; and (3) whether the benefits and deleterious aspects combined are better or worse than a scenario in which the corridor is absent.

Impacts of Edge Effects

Long and narrow corridors can be dominated by conditions related to edge effects, such as noise and light pollution; adverse microclimatic conditions; competition from exotic species; and increased risk of predation, parasitism, or disease (see chap. 3). Species strongly repelled by edge effects are likely to avoid moving through corridors dominated by edge effects (e.g., Lees and Peres 2008), leading to corridors acting as a filter for those species (see below). Edge avoiders are not rare. For example, Laurance (2004) has shown that the majority of Amazonian understory bird species, especially insectivores, respond negatively to edges created by an unpaved road less than 40 meters (231 feet) wide.

As part of a review of potential negative effects of corridors, Haddad et al. (2014) conducted a meta-analysis of the data in twenty-two studies on the impact of edge effects. In about one-third of the studies, edge effects had a negative impact on target species, in one-third there was a positive effect, and in another third there was no effect. Increased predation and reduced abundance or species diversity were documented negative effects of adverse conditions created by edges. Creating or maintaining wide corridors and/or softening the edges between corridors and the matrix are solutions to minimize negative edge effects.

Corridors as Biotic Filters

One of the limitations of corridors is that they generally cannot be used equally readily by all members of the communities constituting the joined patches. This differential permeability of corridors produces a filtering effect that may lead to community patches, identical when isolated, drifting apart in community composition. That possibility is not so much an argument that corridors are a mistake as it is a reason that corridors may not provide complete community connectivity and therefore may fail to solve all problems associated with fragmentation.

To examine further how this filtering process works, we will begin with two situations in which filtering is not a problem. The first is a situation in which corridors are sufficiently broad that entire communities can move through them. Sedentary species and microbiota may take a number of generations to travel even a hundred meters. For long-lived species such as most trees this means that the process may take many decades. A wide corridor with habitats similar to those in the connected patches that is not interrupted by artificial or topographic barriers will provide effective linkages free of biotic drift. Even in that situation, however, one must be cautious about assuming that barriers are absent based on maps alone. Intimate familiarity with the actual corridors is essential.

The second situation where corridors may be completely effective is where large protected areas are connected primarily for the benefit of a few mobile species that require very large home ranges or that need to move seasonally between them. Species in this category would include top carnivores, such as mountain lions, or migratory ungulates, such as deer, elk, and pronghorn in North America. Corridors in these situations could be designed for these target species specifically. This strategy assumes that the rest of the species in the communities involved would not need to be connected, as the individual protected areas would be large and varied enough to meet their needs on a sustainable basis, an assumption that will become complicated as climate change causes many species to shift their ranges to more suitable conditions.

In more typical connectivity projects species filtering issues must be confronted, because they can be anticipated to exclude passage for some species. Patches would thus be disconnected for some portion of the resident species. Unconnected species will generally suffer a greater risk of local patch extinction combined with a reduced chance for rescue by recolonization. Partially connected communities will therefore gradually change in composition as extinctions without colonization proceed over time. Also, while the corridor will facilitate range shifts in response to climate change for some species, others may not be able to respond with movement to the novel conditions.

Differential use of a corridor by species may result from variations in their ability to find a corridor, their inherent vagility, and their responses to whatever edge effects a corridor might manifest. While some species readily cross low-suitability habitat during dispersal or when traveling in the home range (e.g., Keeley et al. 2016), others avoid crossing even seemingly small gaps in suitable habitat. In experimental landscapes, it has been shown that several species of voles (*Microtus*) will readily cross gaps of mowed grass that are less than 4 meters (13 feet) across, are reluctant to cross those between 4 and 9 meters (30 feet), and only rarely will they travel more than 9 meters across inhospitable matrix (Lidicker 1999). Many techniques are now available to study movement and dispersal abilities of species, including GPS collars (Kays et al. 2015), combining location data with cost-distance modeling (e.g., Richard and Armstrong 2010), genetic parentage analysis (e.g., Ismail et al. 2017), and landscape genetic approaches (e.g., Pérez-Méndez et al. 2018). GPS collar studies are especially useful as they are able to analyze and model animal responses to landscape features, identify areas that might provide functional connectivity, or pose higher risks to traveling animals (Tracey et al. 2013; Elliot et al. 2014).

Structural characteristics of corridors can prevent successful passage of some species. A corridor may be too long for some species, it may not have an appropriate width, or it could have gaps or bottlenecks that make passage difficult. It is also possible that specific corridor attributes could select against certain ages or sexes and hence influence the success of corridor passage. Grizzly bear families much preferred using overpasses to cross highways while single individuals were often observed walking through box culverts going under the road (Ford et al. 2017). Thus, highway mitigation measures can act as demographic filters. Finally, some corridors may vary seasonally in their attractiveness for travelers, and so favorable periods may or may not coincide with dispersal pulses of different species.

Species filtering may have effects on the ecosystems connected by a corridor. Even species that are well connected may have strong coactions (interactions) with other species that fail to maintain their connectivity. A predator may lose a prey species in this way and vice versa. Various plant species may lose important herbivores or a pollinator species, and competitors may become more or less abundant. Species in mutualistic (cooperative) coactions may lose out if both partners are not similarly connected. A woodpecker that can easily traverse a corridor may become extinct in a patch if a tree species that is important to it fails to maintain connectivity and becomes extirpated, or even if the tree changes to a younger age structure and thereby fails to provide what the woodpecker needs. Loss of top carnivores through inadequate connectivity may lead to increasing numbers of medium-sized predators that then put prey species at increased risk

(Prugh et al. 2009). So it may come to pass that differential species filtering may lead to a cascade of community disruptions that collectively have a much greater impact over the long term than the corridor filtering does in the first place (see also chap. 2).

In a modeling study, Plotnick and McKinney (1993) conclude that communities will differ in their response to species losses from perturbations, and we can extrapolate this to losses from corridor filtering as well. Specifically, their models suggest that communities in which the component species are strongly connected to each other, both in number of connections and in intensity, will likely experience cascades of extinctions following species loss. On the other hand, the impact on more weakly connected communities that suffer perturbation (or filtering) losses of species may be less likely to include secondary extinction cascades. A corollary of this is Plotnick and McKinney's prediction that species in communities that rarely experience severe perturbations will become strongly connected over time, such as, for example, tropical rain forests or coral reefs. If such a community then suffers a severe impact, it may eventually become devastated. In contrast, members of communities subject to frequent perturbations will not develop numerous intimate coactions and as a result will suffer fewer losses when extinctions do occur.

The message for us here is that species losses caused by corridor filtering may sometimes have major consequences for community integrity, and sometimes the impacts will be minor. A general rule for land planners would be this: The larger the patches to be connected and the more generally passable the connecting links, the less likely it is that problems will arise from community drift.

Facilitation of Invasions

Corridors can facilitate not only the movement of desired species, but also of unwanted species. The effect of corridors on exotic species, deleterious native species, and pathogens and parasites should especially be considered.

Invasions of Exotic Species

Invasions of unwanted species into habitat fragments are perhaps the most widely appreciated of the potential deleterious impacts of corridors. It is generally acknowledged that invasive species are second only to habitat loss and fragmentation in causing the decline and extinction of species worldwide. This is largely a problem generated by humans, because we are responsible for almost all of the introductions of exotic species, either intentionally or accidentally. We have not only transported and released thousands of species outside their natural distributions, but also placed anthropogenically altered communities, in which exotics tend to thrive immediately adjacent to natural ones. Thus, we have set the scene for massive interference in the structure and function of native communities. It is clear that the two most important agents of species endangerment—habitat destruction, and introduction of exotics—are themselves connected in a synergistic fashion. Considerable research effort is directed toward helping us better understand this invasive process (Mooney and Hobbs 2000; Pyšek and Richardson 2010).

Corridors can contribute to this conundrum of negative impacts in two ways. Most important, they usually provide additional edge habitat for exotics living in the matrix to invade natural habitats. They can then interfere with the success of native species whether they are living in the corridor or just passing through. They can do this in multiple ways. First of all, exotics can alter the habitat in corridors by changing cover characteristics or food supplies. Other possibilities are greater predator and parasite pressures and increased interspecific competition. On a longer time scale, exotics living on edges can be selected for improved capability of invading the native communities and thus gradually extend their colonization of the patches as they adapt to the new conditions.

The second way that corridors can be a problem is by allowing introduced species to spread from one patch to another. If an exotic successfully invades a patch of natural habitat, it may then spread to additional patches aided by corridors. Johnson and Cully (2004) describe a likely example of this. They report that colonies of black-tailed prairie dogs (*Cynomys ludovicianus*) that are connected by dry drainage channels, much used by these rodents for dispersal among colonies, are more likely to suffer heavy mortality from sylvatic plague (*Yersinia pestis*), an introduced species in North America, than colonies not so connected. Corridors also increased invasion of a form of nonnative fire ant (*Solenopsis invicta*) that is a weak disperser but forms large colonies that are causing great damage to ecosystems. The other form, not facilitated by corridors, is a strong disperser that forms smaller, less damaging colonies (Resasco et al. 2014). In riparian systems, removal of dams can lead to the spread of invasive species that had been introduced to the reservoir created by a dam (Rahel 2013). Overall however, because most invasive species are strong dispersers corridors likely play a minor, if any role, in facilitating their spread.

Invasions of Deleterious Native Species

Less appreciated is the role that corridors may play in expanding the ranges of native species. Sometimes this entails regional or local range extensions, a process that corridors are often designed to achieve. More commonly, it involves colonization of a patch by native species previously absent from that patch. Such events are mostly not a problem, or indeed a desired outcome if range expansion is an adaptation to climate change, as the invaders are usually normal components of the community type being colonized. They are adapted to patch conditions, and the resident species are adapted to them. Conservation problems do arise occasionally, however, particularly if a target species of concern is negatively affected by the arrival of a new predator, parasite, or competitor. For example, human-induced forest fragmentation has increased wolves near edges and this is contributing to a decline in caribou (*Rangifer tarandus*; Whittington et al. 2011); narrow corridors would also likely be problematic for caribou. In the normal scheme of things, habitat patches that are isolated, either permanently or periodically, may lose species of predators or parasites that are particularly at risk because of the isolation. They may be lost by chance (demographic stochasticity) or by genetic deterioration, especially if the patch is small (chap. 2). Or their prey or host species may become extinct or too rare to support their continued existence in the patch. Prey or host populations may then thrive in the absence of these top consumers. If recolonization of the predator or parasite should occur subsequently, deleterious impacts on the prey or host will ensue. This dynamic pattern is ordinarily not a problem and in fact is one of the processes that is intentionally abetted by connecting patches. Only if a species of conservation concern is so reduced that it survives in only one or a few patches will colonization by a native predator or parasite be potentially disastrous. A poorly conceived corridor construction project may be a bad idea in such a situation.

In assessing specific landscape configurations with the objective of either encouraging or discouraging the persistence of various carnivores, it is important to keep in mind that different predators perceive a patchy environment at different spatial scales. Large and especially more vagile species in some cases may be better equipped to traverse multiple patches and often are more adept at crossing inhospitable matrixes. For them, corridors composed of the same or a similar community type as that of the preferred patch type are not so important. Instead, they may need corridors that allow them to cross human-made barriers. On the other hand, small and less mobile carnivores tend to be tied more closely to specific habitat features and may therefore benefit from corridors that possess the necessary attributes. Orrock and Damschen (2005) report that in patches of clear-cuts imbedded in South Carolina loblolly pine plantations, two species of plants experienced heavier seed predation by rodents in connected than in isolated patches of the same size. Planners contemplating adding corridors to a landscape should consider how the various predators potentially present in the area would be affected by the improved connectivity. Effects could come from changes in the connectivity of the predators or indirectly through resulting changes in their prey base.

Pathogens and Parasites

An area of active research is the ecology of host pathogen or parasite dynamics. Connectivity is an important component of this effort (Plowright et al. 2011). Wildlife species often are hosts of pathogens that can affect humans either directly through infection or indirectly, for example through spread to domestic animals. Emergent diseases are those that have newly appeared in a population or that have been present but are increasing in incidence or geographic range. The spatial configuration of the landscape including the presence or absence of corridors affects how host animals move and come into contact with not-yet-infected conspecifics. A high degree of landscape connectivity, in this context, could be a disadvantage because with uninhibited movements of infected animals disease transmission rate is high and consequently the disease can spread quickly (Meentemeyer et al. 2012). However, there is little evidence that this is indeed occurring (Brearley et al. 2013; Tracey et al. 2014). Inversely, a high degree of fragmentation can lead to eruptions of much more virulent diseases (Quammen 2012). For example, in Australia, deforestation reduced connectivity between populations of flying foxes, which in response moved less between colonies (Plowright et al. 2011). Before landscape fragmentation, bats infected with the Hendra virus regularly moved among the different colonies, quickly reinfecting virus-free colonies. This reduced the number of bats susceptible to the virus. With decreased connectivity, flying fox camps became virus-free for longer periods of time. When the virus then reappeared, it resulted in a short, intense, local epidemic with many individuals succumbing to the disease. With flying foxes thriving on fruit trees in urban and suburban areas, there is an increased risk that humans will come into contact with the disease.

Movements of a pathogen or parasite in a metacommunity can depend on whether transmission is biotic, meaning transported by other organisms (phoresy), or instead depends on abiotically mediated travel. Pathogens and parasites transported by phoresy were found to readily move through a metacommunity and were also present in greater numbers. In contrast, those depending on abiotic transport dispersed with difficulty (Sullivan et al. 2011). Therefore, understanding how animals respond to landscape features can help managers implement strategies to manage diseases. Individual-based models that incorporate functional connectivity and host species density estimates are a powerful approach to assess routes of pathogen transmission, or lack thereof, and inform management strategies (e.g., Root et al. 2009; Marrotte et al. 2017; Tardy et al. 2018).

Demographic Impacts

As we have seen (chap. 3), the matrix community surrounding a habitat fragment may act as a dispersal sink, siphoning off dispersers from the patch and actually keeping the density of a focal species at chronically low levels (fig. 3.11A). Like the matrix, corridors that are habitable can function as dispersal sinks, resulting in populations that cannot persist without continued immigration from source populations (Furrer and Pasinelli 2016; see chap. 3). In fact, they may be even more likely to behave that way because effective corridors are attractive to dispersers from source patches. If the habitat in the corridor is inferior in quality or subject to intensified edge effects, it may also act as a demographic sink. Dispersers may then be enticed to live in the corridors but end up contributing little or nothing to future generations. For example, rates of brood parasitism in southwestern willow flycatcher (*Empidonax traillii extimus*) nests by brown-headed cowbirds (*Molothrus ater*; fig. 6.1) were observed to decrease 1 percent for every meter from the habitat edge. This means that nest success is lower in narrow corridors than in wider ones (Stumpf et al. 2012). High rates of nest parasitism by cowbirds have been shown to cause sites to be population sinks (Rogers et al. 1997). Corridors can also produce demographic sinks if they

FIGURE 6.1 Brown-headed cowbird (*Molothrus ater*) egg in an eastern phoebe (*Sayornis phoebe*) nest. Cowbird chicks tend to be larger than songbirds and can outcompete these other chicks at feeding time. (Photo by Galawebdesign.)

simply lead dispersers to suboptimal habitat patches. Thus, if a corridor is merely a drain on a patch rather than providing connectivity to other favorable patches, its negative effects may outweigh its benefits.

Increased exposure to human depredation is another way in which corridors can function as population sinks. Like predators, humans may find that corridors provide more accessible edge habitat for hunting or allow them to intercept traveling prey more easily. Prey may be further disadvantaged by being confined to a narrow strip of cover, which may offer inadequate protection. Moreover, prey may suffer the disadvantage of traveling outside their normal home ranges and are less familiar with safe hiding places and escape routes. Adina Merenlender observed traps for lemurs in the Masoala Peninsula of Madagascar that were specifically set along narrow corridors of forest. This strategy by local hunters significantly increased the chances that lemurs would be caught. The concentration of waterfowl hunting in migratory flyways is also a well-known phenomenon.

Corridors can influence demographic events such as births, deaths, and dispersal so that they tend to be synchronized among habitat patches. Several studies have been undertaken to determine whether this demographic synchrony increases the risk of extinction because of common responses to perturbations (e.g., Ranta et al. 1995; Kendall et al. 2000; Bunnell et al. 2010). The overall conclusion is that for dispersal to promote demographic synchrony to a degree that would increase extinction risk, dispersal rates would have to be very high. This may happen when distances between patches are short (Haydon and Steen 1997; Kendall et al. 2000). Empirical data also suggest that more abundant species show greater synchrony (Paradis et al. 1999), but more abundant species are less likely to risk extinction. Overall, conditions in which dispersal increases extinction risk are limited to high levels of dispersal where patches are close together and environmental variation is low. Species that would be especially affected would be common, highly mobile species. However, in a system like the North American Great Lakes, it appears that demographic synchrony in an entire lake is reason for concern (Bunnell et al. 2010). It is not known if this is a special case, or if this situation is typical for lakes.

Conservation planners will often face the challenge of balancing the needs for patch connectivity and risky patch synchrony (Allen et al. 1993). However, most landscapes are quite heterogeneous so that this dilemma may not be a concern.

Social Behavior

Earlier (chap. 2), we discussed how a species' social system might influence its ability to travel between habitat fragments. Here we ask whether corridors pose any special problems relating to social behavior. As we will see, there may be reasons why a corridor that seems perfectly adequate to accommodate movements of a focal species may nevertheless fail, with social behavior being the culprit.

In species with complex and perhaps obligatory social groupings, successful dispersal may require that groups of individuals travel together. Corridors will need to be able to accommodate group movements. This may involve better cover and food requirements than would be needed by individual travelers, and perhaps special features such as tree cavities that would be suitable for group resting. Laurance (1995) describes an example in which a highly social arboreal marsupial (*Hemibelideus lemuroides*) never used corridors. Even though this species was common in unfragmented rain forest (Queensland, Australia), it never occurred in habitat fragments. Either the available corridors were not perceived as suitable, or cohesive social forces successfully resisted any long-range dispersal behavior.

Another situation in which social behavior may impede dispersal through a corridor is when it is inhabited by territorial conspecifics. The risk of encountering adults defending their territory can force dispersers out of corridors into the matrix where other risks, including increased encounters with humans, may await them (Keeley et al. 2017). Dispersing young lions (*Panthera leo*) avoided high-quality grass- and shrublands likely to avoid conflict with territorial adults. Instead, they selected for woodlands that constitute low-quality habitats for lions and areas with a high risk of encountering humans (Elliot et al. 2014).

Negative Genetic Effects

The role of fragmentation in influencing the genetic structure of metapopulations has been detailed in chapters 2 and 3. In general, connectivity of demes has favorable genetic consequences. But are there genetic risks implicit in reconnecting landscapes? Potential negative genetic repercussions of corridor construction relate to interbreeding among individuals from populations that have differentiated from each other (fig. 6.2). Such differentiation might result from genetic drift or differential selective pressures operating in different places (chap. 2). Because negative consequences arise from connecting populations with a history of separate evolutionary trajectories, we can anticipate that such consequences will be most likely to occur if populations have been separated for a long time or if large spatial scale connections are developed. For example, corridors that allow organisms to cross major dispersal barriers that have previously allowed subspecific differentiation within a species could lead to hybridization (introgression) and subsequent loss of those subspecific entities. Even taxa at the species level of differentiation can sometimes be at risk. Rare species are particularly vulnerable to this mode of genetic extinction.

There are many examples of North American birds in which eastern and western forms have been brought into hybridization as their distributions have met through urbanization of the Great Plains. Towns with planted trees have provided stepping-stone corridors for expansion of the ranges of forest-adapted species. Three well-known examples are the yellow-shafted and red-shafted flickers (*Colaptes auratus auratus, C. a. cafer*); Baltimore and Bullock's orioles (*Icterus galbula, I. bullockii*); and rosebreasted and black-headed grosbeaks (*Pheuticus ludovicianus, P. melanocephalus*). In all three of these cases, it appears that the hybrid zone has stabilized, so taxon identity may not be lost. In other cases, hybridization can lead to the extinction of native genotypes and enhanced invasiveness of hybrid forms (Meyerson

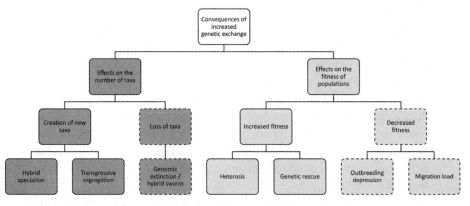

FIGURE 6.2 Flow diagram summarizing the potential consequences of increased genetic exchange. Dark gray boxes contain examples in which gene flow affects the number of taxa; light grey contain examples in which gene flow affects the fitness of populations. Broken box outlines indicate cases in which gene flow has negative consequences for biodiversity, while the solid lines indicate cases in which gene flow has positive consequences for biodiversity. (Modified from Crispo et al. 2011.)

et al. 2010; Rudman and Schluter 2016). Introgression was the cause, or at least a contributing factor, in the extinction of three species of North American fish (Rhymer and Simberloff 1996). A great many plant taxa have disappeared or are threatened by interspecific introgression as well (e.g., Ayres et al. 2008). Often it is an exotic species that genetically swamps a native taxon (e.g., Travis et al. 2010).

In addition to habitat alterations and invasive species leading to contact between formerly spatially separated populations or species, climate change is causing increased genetic exchange (Crispo et al. 2011). As species shift their distributional ranges in response to climate change they are likely to start overlapping with formerly allopatric species. Hybridization and introgression are potential outcomes. Grizzly bears are becoming more common north of their traditional range, and interbreeding with polar bears has been documented. At this point in time it is too early to tell whether this will lead to a widespread breakdown of species barriers in this system (Pongracz et al. 2017).

Disruption of taxonomic entities as depicted in these examples is normally not planned but is an inadvertent result of human activities that connect habitats on a regional or continental scale. Negative genetic impacts are possible on a much more local scale, as well. Where isolation has allowed selection to move populations toward improved local adaptation, enhancement of gene flow may cause the loss of these evolutionary improvements.

This would be an example of outbreeding depression, because mean fitness would decline in the affected demes (Templeton 1986). While outbreeding depression has been observed in several taxa (Waits and Epps 2015), it does not appear to be a widespread problem (Frankham et al. 2011).

On the edges of species distributions, there often are many small demes that tend to harbor novel genetic compositions. This is because in small demes genetically advantageous mutations have a good chance of surviving either through natural selection or by random drift. Thus, small populations can not only improve their adaptation to local conditions, but they also can harbor new genetic innovations. Large populations, on the other hand, show much genetic inertia since mutations, even if advantageous, tend to be swamped out unless the new trait is spectacularly better than the common ones. The dilemma for conservation planning, therefore, is that while small demes are subject to frequent extinction, they are also the main source of new genetic innovations.

With climate change, small populations at the trailing edge of a species' distribution that is shifting poleward or to higher elevations are important because they harbor high genetic diversity and evolutionary potential. Hampe and Petit (2005) argue that improving connectivity should be avoided if it leads to competition with other populations or increases the risk of invasion by nonnative species. However, other studies emphasize the importance of movement between these relict populations that persist in climate refugia to allow repopulation when demes blink out due to stochastic events (Hannah et al. 2014; Morelli et al. 2016; Niskanen et al. 2017).

Although there is clearly a potential for negative genetic impacts with implementation of corridors, the circumstances in which such effects are likely to occur are limited. Particular vulnerabilities are associated with new connections over large spatial scales and where local populations have the potential for local differentiation. It is important to keep in mind that the goal of corridor conservation is to counteract fragmentation, not to create linkages where none have existed before. With climate change forcing species to shift their ranges, previously separated species or populations may nevertheless come into contact, possibly requiring new types of management action.

Conflicting Ecological Objectives

If corridor planners are so fortunate as to be dealing with large blocks of pristine land or aquatic places, there will likely be minimal disagreements over how to proceed. Moreover, protected areas of this type will require minimal maintenance, and connectivity issues will be tractable. However, in this age of environmental onslaught, planners will necessarily be dealing with making the best out of a bad situation. Given this reality, there will inevitably be arguments about priorities, ways to achieve objectives, and even about the best available scientific evidence appropriate for the project at hand (Hodgson et al. 2009; Doerr et al. 2011). A first order disagreement will likely be whether to plan for implementing corridors or enlarging the size of protected areas. What we can be sure about is the following: (1) discussion, compromise, and cooperation among all parties involved will be essential for success; and (2) the more degraded a landscape is to begin with, the more difficult it will be to resolve conflicts and the more extensive maintenance efforts will be required.

Considering the realities of climate change and the uncertainties of its future effects will clearly be an area of considerable concern, and perhaps disagreement, among conservationists. Data will be needed about how species respond to the changing conditions. While a common recommendation for conservation in the face of climate change has been to increase connectivity to allow species to shift their ranges to areas with newly suitable climatic conditions (Heller and Zavaleta 2009), this strategy may not work for species with weak dispersal capabilities or in areas where the climate is changing rapidly (see chap. 8). Instead, other strategies such as increasing climate resilience of existing habitats and increasing protected areas to include a higher diversity of microclimates have been suggested (Schmitz et al. 2015). Focusing on establishing corridors, it could be argued, takes away from focusing on conserving and improving habitat. There is no one answer to the question of how to best balance between these two conservation approaches. It will depend on the landscape, the conservation objectives, the opportunities, and the focal species. Data on future climates and predictions on how species and ecosystems will respond to climate change in an area are important to make informed decisions on the best approach.

Economic Considerations

As discussed in chapter 1, corridors are a tool for establishing effective ecological networks. However, as suggested above, a potential pitfall to avoid in conservation is putting valuable conservation resources into wildlife corridor conservation when initial resources should secure core habitat patches. If it is deemed appropriate to put resources toward wildlife corridors, a related pitfall to avoid is prioritizing corridor conservation based only on ecological criteria. This is because it can lead conservationists to the most expensive solutions. Thus it is critical to also consider cost in optimization models (McRae et al. 2012). One study developed algorithms to optimize corridors for carnivores while minimizing cost. It demonstrated that prioritized corridors with a 75 percent cost savings could fall within the top 15 percent of best corridors from an ecological perspective (Dilkina et al. 2017).

Another cost consideration in securing corridors concerns private lands. The outright purchase price of lands can be prohibitively expensive, but easement expenses (see chap. 10) can be substantial as well. In particular, expenses for monitoring, and dispute costs associated with potential easement violations may occur in addition to regular easement payments. One study suggested that easements may outperform land purchases in long-term cost effectiveness, but only if the rates of disputes and legal challenges remain low (Schuster and Arcese 2015). Another approach to maintaining connectivity on private lands for less money is for communities, or local or regional governments, to require conservation of corridors during the planning phases of land development. If well planned it can ensure that land is set aside and managed to allow for conservation objectives. This approach is likely to require fewer resources as well (Milder 2007; Reed et al. 2014).

In some cases, the restoration of corridors and connectivity can require significant financial resources, especially when particular structures need to be built and maintained. For example, the first two overpasses in Banff National Park cost $1.5 million each, and a proposed overpass for wildlife in Los Angeles over a ten-lane freeway is estimated to cost $55 million (Derworiz 2016). Generally, overpasses and underpasses are expensive due to construction costs and associated fencing (Glista et al. 2009). Likewise, repair and maintenance can be significant. The Trans-Canada Highway that splits Banff National Park is entirely fenced to keep wildlife off the road and funneled to over- and underpasses. The decades old fence along the highway would cost $26 million to replace (Brown and Bell 2016). However, the costs of wildlife collisions by cars are substantial too. One researcher estimated that the average costs for an accident with deer, elk, and moose were $8,015, $17,475, and $28,600, respectively. These costs are offset by development of safe wildlife crossing systems (Huijser et al. 2007). Cost effectiveness of different mitigations measures have also been evaluated and should be considered for any project (Huijser et al. 2009). A number of researchers and groups (e.g., ARC—https://arc-solutions.org) are working to reduce costs of structures and to innovate technologies to lower costs of new overpass designs and alternatives to barrier fencing (Gray 2009). Nevertheless, for any project that requires serious construction or other costs it is important to understand and plan for the immediate and longer-term maintenance costs.

Another economic consideration when restoring or maintaining corridors are costs that affect humans in the matrix beyond the corridor such as those caused by insect pests or wildlife occurring in the corridor that damage agriculture or threaten human security. Incorporating local communities at the planning stage and understanding local people's perceptions of risk is critical in successfully implementing and maintaining corridors in such cases (Treves et al. 2006). Costs of large carnivore impacts on livestock can be enormous. One study estimated that uncompensated costs from 1995–2004 were $222,500 from one cattle allotment in Wyoming (Sommers et al. 2010). Proactively addressing such challenges and developing mitigation measures to significantly decrease costs of predators within and adjacent to corridors will be important if corridors are to function over the long term.

One factor in the economic equation almost universally ignored is the economic benefits that would accrue if a project is successful. Benefits could be aesthetic, educational, recreational, research oriented, health related, improve the quality of life, or result in a multitude of ecosystem services. One reason these benefits are usually ignored is that they are notoriously difficult to convert into monetary terms, in addition to being difficult to quantify and to predict accurately.

In summary, while it is important to be aware of potential pitfalls, most negative impacts of corridors can be effectively mitigated with appropriate planning.

Chapter 7

Identifying, Prioritizing, and Assessing Habitat Connectivity

The focus of this chapter is on planning for connectivity primarily through identifying where habitat corridors currently exist, could be established, or restored. Scale is described in detail because of the influence that it has on identifying and mapping corridors. Then various methods are described that help to identify, prioritize, and assess specific sites that may meet connectivity objectives.

Technological advances in computing power, analysis software, and remote sensing make it possible to view and quantify landscape features more easily than ever before. New modeling approaches are being used to predict where species are likely to be found across the landscape and how they might move through it. Likewise, advances in remote monitoring are allowing researchers to track wildlife better than ever before. The methods discussed in this chapter aid systematic planning for corridor conservation and restoration.

Establishing Collaborations

Conservation scientists have a lot to offer in the planning process to ensure the best possible available scientific information is brought to bear, and that the outcomes will be defensible. However, the science should not be done in a vacuum. In fact, we view initiating a collaborative process as the most important first step in corridor planning and the best way to capture essential local knowledge about the project setting. Collaboration provides a mechanism for incorporating the expertise and desires of local experts and community members into the planning process. Because the sociopolitical realities that communities face are important, especially when it comes to project implementation, stakeholders' diverse opinions should be incorporated early on in the planning process to ensure their long-term commitment throughout the effort. And if the ultimate decisions will be made by certain boards or politically appointed representatives, at least a few of those people should be involved throughout the process as well. If they can voice their thoughts, understand the project goals and constraints, and have confidence in the process, they will be better able to assure other influential decision makers that the best approach is being taken. Input from land managers, wildlife biologists, hydrologists, landscape architects, resource economists, planners, and naturalists is also highly desirable. We discuss the collaborative conservation process in more detail in chapter 10.

Addressing Scale

Scale, both temporal and spatial, comprises extent and resolution. The spatial extent of a connectivity project must be defined early in the planning process, even though, in reality, the geographic extent of an ecological network under consideration often changes over time as additional connections are being protected. Hence, the broader landscape context should be considered even if the current corridor project may be very local in nature. Geographic extent also greatly influences the type of data that will be useful and the outcomes of any prioritization effort.

Generally, for practical purposes we have to sacrifice fine-grain resolution when we increase the extent of a particular study system. Ecosystem variables are usually measured at a smaller scale in the field, and then the information is scaled up to address the larger scope of the project. The opposite is often true for geographic data, which are easier to obtain digitally at a coarse resolution for large geographic areas. However, when applied to smaller areas, it may result in errors due to insufficient resolution of the original data. This is an important point because much of the geographic data used for planning today is collected digitally.

Digital geographic data come in a variety of accuracies and scales and from a variety of sources. It is important to differentiate between local, regional, and global remote imagery and geographic digital data. With drone technology very fine-scale data can be collected for project level planning because they fly close to the ground and therefore can produce topographic maps with 30-centimeter (1 foot) contour lines. Regular airborne photography can usually be acquired at a resolution from 1:200 to 1:10,000 and satellite data are commonly found at 30-meter resolution but are also available at finer resolution (e.g., 0.46 meter). The global scale is exemplified by satellite images from 1:80,000 to 1:1,000,000 resolution. Usually, the larger the geographic extent of these data, the lower the resolution or grain that can be achieved, leading to increased chance for error. While the barriers to using high-resolution information for planning across large expanses of space and time are diminishing as technology advances, there are still tradeoffs between resolution and extent. For example, LiDAR data are captured in the form of three-dimensional clouds of points resulting in billions of points for even a small spatial area. A LiDAR data set for a typical eastern United States county may contain three to six billion points. Furthermore, processing data for a state or province can result in hundreds of gigabytes of data that can tax a single computer using software tools such as ArcGIS (Li et al. 2018). There are Cloud-based solutions to deal with big data such as this, but challenges nevertheless persist.

It is also essential to understand that the minimum mapping unit size for any spatial data coverage may be selected to meet specific project objectives. An example of this is the California Gap Analysis program, which maps land ownership and habitat types using a minimum mapping unit of 100 hectares (247 acres) for uplands and 40 hectares (99 acres) for wetlands (http://www.biogeog.ucsb.edu/research/california-gap-analysis-project). This may be appropriate for state and regional landscape analysis but is too coarse for more local analyses. Extrapolation, the process of estimating information from a given data set, can be problematic when we try to extrapolate from data with a given grain and extent to another area with different dimensions requiring scale transformations. There is a threshold beyond which scale transformation is not effective. For example, classified satellite data often have a limited resolution of 30 by 30 meters (98 by 98 feet) and therefore are not informative at a finer scale. Despite these limitations, it is better to have coarse information for corridor conservation planning than no information.

The scale of a project designed to address connectivity influences the type of digital data that will be useful and the types of analysis that may be appropriate for site identification, assessment, and prioritization. For example, some efforts to reconnect fragmented landscapes are focused on a regional scale, such as within the Greater Yellowstone Ecosystem (Hansen et al. 2002) or across entire countries such as Australia's Wild Country project, which focuses on national-scale landscape connections (McDonald 2004). Other efforts are focused on individual watersheds or even on single barriers to movement such as a highway. It is important to keep in mind that a disconnect between the scale of the ecological processes of interest and the scale of analysis may result in failure to meet conservation objectives.

Identifying Terrestrial Corridors for Conservation and Restoration

Mapping habitat connectivity pathways for the purpose of identifying where to conserve or restore habitat corridors on the landscape greatly influences the eventual outcome. Relevant methods of site selection have evolved in part from research in conservation biology, geographic information system (GIS) science, and systematic conservation planning.

Decisions about which corridors to conserve or where to create new ones may be straightforward in a landscape that consists of clearly delineated patches in an inhospitable matrix where the remaining corridors are an emergent property of the built environment. However, in more complex landscapes modeling the most permeable locations for species movement can help find potentially functioning corridors, and the outcome may differ depending on which species, habitats, or climate benefits are the focus of planning objectives.

While landscapes are often categorized by conservation biologists as patch or matrix, suitable or unsuitable habitat, connected or not connected, more recently, habitat connectivity is viewed as a continuous gradient of permeability. Models estimate the degree of permeability for all grid cells of a landscape based on, for example, their naturalness or their contribution to movement flow (Theobald et al. 2012; McRae et al. 2016). In this section, we give a general overview of the process of modeling corridors for terrestrial wildlife. We divide the process into six steps, starting with the definition of the ecological objectives of the planning effort, modeling where habitat connectivity may be best served, how different corridor options might be prioritized, and ending with assessing the utility of the identified corridors (box 7.1).

Define the Ecological Objectives

At the beginning of the corridor design process it is important to be clear about the ecological objectives. Broadly, the corridors can be planned to improve recovery of specific threatened or endangered species, to support

---

BOX 7.1

STEPS TO IDENTIFY AND PRIORITIZE CORRIDORS
1. Define the ecological objectives.
2. Define what is being connected .
3. Create resistance layer(s).
4. Choose the most appropriate model or algorithm and find existing pathways.
5. Prioritize corridors.
6. Assess the potential utility of the identified corridors.

---

movements of a suite of focal species, to improve connectivity for entire ecosystems that may be increasingly fragmented, or to specifically enable species to shift their ranges with climate change. Strategies to achieve these objectives are to facilitate daily movements or seasonal migrations for focal species, to enable successful dispersal movements to achieve demographic or genetic connectivity for particular species, or to improve overall connectivity in fragmented ecosystems by increasing structural connectivity.

Define What Is Being Connected

Corridor location is largely determined by the endpoints or nodes for what is being connected. There are two options frequently used to delineate nodes: (1) delineated areas such as habitat patches or natural and/or protected areas can be connected to address objectives focused on enabling organisms to move between two or more focal areas; or (2) many points or all cells within a landscape that match certain characteristics can be connected. The objective here is to identify all possible pathways between all locations of a similar nature. For example, pathways between all grid cells with a low human footprint could be identified. In places where many pathways overlap, the connectivity value is high. This approach is often referred to as "node-less" because many similarly defined cells are being connected as compared to a smaller subset of individual endpoints.

To identify which nodes or patches to connect, expert opinion, protected area boundary maps, or empirical occupancy data can be consulted. Using occupancy data, habitat areas can be defined based on actual observed locations, a minimum probability of species occurrence, or polygons drawn around known populations. Alternatively, statistical or subjective rules can be applied to a resistance layer, generally derived from land cover maps, to delineate the nodes. Resistance layers are maps that are intended to provide an estimate of how the landscape affects species' movement. Resistance is high where movement is assumed to be difficult, and low where movement is assumed to be easy. We go into more detail on the importance of resistance in the next section. Options for delineating patches based on a resistance layer include (1) aggregating into patches all cells with a resistance below a threshold value, (2) using a kernel or moving window GIS analysis to group contiguous cells with mean neighborhood resistance values below a threshold, or (3) selecting specific habitat types and special landscape elements that fall above a habitat suitability threshold. Many studies require the nodes to be of a minimum size, which can be based on home range sizes, existing protected area size ranges, or other criteria.

Once the patches have been delineated or nodes selected, a decision needs to be made about the actual endpoints (termini) of the corridors, which have to be specified for the majority of connectivity models. Here, the internal structure of the patch should be considered. Especially when large protected or natural areas are to be connected, suitable habitat for a focal species may only be in small regions within a designated patch. Options on where to place the termini include area centroids, the entire circumference of the patch, points systematically placed within the areas, or in patches of high habitat suitability within the area. If node-less analysis is more desirable, then all cells of a particular characteristic (e.g., native vegetation cover) can be used as termini.

Create Resistance Layers

Resistance layers are spatial rasters, grids of cells organized into rows and columns where each cell contains a value representing information that represent the hypothesized relationships between landscape features (e.g., land cover, topography) and the ability for organisms to move (Spear et al. 2010). Other terms for resistance are "friction," or "cost," referring to the cost of moving through the landscape. We refer to the inverse of resistance as permeability throughout the book to reflect how permeable the landscape is to species movements because this concept is more intuitive.

Estimates of resistance are foundational to most approaches of modeling wildlife corridors and therefore a critical step in the modeling process (Beier et al. 2008; Sawyer et al. 2011; Zeller et al. 2012). If the estimated resistance maps do not accurately reflect actual differences in the probability a species can occupy and/or move through the landscape, then the corridors identified using these maps may not represent highly permeable regions of the landscape. Here we divide estimating resistance into two widely recognized approaches to the problem, one focused on species-specific and the other on structural connectivity approaches (chap. 4).

Species-Specific Estimates of Resistance

Focusing on species-specific connectivity, or what is sometimes referred to as a fine-filter approach, usually requires identifying the type of movement that needs to be facilitated by corridors. For seasonal migration corridors, resistance estimates should reflect how individuals are using the landscape during migrations. For demographic and genetic connectivity, as well as range shift facilitating connectivity, resistance estimates should describe landscape use during dispersal movements. If species have limited dispersal capabilities and can only move through habitats in which they can also persist, resistance estimates should reflect habitat suitability (Wade et al. 2015).

Because habitat suitability is easier to estimate than empirically observing dispersal pathways or developing fully informed movement models, resistance values are commonly derived from habitat suitability models, even for connectivity models intended to facilitate dispersal movements. Low resistance is assigned to highly suitable habitat, medium resistance to habitat of medium suitability, and high resistance to unsuitable habitat. As indicated above and discussed in more detail in the next paragraph, assuming that resistance is the inverse of habitat suitability is not always appropriate. To estimate habitat suitability, expert opinion models are popular because they do not require intensive field data collection (e.g., Larkin et al. 2004; Wikramanayake et al. 2004; Beier et al. 2008). However, some studies that compared resistance estimates from expert opinion and empirical methods found expert opinion to be unreliable (e.g., Shirk et al. 2010, but see Keeley et al. 2016). When feasible, empirical data should be used to inform habitat suitability models. Common field techniques to collect occurrence and movement data of focal species for habitat suitability studies include capture-recapture, GPS telemetry, camera trap, track or fecal survey, and genetic data (Correa Ayram et al. 2016). Depending on available data and ecological objectives, a specific resource selection function is applied to these data to derive habitat suitability (Zeller et al. 2012). Table 7.1 gives an overview of the different resource selection functions used most often.

To estimate resistance from animal movement data there are different options available. Ideally, step or path selection functions are applied to

TABLE 7.1. Resource selection functions for estimating habitat suitability as a basis for resistance estimation (after Zeller et al. [2012] and Wade et al. [2015])

Resource selection function	Correlated variables
Point selection functions	Correlation between presence data and ecological variables
Home range selection functions	Correlation between home range data (generally based on telemetry data) and ecological variables
Matrix selection functions	Correlation between distance (genetic or individual occurrence locations) and ecological variables without assuming the actual movement paths between locations
Step selection functions	Correlation between observed steps from GPS movement data and ecological variables
Path selection functions	Correlation between observed paths from GPS movement data and ecological variables

movements documented by GPS telemetry (Blazquez-Cabrera et al. 2016). Because this kind of data is rarely available, another option is to characterize the behavioral state of all movement steps and use only those for parameterizing the resource selection function during which the animal moved in a fast, directed way (Abrahms et al. 2017). Recognizing that knowledge of habitat suitability is already available or relatively easy to obtain, Keeley et al. (2016) tested linear and nonlinear functions to convert habitat suitability estimates into resistance estimates. After evaluating the resulting resistance maps on a set of documented movements they recommend transforming habitat suitability into resistance with a negative exponential function to reflect that during dispersal animals are often able to move through habitat of medium and even low suitability (fig. 7.1).

Genetic data that reflect the relative degree of relatedness among fragmented populations and hence provide an overall picture of genetic structure for a particular species across the landscape is sometimes used to estimate how connected different habitat patches are for the focal species. By exploring the relationship between population genetic structure and environmental variables, a resistance surface can be developed (Cushman et al. 2006; Shirk et al. 2010; Epps et al. 2007). A comparative study of using habitat suitability versus genetic data for resistance estimation for brown bears in Spain concluded that habitat suitability models tend to overestimate resistance to movement through nonoptimal areas. In other words, in this case

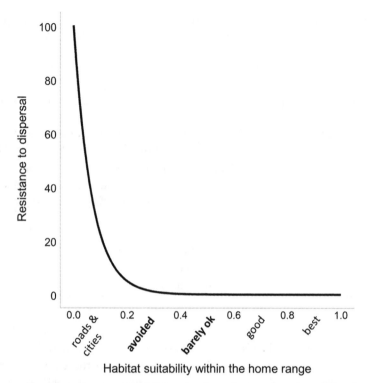

FIGURE 7.1 A negative exponential function best characterizes the relationship between habitat suitability within the home range and resistance to dispersal for mobile mammals.

the brown bears had more gene flow among the habitat patches than might be expected using estimates of movement between patches based on habitat suitability alone (Mateo-Sanchez et al. 2015). For more detailed information on estimating resistance, see Zeller et al. (2012) and Wade et al. (2015).

Estimates of Resistance Based on Matrix Structure

Structural connectivity modeling is sometimes referred to as a coarse-filter approach aimed at identifying areas through which a diversity of species may be able to move (see chap. 4). Resistance in these models implicitly attempts to reflect the relationship between landscape elements and resistance to movement, risk of mortality, or behavioral avoidance. Several studies use naturalness (the degree of human modification) as a proxy for resistance, arguing that it likely represents the potential costs of movement, especially for species that are sensitive to human disturbance. Resistance layers that are based on naturalness assume, for example, that species avoid the built environment and areas of intense agriculture. Some studies also add elevation, slope, and large rivers into the resistance calculation to account for the tendency of species to avoid steep terrain and crossing large rivers (Dickson et al. 2016).

A different approach is taken in the land facet corridor modeling approach that aims to maximize corridor pathways between similar types of landscape units, which are generally defined by elevation, slope, and soil type, found in the neighboring natural areas, thereby providing a conduit for organisms adapted to these conditions to move among these similarly defined regions (Beier and Brost 2010; Brost and Beier 2012; see also chap. 8). Resistance for this approach represents the dissimilarity of a landscape cell to the focal facet type. For example, if the focal land facet is steep, north-facing slopes, cells with this topography would be assigned a low resistance value; gentle, north-facing slopes would have medium resistance and flat areas, high resistance. The assumption is that species adapted to certain land facets, for example, steep, north-facing slopes, will always be able to move between protected areas through corridors that contain this topography or areas most similar to it.

Choose the Model or Algorithm to Identify Pathways

In this section we describe five approaches to modeling connectivity (box 7.2). All of them are performed in a geographic information system (GIS), and all but individual-based modeling need a resistance surface as input. The approaches vary in their objectives, data requirements, and output. Several software packages are freely available online to implement these algorithms, and a good place to look for tools to get started is corridordesigner.org.

Least-cost analysis is focused on determining the least resistant route between two termini by minimizing the cumulative cost of all the cells that make up the entire pathway (fig. 7.2). Increased distance of a path option is also considered more costly, which is important to note because many species may not be able to consider total corridor length when utilizing a particular pathway for movement between patches of suitable habitat. The very lowest cost path exists only for a one-cell-wide corridor, which is not wide enough to function as a wildlife corridor. Therefore, the algorithm has been modified to generate least-cost corridors, either by buffering the least-cost

BOX 7.2

OVERVIEW OF COMMON METHODS FOR IDENTIFYING AND
MAPPING HABITAT CORRIDORS

Model Type	Brief Explanation
Least-cost path analysis	Seeks the path of least cost that a species uses from one point to another or between two to many nodes.
Circuit theory	Adapted from electrical current theory; identifies pathways of least resistance given multiple pathway options between nodes.
Omniscape	Builds on circuit theory to measure current flow from all grid cells within a window to the central target cell, with the window moving systematically across the landscape to measure current flow from all like-cell types to all target cells on the landscape.
Resistant kernel	Calculates the expected relative density of dispersing individuals around source habitats; based on least-cost dispersal from a defined set of source habitats.
Individual-based modeling	Simulates movement paths of individuals by following movement rules to map the estimated relative frequency of use for movement across landscape.

path or by taking the top n^{th} percentile of least-cost paths. Least-cost path analysis has computationally expanded with advances in computing power and it is now possible to calculate least-cost paths between a multitude of patches or cells.

Circuit theory has its foundation in electrical circuitry (McRae et al. 2008). A set of nodes, or cells in a raster landscape, is connected by edges or resistors that reflect the ease of movement. When a current is applied to a node it will flow to other nodes that are electrically grounded. Just like each element in an electric circuit can transmit a current, each cell in a landscape can support movement. More electrical current or movement will flow through elements with low resistance. Current flow in connectivity modeling can also be described through the metaphor of a random walker that will move from one cell to the next, indicating that the model assumes that

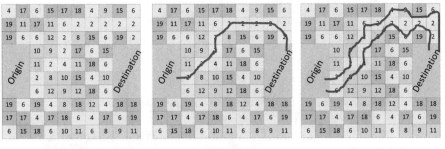

FIGURE 7.2 Least-cost path analysis with values noted from a resistance surface. (Modified from Rudnick et al. 2012.)

individuals moving through the landscape do not have perfect knowledge of the landscape as is the case for least-cost path models between distant termini, but rather make decisions based only on the immediate surroundings.

If most of the landscape between patches is available for movement, the overall current flow will be low, because there are so many options that the flow remains unconstrained and therefore running through at a lower magnitude (fig. 7.3). If there are only few options for movement, the current is constrained, and current flow will be concentrated at a higher level through delineated pathways. These areas of concentrated current flow can be used to designate corridors that may be a priority for protection between habitat patches or nodes.

Omniscape is a modeling approach that extends circuit theory to model connectivity for an entire (node-less) landscape, and does not focus on connectivity between previously and often subjectively defined habitat patches (fig. 7.4). In this case, current flows from all grid cells within a circular window to the central target cell. By systematically moving the window of attention across the landscape, current flow is measured to all target cells on the landscape. The flow maps are then added up to produce one landscape-wide current flow map. The radius of the moving window determines the scale at which landscape connectivity is being considered. As with the previous approaches, flow through cells is affected by landscape resistance. Source and target cells can be weighted by the probability of movement emanating from and traveling to them. For example, higher flow may originate from and go to cells with a higher degree of naturalness, or a higher degree of habitat suitability.

The *resistant kernel approach* models organism dispersal from different source areas by taking into account the resistance of the surrounding landscape to movement, dispersal capabilities, and an assumed dispersal

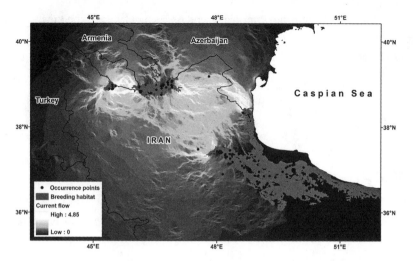

FIGURE 7.3 Model of cumulative current flow used to evaluate habitat connectivity for Persian leopard (*Panthera pardus saxicolor*) across the Caucasus ecoregion (Farhadinia et al. 2015).

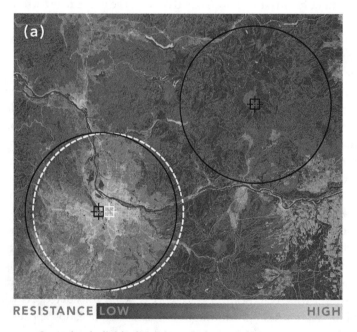

FIGURE 7.4 Summing individual moving window results to create a seamless current flow map. The top panel (directly above) shows locations for two 50-kilometer-radius moving windows (the left window centered on a city, the right window located in a less-human-modified area). The white, broken-line circle illustrates the concept of the moving window operation with "snapshots" taken every time the window is moved. The middle panel (top, opposing page) shows results for the two window snapshots. In both windows, current concentrates toward the center of

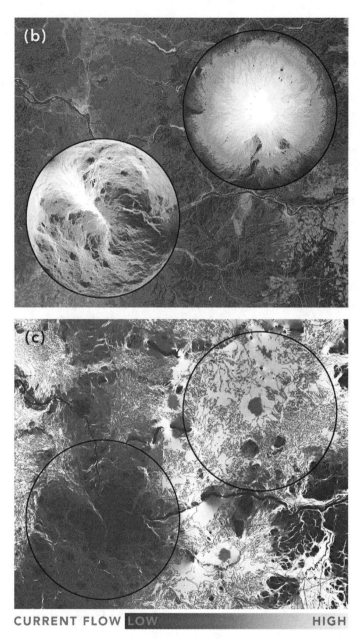

CURRENT FLOW LOW   HIGH

the window. But flow is less constrained, and thus more evenly spread, throughout the less-human-modified area. The bottom panel (directly above) shows the same subset of the study area, with summed current flow from moving windows passed over the entire study area. Flow is lower in heavily modified areas because (1) high resistance causes flow to divert around them when other routes are available; and (2) there are fewer natural areas to connect within 50 kilometers, and thus current sources and targets are fewer and weaker (McRae et al. 2016).

function (i.e., how likely the focal species is to disperse short, medium, and long distances) (Compton et al. 2007). A cost-distance algorithm is used to calculate the expected density of dispersing individuals for each raster cell around the source population. For example, imagine a pond with breeding frogs. When hundreds of young frogs are dispersing, there will be many froglets in suitable habitat close to the pond, but fewer in unsuitable habitat close to the pond and of course far fewer far away from the pond. Summing up the expected densities of dispersing individuals around multiple source populations results in a kernel map for the entire study landscape. This approach can be used to estimate the influence of dispersal patterns across the landscape but may not necessarily reveal specific movement corridors between two nodes.

While all the connectivity modeling approaches described above are based on resistance surfaces, individual-based modeling extracts movement characteristics from empirical movement data. A different model is built to simulate animal movement using species-specific behavioral rules (e.g., movement angles) and sometimes mortality risks in response to landscape elements (Horne et al. 2007; Ament et al. 2014; Allen et al. 2016). Because the simulations are often based on a random walk algorithm, the model assumes that individuals only perceive the immediate surrounding. By generating many simulated movement paths, a map can emerge of the relative frequency of use for movement.

With fast-improving technologies, obtaining animal movement data is becoming easier, even for small, light-weight species (Kays et al. 2015). Consequently, collecting extensive movement data for many species will become possible and building sophisticated movement models or estimating corridors directly from movement paths will become feasible for the first time (LaPoint et al. 2013; Tracey et al. 2013; Bastille-Rousseau, Douglas-Hamilton et al. 2018; Bastille-Rousseau, Wall et al. 2018; Doherty and Driscoll 2018). If based on large amounts of empirical data, these approaches may provide a more realistic representation of how animals perceive and move through the landscape than approaches to identify corridors that are based on resistance maps that represent hypothesized relationships between land cover and species movement.

Prioritization

The approaches outlined above for identifying and mapping corridors as part of the planning process often produce more options for corridor conservation than can be implemented in the near term. The final decision on where to act depends on a wide variety of issues, some of which can be analyzed a priori. The most important components of corridor prioritization are the importance for species conservation, economic costs of the conservation action needed, probability that existing connectivity areas will be lost if no action is taken, and the benefit for climate resilient landscapes.

In addition to focusing on which corridors will provide the most ecological benefit, expertise should be sought to determine the relative costs and threats of potential sites. That way, time and money can be invested in sites that require investment, so that they can continue to provide ecological benefits, and thereby avoid investing in sites that by default will continue to support biodiversity. It is important to estimate only the relative probability that a site will be lost and the relative cost of conserving it compared to other sites. This means that exact threat levels are not as important as how the various sites are ranked from low to high. Failure to consider vulnerability or cost in prioritization efforts will result in suboptimal targeting (Newburn et al. 2005). Economic costs of alternative strategies need to be included in order to address opportunity costs, that is, costs of alternatives that must be forgone in order to pursue a certain action (Faith and Walker 2002).

Approaches from operations research have been used to optimally select the most desirable protected area design from complex options (Williams et al. 2004). There are software packages that use iterative heuristic algorithms to solve these problems based on the concepts of systematic protected area design for conservation planning focused on maximizing biological benefits of site selection while minimizing cost (Moilanen et al. 2009). Several of these programs rely on stochastic optimization routines to prioritize spatial protected area systems that achieve biodiversity or other benefits such as balanced representation of habitat types or connectivity goals with reasonable optimality.

Graph Theory

Most corridor plans are based on static maps that attempt to connect all the potential core protected areas or wildlands. Graph theory is often used to prioritize linkages based on protected area network geometry. The theory is to reduce the number of priority linkages required to connect a multinode network by evaluating the consequences of losing connections on the whole network (Urban and Keitt 2001). Graph theory is the study of the

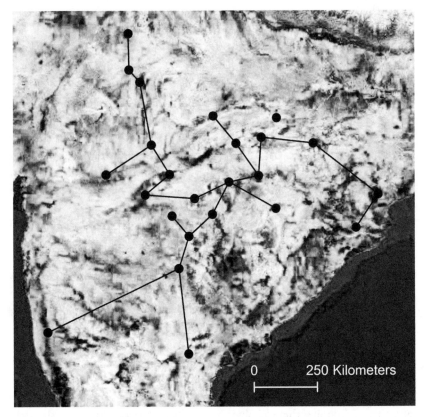

FIGURE 7.5 A feasible tiger corridor network in central India, given by a minimum spanning tree overlaid on the terrain map of the focal landscape. Circles indicate the location of modeled habitat patches; lines are modeled corridor connections. (Modified from Shanu et al. 2016.)

relationships between points (also called nodes) and the lines connecting them (also known as edges) and can be applied to prioritize corridors or protected areas. We can evaluate nodes (habitat patches) and edges (modeled corridors) for their overall contribution to the network, for example by assessing redundancy in connections. Using graph theory to identify the minimum spanning tree can reveal the most efficient geometric solution to identifying the fewest number of corridors required to connect all nodes in a protected area network with the minimum number of edges. See figure 7.5 for an example of a minimum spanning tree linking protected areas for tigers in India. Also, by combining graphs of patches and corridors with a resistance model of the landscape, we can consider, for example, the importance of the corridors specifically for genetic connectivity or de-mographic connectivity (Creech et al. 2014). The approach is valuable for understanding the cohesion of large protected area networks, identifying isolated protected areas, recognizing weak links, and evaluating the conse-quences of restoring or eliminating corridors or patches.

Species Persistence

The main raison d'être for protecting habitat corridors is to increase the likelihood that species and ecosystems will persist into the future, thereby preventing extinctions. Yet, few connectivity studies directly measure the outcomes of implementing corridor conservation on species persistence. Persistence is an important concept for assessing which connections are most important to maintain. Incorporating species persistence into pro-tected area network design relates back to metapopulation theory, where groups of subpopulations need to remain linked to facilitate dispersal be-tween patches and maintain some gene flow. Persistence metrics, such as mean time to extinction or probability of extinction within a given time-frame, are useful for determining the benefits inherent in different connec-tivity configurations. This approach generally relies on modeling species viability based on species-specific life-history information. The goal is to prioritize corridors based on their importance as dispersal pathways be-tween patches to lowering species extinction probabilities.

Population viability analyses take advantage of life-history, demogra-phy, and ecological information to quantify extinction risk (Shaffer 1981; Beissinger and McCullough 2002). These are stochastic models (i.e., they include random variables) of population dynamics, in which population vi-ability is an estimate of the probability of a species going extinct over a spec-ified time period. These models always include species population dynamics based on life-history characteristics such as birth, death, immigration, and emigration rates. Genetic processes, including inbreeding and genetic drift, can be important in small populations (see chaps. 2 and 3) and are some-times included in population viability analyses. Environmental stochasticity is another important determining factor that will influence species persis-tence and can be modeled within a population viability analysis.

The most useful population viability analysis models for corridor plan-ning are spatially explicit and attempt to estimate the population dynamics at different locations in space. This approach combines traditional popu-lation viability analyses with a GIS that includes attributes that influence the parameters used in the population viability analysis, such as the impact of habitat type on reproductive and dispersal rates. Linkie et al. (2006) identified and mapped tiger subpopulation in a region in Sumatra, Indonesia, and used a population viability analysis to assess the importance of corridors for the persistence of the populations under different levels of poaching pressure. The authors concluded that maintaining connectivity between the largest core area and a smaller core area would greatly improve the long-term persistence of tigers in the region.

However, the construction of complete population viability models can require a good deal of data for a single species, making these methods infeasible for applications to large numbers of species. An important way to address this constraint is to develop spatially explicit, stochastic, demographic metapopulation models that can be parameterized for many species. Simulation models can be used to compare how the subtraction of each patch and linkage in a complete network influences mean time to extinction.

Maximizing persistence using a persistence-like index based on the probability of occurrence and colonization was used for European mink (*Mustela lutreola*) and a water clover (*Marsilea quadrifoliar*) as one approach to evaluating spatial linkage design outcomes (Alagador et al. 2016). The influence on species persistence was estimated for 2,500 unique corridors, where approximately 500 of the corridors received a persistence score of 0.01 or greater out of a maximum score of 0.04, revealing that a large number of corridors had little impact on persistence. The need for tradeoffs between maximizing species persistence and the cost of conserving corridors is stronger when less money is available; however, with larger budgets maximizing persistence provides the best performing model for designing ecological networks for conservation.

Financial Investment or Cost

Effective incorporation of economics remains relatively rare in the contemporary protected-area planning literature, but we highlight a few good examples. A study done for Papua New Guinea demonstrated the importance of including opportunity cost into conservation priority-setting algorithms. This was done by illustrating the selected protected areas that met biodiversity targets, minimized costs, and took into account that some sites were unavailable due to conversion to agricultural land (Faith and Walker 2002). In South Africa, scientists evaluated the cost of various acquisition strategies to conserve targeted lands and demonstrated that costs vary depending on the tools used to conserve private land (Pence et al. 2003).

While the price of conserving corridors is certainly accounted for in the implementation phase it is not generally found in connectivity planning prioritization efforts. This is unfortunate as cost is important to consider when prioritizing conservation investments, which was aptly demonstrated by Dilkina et al. (2017). In this study, when corridor options for connectivity for wolverines (*Gulo gulo*) and grizzly bears (*Ursus arctos*) in western Montana were constrained by a fixed budget, connectivity benefits only slightly lower than optimal (11–14 percent lower, respectively) could be achieved for 25 percent of the cost required to protect the maximum connectivity options. Also, as expected, optimizing for multiple species is more cost effective than separately investing in each species alone. This example demonstrates that including costs in corridor conservation planning can improve the efficiency of investments to enhance habitat connectivity for multiple species.

It is clear that more can be done for conservation with less if cost is incorporated at the onset. However, to address cost explicitly, it is useful to employ spatially accurate land valuation models, which are increasingly feasible to develop owing to the availability of parcel databases and advancements in GIS technology. For example, the *hedonic approach* is often employed and uses observed market transactions to infer the market value of parcel characteristics (Rosen 1974). These characteristics may include physical land quality (e.g., slope), location attributes (e.g., proximity to urban centers), and land-use regulations along with other factors influencing the returns to land (e.g., zoning). In many places, parcel records, collected for tax assessment purposes by local and state governments, can provide detailed information on property sales, existing-use value assessments, land use, and other characteristics.

Land-Use Change

The likelihood of future land-use change due to development resulting in conversion and habitat loss also needs to be considered when prioritizing which connectivity areas require protection and over what time horizon. Some areas of the landscape will remain protected because they are not suitable for development or may be at risk in a long time horizon compared to other sites. If only conservation benefits and cost were considered, then it would be assumed that all sites are equally likely to be lost in the future, which is rarely the case. The expected probability of land-use conversion often increases as a function of the value of developable land. This is usually true when the external threat to the site is development, agriculture, logging, or mining. The positive correlation between the probability that a site will be converted and its cost means that low-cost parcels typically have a low likelihood of future conversion. If the relationship between vulnerability and cost is not accounted for, low-cost sites will be selected even if they are not threatened. For that reason, it is important to consider whether a site needs investment to protect or restore it, in addition to how much it would cost to do so. Therefore, it is important to estimate the probability that a site will be vulnerable to land-use change. Typically, ad hoc ranking or rule-based classification is used to formulate a proxy for vulnerability (Abbitt et al. 2000; Pressey and Taffs 2001).

A better approach for finer-scale decision making is to actually estimate land-use change as a function of the underlying biophysical and socioeconomic characteristics, based on sites that have previously been converted (Wilson et al. 2005). Consider a simple land-use change model constructed with respect to a set of developable parcels observed at two time periods. For each developable parcel, there is a binary outcome: remain in the initial developable land use (e.g., forest habitat) or be converted to a more intensive type of land use (e.g., housing). Mapped biophysical and socioeconomic characteristics derived from a GIS serve as explanatory variables in a logistic regression to estimate the relative probability of each land-use alternative. For example, forest conversion to agricultural use will be more likely on areas with suitable soil quality, slope, access to water or precipitation, and access to markets. Coefficients from the logistic regression then may be used to predict the relative probability of land-use conversion for remaining developable sites. The important point is that land-use change models are a better way to examine vulnerability compared to more ad hoc estimates.

There is wide recognition that land use can impact local climate with the classic example being local heat islands in heavily urbanized areas (Taha 1997). In addition, climate change is influencing patterns of land-use change. For example, agricultural development will shift according to climatic suitability, and these changes can threaten natural areas. In fact, models predicting a shift in vineyard development into newly suitable areas is expected to impact native species conservation in California (Roehrdanz and Hannah 2016). With these pending changes in mind, it is important to consider existing land-use change trends as well as how climate change may influence future land conversion in order to fully assess the threats existing connectivity areas may be facing and the relative urgency for protection.

Climate Resilience Benefits

The relative climate benefits that different options for corridor conservation may offer can be an important consideration in prioritizing eventual implementation. One way to approach quantification of differences among climate benefits across various mapped corridor options is to estimate which corridors will facilitate movement of species to cooler climates in response to climate change. The assumption is that a network of connected protected areas may provide the opportunity for range shifts if future climatic conditions at some sites are no longer suitable. Thus, connecting neighboring protected areas with a linkage may provide a temperature benefit in the form of access to cooler locations.

To prioritize corridors in this way the temperature benefit added to each patch or node by maintaining a linkage with a neighboring patch can be calculated using net cooling as an indicator of resilience to climate change. Net cooling values are derived by calculating the difference between the lowest grid cell values for a particular climate metric (such as summer maximum or winter minimum temperatures) based on historic climate data within two connected patches. This value represents the net cooling the corridor provides for any one patch in the network. This value is then assigned to the corridor to represent the added benefit of the corridor in maintaining cooler winter minimum temperatures for example. The corridors offering the greatest climatic benefits to the protected area network can be determined by mapping the climate benefit for each various climate variables into the future. Other climate metrics can also be assessed for different core and corridor networks such as assessing climate diversity, which is often higher in areas with steep elevational gradients. These and other climate considerations are discussed further in chapter 8.

Finally, it is important to remember that the mind is a wonderful tool for integrating information about complex systems and coming up with informed decisions. A new development in connectivity planning is that instead of one conservation planner or a team of planners modeling and prioritizing the corridors for a region, maps are made available on public or private repositories (e.g., databasin.org; Morgan Gray, John Gallo, pers. comm., Jan. 25, 2018), displaying all corridor options and data layers that help multiple groups with their prioritization goals. Also, interactive corridor design tools that allow stakeholders to choose the underlying resistance surface and view and discuss the resulting corridors are highly desirable. Stakeholders in an area can use these data and tools when developing their own conservation plans. This approach fosters the co-design of corridors by those who are involved in implementing corridors and a wide variety of stakeholders.

Assessing Corridors

The type of corridor modeling discussed above produces maps of potential corridors based on geographic map layers. These models contain a good deal of uncertainties due to errors in the digital data and problems with the assumptions inherent in the modeling effort. Therefore, prior to any conservation action in an area it is important to conduct field visits and talk to local experts. The type of data that may be useful to collect at a particular site to estimate its utility as a potential corridor will depend on the initial goals for the site, resources, time available, and site access. Here we review some methods that can be useful for quantifying site characteristics of various corridor options.

Baseline information about potential corridor locations should include social, physical, and biological information. Social information that should be obtained includes land ownership, and possibly the risk of future urban development in the corridor. Identifying the level of interest in conservation and tolerance toward coexisting with wildlife may be useful for understanding the feasibility of implementing corridor plans. Some physical information, such as the location of the site within a particular drainage area, can often be determined from existing maps that may already be part of the project GIS. However, some mapped information may be too coarse to accurately portray a particular site. Therefore, site characteristics such as roads, canals, and fences should be evaluated and existing crossing structures noted during field visits (e.g., Beier et al. 2008). Particular attention needs to be paid to identify any barriers to movement on the site.

Slope, aspect, geology, and soil types may need to be confirmed with field measurements. These variables influence the type of vegetation that may currently or historically be found at the site. If the site is large, it may be necessary to select a set of sampling points where data will be collected. It is important that the sample sites represent the breadth of variation found across the area under consideration. The best way to ensure this is to select sampling sites based on the mapped information, even if it is coarse information. For example, if multiple soil types and changes in stream gradient occur, randomly selecting sample locations evenly across the range of soil types and classes of stream gradient would be best. This is referred to as stratified random sampling and is often used in ecological studies (Lookingbill and Urban 2004).

While digital maps of land cover can sometimes provide coarse-scale information on vegetation types, that information is never completely accurate. Therefore, it may be useful to determine the habitat type of a particular site in the field. This is often done using existing classifications, which define habitats according to dominant plant and associated species. There is much debate in the literature about the utility of habitat classifications that rely on defined climax community types (Cook 1996) because ecosystems are dynamic over time and may not reach equilibrium at a single climax community type due to perturbations and site and changing global conditions. However, it is helpful to obtain existing plant community classifications and determine which community types are represented within the corridor sites of interest.

It is also important to describe how well a corridor is anticipated to function for the focal species by assessing habitat quality and surveying potential corridors for the number of road crossings, the number and severity of bottlenecks, and the distances between species-specific habitats in the corridor in relation to focal species' dispersal distances (Beier et al. 2008; Larkin et al. 2004; Jenness et al. 2011).

There are now many options to determine whether model-determined corridors are being used by focal species. Camera traps are quickly becoming a tool of choice for detecting medium- to large-sized animals. They offer several important advantages, particularly in narrowly vegetated corridors (Hilty and Merenlender 2004; LaPoint et al. 2013; Olsson et al. 2008) and crossing structures (Ng et al. 2004), where they do not require baiting and therefore represent a passive monitoring method. Camera traps combined with occupancy models can also be used to monitor species richness in wide corridors (Cove et al. 2013). Photographed animals can easily be identified by species, and sometimes individuals can be distinguished from one another even by nonexperts (Karanth and Nichols 1998).

Roadkill distributions, winter snow or sand tracking transects, track plates, scat surveys with or without scat-detecting dogs, and bird surveys are ways wildlife biologists or local naturalists can determine whether focal species are using a suggested corridor (Merenlender et al. 1998; Poulsen and Clark 2004; Singleton and Lehmkuhl 1999). For species that are easy to identify, interviews combined with occupancy modeling can be employed to verify the presence of species in a corridor (Zeller et al. 2011). Animal movement data, obtained with GPS collars are another, powerful way to verify animal use of proposed corridors. Documenting movements of mountain lions has been fundamental for promoting and achieving corridor implementation in Southern California (e.g., Morrison and Boyce 2009; Schlotterbeck 2012). In Arizona, GPS movement studies of desert bighorn sheep resulted in three wildlife overpasses being built (Gagnon et al. 2014). As GIS tracking systems become more affordable and easier to deploy, we suggest extensive use of movement data to identify and justify corridors. Direct observations of animal movement paths are powerful but may not be available for a specific study site. Applying statistical models instead to animal movement data from nearby can be a good way to link connectivity evaluations to empirical data. Tracey et al. (2013) developed maps representing movement responses of bobcats to landscape features, which help to evaluate the functional connectivity of proposed corridors for the focal species.

Potential corridors should also be scrutinized for their resilience to climate change. For example, due to increased flooding, culverts may lose their connectivity potential, unless ledges or walkways are built in (fig. 7.6). Considering whether the plant community in the corridor will be resilient to climate change and will provide resources to wildlife during all seasons can inform necessary restoration or management actions (Parodi et al. 2014).

While most of this book is focused on terrestrial connectivity, without increased attention to hydrological connectivity, we stand to lose a good deal of aquatic-dependent biodiversity. As discussed in chapter 5, hydrological habitat connectivity is the explicit coupling of species habitat requirements with hydrological connectivity analysis to provide guidance for the maintenance and restoration of functional freshwater-stream connectivity. The merging of species functional response models with spatially explicit hydrological modeling provides a useful approach to understanding the potential consequences of flow alterations and impaired connectivity on freshwater communities. For example, freshwater removal for human use facilitated by water projects on the Iberian Peninsula designed to mitigate flood risk have led to reductions in the distribution of nine native fishes and the expansion of eighteen introduced freshwater fishes (Aparicio et al. 2000).

The modeling framework we describe below is designed to illustrate recommended spatially explicit, hydrological and ecological modeling components essential to quantify hydrological habitat connectivity. This allows the evaluation of the impacts channel and flow alterations will have

FIGURE 7.6 Culverts with ledges or walkways will retain functionality for connectivity even if flooded. (From Jaeger et al. 2017.)

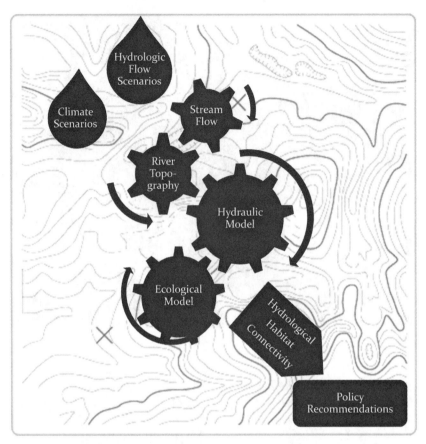

FIGURE 7.7 This integrated hydrological habitat connectivity modeling framework includes: (1) incorporation of climate change models at a fine temporal scale; (2) proposed alterations to the channel morphology at a fine spatial scale; (3) fine-scale empirical hydrology; (4) fine-scale empirical topography data; (5) spatially explicit hydraulic modeling; and (6) species and life-stage-specific ecological models; (7) to produce quantitative ecological outcomes such as changes in habitat availability; and (8) decision support for policy and management adoption. (Modified from Merenlender and Matella 2013.)

on the habitat quality of individual species. It is based on coupling physical and ecological models and is specifically focused on quantifying impacts to hydrological habitat connectivity associated with topographic alterations, flow restoration, or climate change scenarios. Previous research, including a good deal of advances in GIS science, hydrological modeling, and habitat suitability analysis, provides the fundamental components incorporated into the framework illustrated in figure 7.7 (Merenlender and Matella 2013).

This integrated modeling framework is useful to examine hydrological habitat connectivity for aquatic species by evaluating different flow and climate scenarios (step 1) as well as changes in the physical landscape (step 2). Examination of differences in year-to-year changes in stream flow (step 1) is particularly important for many regions where interannual variation in rainfall is high, such as in Mediterranean and desert climates. Forecasted changes in climate introduce another source of uncertainty with respect to the amount and timing of rainfall, which influences hydrological habitat connectivity. Global climate change models can be integrated with hydrological flow estimates to examine the influence of climate change on future stream flow patterns and hydrological connectivity.

Hydrological rerouting, roads, and other physical alterations to the stream channel that can reduce connectivity are common in highly managed watersheds. These alterations can obstruct natural hydrological processes and connectivity by controlling the intensity and frequency of flooding (Bunn and Arthington 2002). Hence, the restoration of stream connectivity often requires physical restoration treatments such as the removal of dams and levees, which changes the physical landscape (step 2) and the resulting hydrological routing of water. This improves lateral connectivity between river channels and adjacent floodplains. Once hydro-climate and physical restoration scenarios have been selected, changes to stream flow and topography caused by the proposed physical alterations must be assessed and processed as the gears in steps 3 and 4 suggest (fig. 7.7). Stream flow records influence the hydrological dynamics (step 3), and can mirror historic periods, future predictions such as climate changes, or proposed changes in managed stream flow. In addition, a physical representation of the channel-floodplain topography (step 4) must be specified in order to examine longitudinal, lateral, or vertical hydrologic connectivity under alternative flow scenarios.

Step 5 takes advantage of advances in hydraulic modeling. For example, using cross sections perpendicular to the flow direction in a standard riverine numerical model can be based on one-dimensional (1D) finite difference solutions of the full Saint–Venant equations using programs such as MIKE-11 and HEC-RAS (Bates and De Roo 2000). Two-dimensional (2D) models using depth-averaged Saint–Venant equations are increasingly used in aquatic biology and geomorphology because they provide finer-scale distributions of velocity vectors with lateral components instead of cross-sectional average downstream speeds that result from a 1D model

(Pasternack et al. 2004). Increasingly efficient 2D models of shallow water hydraulics are also available for creating more realistic representations of floodplain flows (Bates et al. 2006). Three-dimensional (3D) models of fluvial dynamics offer a more complex representation of flows, including a vertical component, but require more computational power and sophisticated assumptions to parameterize a model (Lane et al. 1999). For an example, see Moussa and Bocquillon (2009): a reduced complexity 2D hydraulic model was used to model effects of fourteen construction scenarios of dams, embankments, or both in Southern France.

The connection between hydraulic results and ecological outcomes requires an integrated ecological model (step 6) that stipulates relationships between flow regime characteristics and species responses. For example, examining environmental flow requirements established using generalized flow-ecologic relationships provide the basis for recovering natural flow regimes into river management practice across biogeographic regions (Poff et al. 2010). Unfortunately, very few quantitative ecological response models are available for freshwater species because they require an in-depth understanding of species biology and responses to different hydrological conditions. Hence there is a need for quantifying ecological responses across well-defined gradients of flow regime alterations to support the type of ecological models we propose (Poff and Zimmerman 2010).

A similar framework for evaluating wetland eco-hydrological responses to climate change specific to climate models and static physical landscapes in the United Kingdom can be found in Acreman et al. (2009). They emphasize the need to run models to define relevant ecosystem variables. These can be used to assess the potential impacts of climate on the habitat requirements of the species and communities of interest. Shafroth et al. (2010) used similar approaches to estimate riparian tree seedling establishment, and to evaluate the response of these seedlings to floods based on a 1D HEC-RAS hydraulic model. In addition, they employed a MODFLOW model to assess groundwater interactions contributing to the reemergence of surface flow downstream.

In summary, the steps of the integrated modeling framework (fig. 7.7) currently require the use of an individual model chosen by a user for each major step. This provides flexibility in choice of hydraulic model, ecological model, and data management and spatial tools to evaluate model output. Selection of component models must be made wisely, with a clear understanding of project purpose, data availability, and computational limitations. The final outcome of this integrated modeling framework quantifies the potential ecologic response to changes in hydrological connectivity under the various hydrologic and/or physical alteration scenarios. The results from these integrated modeling scenarios can inform policymakers considering options in a given study area (step 8).

Caveats

There are many caveats associated with constructing GIS models to estimate landscape permeability and determining the best possible pathways for organism movement because in-depth understanding of species dispersal is rare. Models by definition are a simplification of reality and when they differ from truth it can cause consternation. Uncertainty can build due to imperfect species' information, unsubstantiated theoretical basis for model development, and overconfident data extrapolation beyond what is actually supported. For example, if connectivity is modeled based on habitat suitability but the habitat requirements of the designated focal species are not well understood, and the resistance surface is therefore not accurately parameterized, the results will be fraught with errors. Another place where high levels of error can occur is when models are used for extrapolation, that is, models are run outside of the originally observed conditions. For example, species data for resistance estimates should be collected in the landscape or in a similar landscape and for the focal species that is used in the models. Explicitly noting uncertainties allows for honest interpretation of the results and identification of weaknesses or missing data in the modeling process can point to future directions for research, monitoring, and evaluation.

These uncertainties should not diminish the utility of using modeling tools to help plan for habitat connectivity. The models can indeed be very useful as planning tools especially when applied by experts familiar with the local landscape. But it is worth noting the following limitations of some of the most commonly used modeling approaches. The least-cost algorithm assumes that the organism has perfect knowledge of the landscape and purposefully moves to its destination while minimizing travel cost. The results of circuit theory are hard to predict due to the black box nature of the methods used to estimate current flows, and can result in current flows across known barriers in highly constrained or high-resistance areas. This creates challenges for interpretation. Also, this method does not provide much insight for regions with high levels of permeability with, for example, a seminatural matrix. Dispersal kernel methods, as mentioned above, do not result in linear connections between termini and so may need to compliment another method for corridor identification.

Corridor models are simplifications of reality and a tool for evaluating options. They do not necessarily provide the final answer to the question of where to best maintain or reestablish connectivity. Field verification may reveal that a physical barrier such as a dam, road, or natural cliff may prevent animal movement through a particular corridor. Or a particular invasive species or urban adaptor may prevent native species from using or reproducing in a corridor (see chap. 6). The sociopolitical environment and economics are also key in determining how to move from models to implementation. Even with sound corridor models based on the best data, successful implementation will depend on factors such as the willingness of the people to coexist with wildlife, the support of the community for conservation action, and existing policies. For example, public support was strong for building a crossing structure to facilitate wildlife moving across a busy highway, because it simultaneously reduced vehicle collisions with wildlife and made the road safer for motorists (Nancy Siepel, pers. comm., 2016). If private landowners do not need financial benefits for conservation action and/or they are skeptical of land trusts making such proposals, but easements or acquisitions are important to achieving corridor conservation, it may be important to look at alternative pathways where the social dynamics may yield more success. In Australia, the conservation community developed a shared vision of connecting landscapes to conserve biodiversity as a way to increase resilience against climate change and fostered a social movement behind this vision (Pulsford et al. 2012). The National Wildlife Corridors Plan was released for public exhibition in 2012. However, a final plan was not adopted by the Federal Labor Government before the 2013 election. The incoming Abbot Liberal/National Coalition Government shelved it (I. Pulsford, pers. comm., 2018). While connectivity conservation is still proceeding at a local level, high-level national support is lacking.

In summary, determining the type, quality, and potential utility of possible corridors, as well as prioritizing sites for conservation, should be done before long-term investments in conservation and restoration are made. Measuring these factors will also provide baseline information for continued monitoring and adaptive management of the corridor or network. A good safety measure in a world filled with uncertainty is to plan for increased connectivity, and to conserve existing corridors to account for changing landscape conditions and threats.

Chapter 8

Climate-Wise Connectivity

Climate change has added a new dimension to connectivity conservation. Because the need for counteracting fragmentation caused by diverse human land uses has been recognized for many decades, conservation efforts have focused on ensuring that protected areas stay connected so that populations occurring in these areas do not suffer from inbreeding and local extinctions due to stochastic processes. Climate change offers an additional impetus for engaging in linking landscapes for biodiversity conservation. As the climate warms, current habitats may become unsuitable or newly suitable, and many species may be able to respond by shifting their ranges. For connectivity conservation this means that planners cannot use historical ecological data as a reference point for desired outcomes. Connectivity conservation may need to be planned for species that are predicted to move into the area in the future (Lawler et al. 2015). From a focal species perspective, areas established to protect particular species may become obsolete and the need to protect these species in other places will arise (Alagador et al. 2014).

In this chapter, we delve into principles of climate space that influence how species will move in response to changes in climate: climate velocity, climate analogs, climate refugia, and range dynamics at the trailing and leading edges of species' distributions. We then explore the two primary approaches to designing climate-wise connectivity. As discussed in chapter 4, structural connectivity approaches use physical features as surrogates for biodiversity and design corridors to maximize presence of these physical features in the corridors. The focus may be, for example, on land facets, climate gradients, or areas of low human impact. Focal species–based approaches, in contrast, rely on modeling current and future species distributions and movements to identify corridors that promise to facilitate movement even under changing climatic conditions, or help species move to regions that will become suitable.

Principles of Climate Space

Climate Velocity

Ecologists like to work with indices that describe environmental conditions. These indices combine different sources of information to characterize an attribute of interest and enable comparisons between similar environmental systems. Climate velocity is such an index (Loarie et al. 2009; fig. 8.1). It describes the rate at which species have to move to maintain constant climate conditions for every grid cell in the landscape. To compute the index, the change in temperature from current to future conditions is calculated for each grid cell as degrees Celsius per year. This temporal gradient is divided by a spatial gradient: the change in temperature (under current conditions) from a grid cell to its surrounding cells, which has the unit °C/kilometer. The resulting unit is kilometer/year.

A hypothetical example illustrates the results for different landscape features: If we assume a steady 4°C temperature increase over 100 years, it would be an increase of 0.04°C over one year. On a mountain slope, the temperature difference across 1 kilometer may be 0.5°C/km. The climate velocity on this mountain slope comes out to be 0.04/0.5 = 0.08 km/year. To keep up with the changing climates, species would have to move 80 meters/year. However, in a flat landscape, the temperature difference across 1 kilometer is very low—let's assume 0.005°C/km. The climate velocity in this landscape would be 0.4/0.005 = 8 km/year. There are not many terrestrial species that can move and settle in a new location this far away in a single year.

The advantage of this index of climate velocity is that it is simple, easy to interpret, and can be applied globally. It does not make any inferences about the biological response of organisms to climate change, but it is useful for thinking about the effect of climate change on populations in a spatially explicit way. Because all organisms are adapted to certain climatic conditions, climate velocity gives a general idea of how fast populations need to move to track similar conditions over time. However, the effect of climate velocity on a particular species, especially terrestrial species, depends on how tolerant this species is to different climatic conditions, and how well it can disperse to newly suitable conditions (Sandel et al. 2011; Carroll et al.

FIGURE 8.1 Velocity of change for mean annual temperature in the San Francisco Bay Area, California. Velocity of change is calculated as the rate of change per time divided by the spatial gradient of change, and ranges from approximately 0.01 km/year to 5 km/year (Ackerly et al. 2012).

2015). Strong dispersers (many bird species) are less negatively impacted by high velocity than slow dispersers (amphibians; Sandel et al. 2011). For marine species, climate velocity surprisingly explained the magnitude and direction of range shifts very well; knowing a species' characteristics did not improve the predictions of the range shifts much at all (Pinsky et al. 2013). Dispersal through ocean currents and lack of barriers to dispersal movements may explain why marine species track the shifting climate more closely than terrestrial species that often lag behind (Pinsky et al. 2013; Zhu et al. 2012; Moritz et al. 2008).

Since Loarie et al. (2009) introduced the concept of climate velocity, it has been applied to study a variety of questions including its effect on species distributions (Burrows et al. 2014; Pinsky et al. 2013), and the location of climate refugia in relation to species ranges (Roberts and Hamann 2016). By incorporating other climate variables such as mean annual precipitation and mean temperature of the warmest month in addition to mean annual temperature we can examine changes in climate space that may influence species response to climate change (Hamann et al. 2015). This may improve the models because species' distributions are not only determined by the mean annual temperature. However, including precipitation variables adds to uncertainty because climate models vary widely with respect to forecasted changes in rainfall.

One study noted that the original climate velocity index can dramatically overestimate velocity in flat terrain, and underestimate velocity on mountain tops (Hamann et al. 2015). To remedy this situation, the authors suggested an alternative approach to calculating climate velocity. They devised a way to find the shortest geographic distance between a cell and its climatically closest match or analog. They divided this distance by the number of years between the current and future climate to find the velocity required for migration. This new distance-based method can be used to calculate forward and reverse velocity. Forward velocity takes the perspective of species' populations and gives an indication of where species need to move to keep up with climate change. Reverse velocity, on the other hand, takes the perspective of a specific location and asks where new species will be coming from as the climate changes.

Climate velocity is an important concept in corridor ecology, because it can inform connectivity designs. Minimizing velocity along a corridor increases the likelihood that species will be able to keep up with the changing climate (Anderson et al. 2014; Anderson et al. 2016; Heller et al. 2015). Examining maps depicting forward velocity allows resource managers to consider conservation actions to facilitate dispersal of focal species; reverse velocity maps can help protected area managers decide which areas to prioritize for connectivity conservation to ensure that new species can reach a particular reserve (Dobrowski and Parks 2016). Also of interest to reserve managers is the use of climate velocity for characterizing the potential of sites to serve as climate refugia, or whether the sites will likely experience loss of species and the resultant changes in ecosystem processes (Carroll et al. 2015). Finally, the concept of climate velocity is important in the debate about managed relocation: species with low dispersal capabilities in areas of high velocity may not be able to successfully track suitable conditions and therefore may be prime candidates for intense human intervention.

Climate Analogs

Species are adapted to certain climate conditions. While some may be climate-generalists and have a wide distribution spanning several climate zones (e.g., gray wolf [*Canis lupus*], and European roe deer [*Capreolus capreolus*]), many others are climate-specialists adapted to live only in a narrow band of suitable climate and will need to either adapt to a novel climate type or move to areas that will provide suitable climates in the future. To assess the impact of climate change, a species' distribution can be projected onto a future climate map if the current species distribution can be effectively modeled based on climate and other environmental variables. The resulting current and predicted species distribution maps indicate where a species might be able to move to in order to maintain suitable climate conditions.

Along these same lines, sites with today's climate types can be matched to sites that may present a similar climate regime in the future, making the two geographically separated sites analogs to one another. Ensuring habitat connectivity between climate analogs also does not make any assumption about the effect of climate change on species responses. Therefore, increasing connectivity between climate analog sites can be a useful strategy for climate-wise structural connectivity designs intended to facilitate movement for entire communities made up of a large number of species that may be restricted to a particular climate type (Nuñez et al. 2013; McGuire et al. 2016; Littlefield et al. 2017).

Climate Refugia

The concept of refugia was first explored in paleoecology. During the Pleistocene, refugia were identified as areas buffered from climate change that enabled species to persist despite overall climatically adverse conditions. Because of their long-term isolation, new species evolved in the refugia. When the climate began warming, populations expanded and colonized the surrounding landscape (Keppel et al. 2012; Morelli et al. 2017). Similarly, under today's changing climate we expect populations to experience varying rates of climate change and those in places slower to change might provide source populations that can more easily adapt to change and possibly expand into novel climate space over time. Hence, in thinking about climate-wise connectivity, we refer to climate refugia as places of lower climate velocity relative to the surrounding landscape.

There are several physical landscape attributes that can buffer an area from a warming climate (fig. 8.2), including north- or south-facing slopes, areas adjacent to deep lakes or oceans, deep valleys that harbor cold air, streams fed by cold groundwater from deep aquifers, dense canopy cover, and topographically complex terrain (Morelli et al. 2016). The latter, topographically complex terrain, provides reprieve from changing climate conditions because it offers a diversity of microclimates that allows individuals to make small shifts in location but persist in the region (Anderson et al. 2016). Two other terms that we need to introduce here are "in-situ refugia" and "ex-situ refugia." In-situ ("in its original place") refugia are locations that will remain, at least temporarily, suitable for a species (i.e., the type of climate refugia discussed above); ex-situ ("off-site") refugia, on the other hand, are sites that are currently unsuitable for a species but will become suitable in the future due to forecasted climatic shifts (Ashcroft 2010). Populations are expected to persist longer in these refugia than in the surrounding areas. It is important to note that with rapid climate change, populations may not always persist until the climate will cool again; however, refugia could provide some species more time to adapt (Heller et al. 2015), and allow for dispersal opportunities to climatically suitable areas. Inclusion of climate refugia is now a recommended approach for planning of climate-resilient protected area networks (Keppel et al. 2012; Hannah et al. 2014; Keppel et al. 2015; Keppel and Wardell-Johnson 2015).

Support for the importance of refugia for population persistence comes from a study of Belding's ground squirrels (*Urocitellus beldingi*) in California that live in mountain meadows at mid to high elevations. Ground squirrels are more likely to occupy meadows that have experienced less change over the last century with respect to several temperature and precipitation indices (Morelli et al. 2017). Also, the genetic variability was highest within the coolest sites, which indicates that the metapopulations in these sites are large and have persisted for a long time. Taking advantage of a previous study that characterized connectivity between the meadows (Maher et al. 2017), researchers found that well-connected meadows were more likely to harbor a population of ground squirrels and that these populations had higher genetic diversity than more isolated meadows demonstrating the importance of habitat connectivity for species persistence.

Range Dynamics: Trailing and Leading Edges

Species and populations can respond to changes in climate in four different ways. They can go extinct; they can genetically adapt to the new conditions; they can slightly shift their locations to different microhabitats (e.g., from south-facing to north-facing slopes, into a canyon, or upslope); or they can shift their geographic distribution at a large-scale, usually poleward, upslope, or toward the coastline of oceans or large lakes (Jackson and Overpeck 2000; Ackerly 2003).

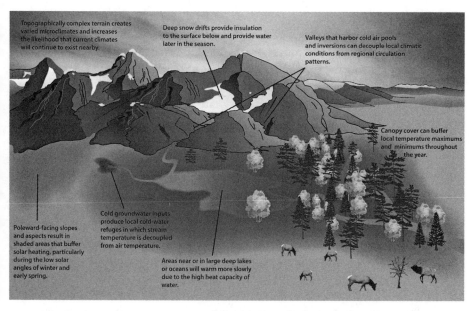

Topographically complex terrain creates varied microclimates and increases the likelihood that current climates will continue to exist nearby.

Deep snow drifts provide insulation to the surface below and provide water later in the season.

Valleys that harbor cold air pools and inversions can decouple local climatic conditions from regional circulation patterns.

Canopy cover can buffer local temperature maximums and minimums throughout the year.

Poleward-facing slopes and aspects result in shaded areas that buffer solar heating, particularly during the low solar angles of winter and early spring.

Cold groundwater inputs produce local cold-water refuges in which stream temperature is decoupled from air temperature.

Areas near or in large deep lakes or oceans will warm more slowly due to the high heat capacity of water.

FIGURE 8.2 Examples of the physical basis for geographic locations likely to experience reduced rates of climate change (Morelli et al. 2016).

When species shift their ranges, there are leading edges (also called expanding edges or the colonizing front), where improving conditions allow range expansion, and trailing edges (or rear edges) with declining conditions that cause range retraction. It is important to understand that in topographically homogeneous terrestrial landscapes, the edge of the shifting ranges may be at the northern and southern ends of the distribution, but in heterogeneous landscapes, leading and trailing edges may occur throughout a species' range, for example, on mountain slopes. The "leading edge model of colonization" explains that populations at the leading edge are most important for range expansions (Hewitt 2000; Hewitt 1993). Four processes—reproduction, dispersal, recruitment, and population growth—determine the speed of range shifts (Thuiller et al. 2008). Long-distance dispersal events into previously uninhabited areas with subsequent exponential population growth is what drives colonization of new areas (Hewitt 1993; Hampe 2005), but species interactions, such as intra- and interspecific competition, parasites, and pathogens affect the successful establishment of new populations at the leading edge.

Many marine organisms produce large numbers of propagules, which are dispersed by ocean currents, resulting in rates of range expansion a magnitude greater than any observed on land. Poloczanska et al. (2013) report a mean rate of expansion at the leading edges for marine species of about 70 kilometers (44 miles) per decade, a rate that has not been documented for terrestrial species most likely due to their more limited dispersal ability, lower fecundity, and barriers associated with land-use change.

Genetic variability is the key for adaptation to novel conditions. At the trailing edge many populations are likely to disappear over time. Successful dispersal throughout a species' range is critical for adaptation through range expansion (Davis and Shaw 2001). However, if populations inhabit areas that are climate refugia, areas where conditions remain comparatively stable through time due to a heterogeneous topography, they can be key for species persistence because they tend to be genetically diverse and may contain traits adaptive to new conditions (Hampe and Petit 2005). These stable populations are typically spatially isolated. While conserving these populations for their genetic potential is vital, improving landscape connectivity may not be possible or even counterproductive because of potential invasions by competitors and loss of genetic diversity.

In conclusion, to facilitate range shifts and retain evolutionary potential, it is important to improve connectivity between current and future suitable habitat, as well as within the current range, including for populations at the trailing edge. However, for species currently occupying climate refugia, increasing connectivity may not be necessary and could even be counterproductive.

Strategies to Improve Range Expansion

Because anthropogenically influenced climate change is still relatively new, there are only few empirical studies testing the effect of corridors and stepping-stones on range expansion. One of them looked at a long-term data set on butterflies in Great Britain that started in 1970. Researchers determined the baseline distribution of five butterfly species in an urbanized landscape and identified a range expansion period from 1983 to 2004. Of the five species, two generalist species were able to move through the matrix just fine; another two used habitat patches that acted as stepping-stones and corridors; and one, a habitat specialist, only expanded in well-connected woodland habitat (Gilchrist et al. 2016).

Because field studies of this type are rare and it is difficult to control for amount of habitat and various landscape configurations in real landscapes, simulation studies are sometimes used for examining the effects of the landscape and species characteristics on range shift patterns. A common conclusion from these studies is that increasing the amount of habitat throughout the landscape is one of the most effective strategies for facilitating range shifts (Keeley, Ackerly et al. 2018; table 8.1). However, concentrating habitat in few large areas reduces the capacity for rapid range shifts, because distances between suitable habitats tend to be too far (Hodgson et al. 2012). Adding corridors between natural or protected areas can be an effective strategy to facilitate range expansion, but the effectiveness depends on many factors discussed throughout this book, such as the size and the elevational gradient in the corridor; the degree of landscape fragmentation; the amount of available habitat; climate velocity; species' dispersal ability; and habitat preferences.

In more intact landscapes, the quantity of suitable habitat for strong dispersers may be more important than the spatial arrangement of suitable habitat for determining successful dispersal. However, in highly fragmented landscapes where less than 20 percent of the habitat remains, plant species will not be able to shift their ranges even if they have strong dispersal capabilities and climate change is kept at bay (Renton et al. 2013). Conserving or restoring connectivity between suitable habitats to facilitate range expansion is most effective for species with medium dispersal capabilities in moderately fragmented landscapes with lower climate velocity. If habitats are naturally isolated (e.g., vernal pools or microrefugia), they need to exist at sufficient density that distances between them are close enough to enable dispersal.

Corridors that cover large areas and have high altitudinal gradients were found to benefit the greatest number of species, whereas small stepping-stones embedded in the matrix are beneficial only to a few species. The effectiveness of both—increasing the number of protected areas in the landscape and adding small but critical corridors—often creates a tradeoff in conservation strategies because, in theory, financial resources can go toward either new protected areas or corridors.

Other strategies to improve landscape connectivity include specifically increasing the size of existing protected areas and enhancing and diversifying the matrix. Simulation studies designed to compare the effectiveness of these different strategies in facilitating range shifts differ in their conclusions. One modeling study concluded that creation of new habitat adjacent to existing small patches gives the most consistent benefit across species (Synes et al. 2015). Another found that adding new habitat to cells chosen at random and to cells with high dispersion and low connectivity provided the most consistent increases in the speed of predicted range expansion

TABLE 8.1. Advantages and disadvantages of strategies to improve climate-wise connectivity (from Keeley, Ackerly et al. 2018)

Strategy	Advantages	Disadvantages
Increasing the amount of conserved habitat throughout the landscape	Increases speed of range shifts in fragmented landscapes; benefits most species	
Concentrating habitat in few, large areas	Increases species persistence for some species	Slows speed of range shifts
Adding corridors between natural or protected areas	Increases speed of range shifts in fragmented landscapes	Potential tradeoff with increasing protected area system; most effective for species with medium dispersal capabilities in moderately fragmented landscapes with lower climate velocity
Creating small stepping-stones embedded in the matrix	Increases speed of range shifts in fragmented landscapes	Benefits fewer species
Increasing the size of existing protected areas	Increases species persistence; improves temporal connectivity for some species	
Improving the permeability of the matrix	Increases speed of range shifts in fragmented landscapes; benefits many species	Unlikely to serve specialist species unless significant habitat restoration is undertaken
Maintaining naturally isolated habitats at a density that permits exchange between habitats	Enables dispersal; ensures species persistence; creates genetic refugia	

(Hodgson et al. 2011). Because different species have different ecological requirements with respect to landscape configuration, landscapes containing large protected areas connected through corridors, or stepping-stones embedded in a permeable matrix, will promote population persistence and facilitate range expansion at the leading edge for the greatest number of species.

Designing Climate-Wise Connectivity

Structural Connectivity

One estimate puts the number of species on Earth (not including bacteria and viruses) at 8.7 million, of which over 85 percent have not yet been described (Mora et al. 2011). Good knowledge of habitat preferences and dispersal abilities has been obtained for only very few species. To design corridors for the majority of species, researchers are using physical features of the landscape as surrogates for biodiversity (Beier et al. 2015). Abiotic surrogates are easily observed landscape characteristics that represent patterns of biodiversity. Surrogates may be land-cover or land-use patterns, topography, geodiversity, or measures of climate space. Below we describe several approaches to structural connectivity modeling.

Riparian Corridors

Riparian corridors tend to span climatic gradients as they are oriented along elevational gradients, and often serve as refugia, especially in more arid climates, because they provide cooler and moister microclimates than the sites immediately surrounding streams and wetlands. Riparian corridors are also commonly used as movement corridors by many species of animals and plants (including terrestrial and aquatic species). Other reasons that make riparian corridors attractive for climate-wise connectivity are that humans often support their conservation such as for water quality and recreation benefits, and they do not require modeling, making them easy to utilize for community conservation efforts (Townsend and Masters 2015). In many places, riparian zones already have some legal protection (Fremier et al. 2015), though the legal requirements may not be wide enough to support a full suite of species that could potentially benefit from the corridors.

For these reasons, riparian corridors are often a priority for climate change resiliency. Applying fixed buffers around riparian areas that connect desired termini has been suggested as a simple method to design riparian corridors (Rouget et al. 2003; Brost and Beier 2012). In cases where no specific areas need to be connected, a method has been developed to decide which riparian areas are best for climate adaptation. It is based on the temperature gradient the river spans over its length; the width of the riparian area; and the levels of canopy cover, solar insolation, and human modification (Krosby et al. 2014). This information is combined in an index of climate-corridor quality to estimate the climate adaptation potential for each of the different segments from the headwaters to downstream reaches.

Lattice-Work Corridors

The lattice-work corridor approach builds on the concept of riparian corridors protecting elevational connectivity, which can enable range shifts to higher elevations, but also delineates perpendicular elevational bands for biodiversity conservation (Townsend and Masters 2015; fig. 8.3). One advantage of this approach is that it provides comprehensive, climate-wise connectivity without the need for developing complex models, which makes involvement by the local communities easier. Transparent, comprehensible conservation planning that includes the stakeholders greatly increases the likelihood of successful connectivity conservation implementation.

Climate-Gradient Corridors

Species with limited dispersal abilities may not be able to cross large expanses of unsuitable climate space. Therefore, one corridor design technique is to plan corridors that follow locations with similar temperature and precipitation patterns, avoiding reversals and abrupt changes in climatic conditions and the built environment (Nuñez et al. 2013; McGuire et al. 2016). Software is available to map these climate-gradient corridors (Climate Linkage Mapper, http://www.circuitscape.org/linkagemapper). Climate-gradient corridors can be more readily achieved in topographically diverse areas and may not be achievable in large flat landscapes.

Naturalness-Based Corridors

The paleo-ecological record suggests that species responded to previous climate changes (e.g., during the Pleistocene) with extensive range shifts, while few species extinctions have been recorded (Davis and Shaw 2001). One of the greatest differences to today's world is the human footprint: extensive areas have been converted to agriculture, cities and towns; and an immense network of roads is fragmenting the landscape, all of which create barriers to species movement. Based on the idea that areas with a low

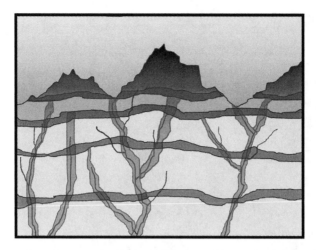

FIGURE 8.3 Schematic example of a lattice-work corridor system. Elevational connectivity along rivers (light gray bands) facilitates dispersal to higher elevations. Connectivity within an elevational band (dark gray bands) facilitates population persistence (Townsend and Masters 2015).

human footprint will facilitate range shifts of species that are sensitive to human disturbance, naturalness-based corridor models prioritize corridors in areas with the least amount of human development, the highest index of wildness, ecological integrity, or ecosystem representation (Belote et al. 2016; Belote et al. 2017). Connectivity can be modeled either between protected areas or on a continuous landscape. To account for the tendency of species to move toward areas that will provide suitable climates in the future, naturalness-based corridors can be prioritized to lead to climate analog sites. To ensure inclusion of high microclimate diversity, sites with high topographic diversity can be prioritized (Schloss et al., forthcoming). In areas where habitat loss and fragmentation are occurring at a rapid rate (e.g., Cho et al. 2014; Cameron et al. 2014), it is also important to take into account predictions of future land-use change when prioritizing naturalness corridors.

Land Facet Corridors

Land units with different topographic and soil characteristics are called "land facets," "enduring features," "geophysical settings," or "ecological

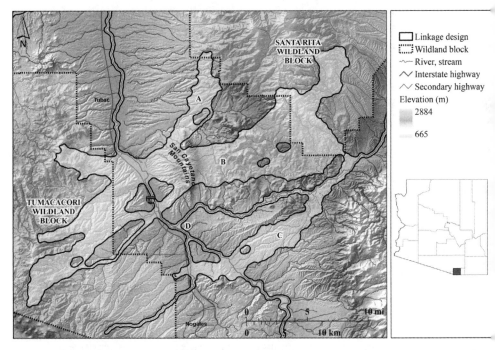

FIGURE 8.4 Map of the land facets linkage design (outlined in black) for the Santa Rita-Tumacacori planning area in Santa Cruz County, Arizona. Linkage strands consisted of corridors for (A) high-elevation, steep canyon bottoms; (B) low-and mid-elevation, gentle canyon bottoms and ridges; mid-elevation, steep canyon bottoms and ridges, and steep, cool and hot slopes; high-elevation, steep ridges, and gentle and hot slopes; and high diversity of land facets; (C) low-elevation, gentle, warm slopes; and (D) riparian habitat. Inset shows location within Arizona, USA. (Permission to use by Brian Brost.)

land units." These features greatly influence the distribution of plants and animals. A different suite of species is associated with a sunny, flat area with fine sediments rather than with a sunny, flat area with coarse sediment, or even a north-facing slope of granite rocks. The diversity of the land units is described by the term "geodiversity." This concept, applied in corridor design and systematic reserve network planning, is commonly referred to as "Conserving Nature's Stage" (Beier et al. 2015; Lawler et al. 2015), from an understanding that these physical factors, that remain relatively constant over time, greatly influence the observed patterning of community types. In land facet corridor design, the objective is to maximize the continuity of land facets among neighboring protected areas, and thereby support movements of organisms adapted to these conditions (Beier and Brost 2010; Brost and Beier 2012; fig. 8.4).

Corridors in Conservation Network Planning

Several approaches have been developed to incorporate climate-wise connectivity into systematic conservation planning. One study took advantage of the innate connective properties of streams (see discussion above) and prioritized land units that are located near streams, while balancing inclusion with acquisition cost (Klein et al. 2009). Applying the concept of Conserving Nature's Stage, a team of sixty scientists developed a comprehensive landscape plan for the eastern United States that seeks to make the landscape climate resilient. Climate-wise connectivity was included by prioritizing microclimatically diverse and locally connected grid cells in each geophysical category. Also, the scientists prioritized regional movement pathways that increase in altitude and latitude (see Anderson et al. 2016 for a in-depth discussion on this method).

Systematic reserve designs that use the conservation planning software Marxan, which optimizes spatial reserve systems to achieve defined biodiversity goals, have found ways to incorporate climate-wise connectivity. One study required that the final site selection be clustered in the end to minimize dissimilarities in topography, soil, and climate in adjacent areas (Game et al. 2011). Another study focused on maximizing hydro-climate diversity in the reserve network thereby capturing the diversity of climate types in the planning region (Heller et al. 2015). Alternatively, climate-wise corridors can be incorporated into systematic conservation plans by identifying and delineating climate gradient corridors and including them as conservation targets (Rouget et al. 2006).

Carbon-Stock Corridors

While all the connectivity design approaches so far address climate adaptation, carbon-stock corridors, aiming to maximize the amount of biomass contained in the corridor, are a climate mitigation strategy aimed at increasing carbon sequestration (Jantz et al. 2014). Implementation of carbon-stock corridors may qualify for funding through climate mitigation programs such as the Reducing Emissions from Deforestation and Forest Degradation (REDD+) and Cap-and-Trade programs. Carbon-stock corridors help reduce emissions and contribute to biodiversity conservation by increasing landscape connectivity.

Including Refugia in Climate-Wise Connectivity Design

Because climate refugia are an important strategy for climate-wise connectivity, we want to pay special attention to how they can be considered in climate-wise connectivity designs. There are four different ways: (1) A set of climate refugia can be identified and included as targets in systematic ecological network planning efforts. In Australia, drought refugia were characterized in arid and semiarid regions as areas of relatively high plant productivity estimated from satellite data. These sites were then prioritized for inclusion in the existing reserve system (Klein et al. 2009). (2) Refugia can serve as the start and end points for corridor selection. Wolverines (*Gulo gulo*) depend on persistent spring snowpack for successful reproduction. McKelvey et al. (2011) used climate models to predict areas where spring snowpack is likely to remain high under climate change, and then modeled the least-cost paths between these areas. (3) Climatically more stable areas can be included in the corridors themselves by using a resistance map that is based on the vulnerability of cells to climate change. Vulnerability to climate change has been determined by measuring topographic diversity and elevation gradients (Anderson et al. 2014) or by assessing the change and variability of climate variables over the past decades (Coristine et al. 2016). Areas with high levels of topographic heterogeneity are also considered refugia because of slower climate velocity and can serve as endpoints for corridor selection (Gallo and Greene 2018). (4) For focal species, ex-situ refugia can be connected to current habitat. As mentioned above, ex-situ refugia are sites that will become suitable for a species in the future. Species distribution models can predict where a species is likely to occur under future climates. By taking into account information on dispersal, connections between current habitat and ex-situ refugia can be modeled. Maintaining or increasing these corridors will facilitate colonization of new habitat (Vos et al. 2008; Pellatt et al. 2012; Brambilla et al. 2017).

Estimating Range Shifts Using Species Distribution Modeling

Biodiversity conservation has long been species-centered, which is reflected in environmental policies (e.g., the IUCN red list of threatened species; the

US Endangered Species Act), supporting science, and management of biological reserves. With climate change, the need for conservation approaches that go beyond individual species and focus on overall biodiversity conservation is recognized. This is reflected, for example, in the focus on structural climate-wise connectivity planning. However, climate-wise species-specific corridor models have their place for several reasons. Humans connect with individual species conservation more easily than with more abstract configurations of land facets or ecosystems, making species-focused corridor plans under some circumstances more effective. Selected suites of species or large species groups may also serve as umbrellas for the majority of species. Often, species-focused policies direct attention to threatened and endangered species, and the connectivity needs of some species may not be well represented using structural or coarse-filter approaches discussed above. If current and predicted species distributions or suitable habitats are spatially disconnected, corridor models can identify the best connections between them, taking into account the species' dispersal capabilities using methods described in chapter 7.

Climate-wise focal species connectivity models design or prioritize corridors based on current and predicted species distribution models. Current species distribution models combine species observation data with environmental variables, mostly topographic and climatic attributes (e.g., elevation, slope, mean annual temperature, precipitation in wettest month) and sometimes land use, to map currently suitable habitat across the landscape. To predict species distributions, mid- to late-century modeled estimates of future climatic conditions are used to estimate where suitable habitat may exist in the future. The final maps show connections between current and likely future suitable locations for individual species.

These models can be run at the local scale for individual, often endangered, species (e.g., Dilts et al. 2016), or at a continental scale to provide a conceptual awareness of which regions will be most important for facilitating range shifts. One study modeled current and predicted species distributions for almost 3,000 vertebrate species in the Americas and found areas that will be especially important for range shifting movements (Lawler et al. 2013).

Recognizing that species will not suddenly, in one step shift their ranges from currently suitable habitat to areas that will be suitable toward the end of the century, but likely will occupy areas with increasing distance to the current range, a set of studies modeled future suitability for every decade. By predicting suitable habitat for every decade of the first half or the entire twenty-first century and modeling corridors between the temporally and spatially advancing habitats, this approach ensures that suitable habitat is connected through time and space allowing species to shift their ranges gradually over time (Williams et al. 2005; Phillips et al. 2008; Rose and Burton 2009; Hannah et al. 2012; Pellatt et al. 2012; Fleishman et al. 2014). It therefore considers temporal connectivity in addition to spatial connectivity

Paleo-Connections

Instead of modeling connectivity areas based on current and future species distributions, the paleo-connections approach identifies regions that likely functioned as biodiversity corridors under past climates. There is a large body of literature about pathways of migration following the ice ages (e.g., Hewitt 2000; McLachlan et al. 2005), and some researchers argue that the areas that connected populations under past climate regimes will also be important under future climate changes (Wu et al. 2017; Fan et al. 2017; Mokany et al. 2017). To identify important areas for past range shifts, one study used current bird distribution data to describe current patterns of diversity and compared these patterns with simulated species richness patterns under paleoclimate models (Wu et al. 2017). This comparison allowed them to assess changes in species richness over time and delineate areas that bridged major biotas in the past.

Assumptions

Studies modeling species distributions into the future make several assumptions. Current distributions of species are assumed to be in equilibrium with climate, and species are not expected to evolve to tolerate new climates (Razgour 2015), even though empirical studies indicate that some species can adapt to the changing climate (Parmesan 2006). Modelers also anticipate that species will move directionally toward locations with analogous climates and that species will be able to disperse from the leading edge of their range, and survive and reproduce in newly suitable habitat (Hodgson et al. 2016). Species interactions are usually not considered, and the lack of this type of realistic biological information decreases model accuracy. However, the species-specific approach is increasingly taking advantage of newly available, fine-scale climate data and georeferenced individual observations for a wide variety of species. These data sets allow for very detailed modeling of the relationships between physical variables and species occurrences that could be used to provide detailed guidance for species conservation and recovery planning (Midgley et al. 2010; Schumaker et al. 2014; Pérez-García et al. 2017).

Model Uncertainty

Many, although not all, of the approaches to climate-wise connectivity planning include forecasting future climates using global climate models. Projections of future climates depend on the rate of continued carbon emissions, how the atmosphere and oceans respond to these emissions, and different model parameters. Also, empirical data, of varying accuracy, are used for downscaling global predictions to a finer geographic scale. Many of the studies that apply these projections deal with the uncertainty by looking at model results based on an assemblage of models, or by bracketing the results using the extreme models on both ends. For example, two of the carbon emission scenarios developed by the Intergovernmental Panel on Climate Change (Pachauri et al. 2014) that are frequently applied to test optimistic and pessimistic conditions are "RCP 4.5," which assumes that emissions will peak around 2040, and "RCP 8.5," which assumes that emissions continue to rise through the twenty-first century (RCP stands for Representative Concentration Pathways).

Uncertainty in examining climate-wise connectivity also comes from land-cover change projections that can vary wildly, uncertain biological information on how species will respond to different aspects of climate, and species' dispersal abilities, which usually are not well known (Rudnick et al. 2012). Perhaps the most challenging aspect of these models is that, under climate change, novel types of climates are predicted; how suitable these novel climates will be for existing species cannot be reliably determined (Capinha et al. 2014). Because the models are predicting species' potential climate space or refugia in the future, it is not possible to validate modeled events by empirical tests. Also, inherent to most models is that they cannot account for all factors driving the response. For example, even models designing corridors for individual plant species to shift their ranges do not account for specific factors such as soil type, seasonally varying soil properties like wetness (Pellatt et al. 2012), or the habitat requirements and dispersal characteristics of animals that disperse the plants.

Recommendations

Selecting the best methods for connectivity design that incorporates climate change depends on the objectives, available data, and the landscape. If the conservation objective is to connect protected areas, we suggest starting with structural connectivity designs. In regions with high topographic diversity, land facet corridors or environmental gradients are two good options. While the land facet corridor approach usually yields several corridor strands between two protected areas offering suitable habitat to species adapted to the different land facets, environmental gradient corridors are limited to one best option. In regions with low topographic diversity, micro-climatically diverse corridors that connect protected areas with an altitudinal or latitudinal differential should be prioritized. If, for public relations reasons, it is critical to base planning on focal species, we recommend selecting a suite of species that represents different life-history strategies, and design climate-wise corridors based on the connectivity needs of these species.

When selecting new sites for protection or to plan a conservation network, we recommend maximizing representation of geophysical settings, and prioritizing micro-climatically diverse and locally connected grid cells in each geophysical category; regional movement pathways should increase in altitude and latitude. In topographically complex regions, naturalness models that include information on climate analog sites and prioritize topo-climatically diverse cells are an alternative option.

For conservation action designed to address the conservation of species of concern within their current range, we recommend finding corridors that will retain function under predictable future climate conditions. At the leading edge of the range, corridors should be designed that will connect current to future habitat. For poor dispersers modeling temporal connectivity will be important; for good dispersers modeling only one time step may be sufficient.

By combining results from structural connectivity and species-focused approaches, complex ecosystems can be addressed as well as particular focal species of interest. Riparian corridors should be included in all connectivity plans because of their importance as natural movement corridors, climate gradients, and refugia (Beier 2012). It is important to provide resident habitat in the corridors when at all possible (Beier et al. 2008; Mackey et al. 2008), implying that wide landscape linkages (e.g., > 1 kilometer) will be more functional than narrow corridors. Making corridors as wide as possible is a simple way to ensure that they contain a diverse topography that provides micro-refugial sites for species persistence (Jewitt et al. 2017). Quantifying the impact of natural and anthropogenic barriers on possible range shifts could inform management strategies within corridors.

In summary, today it is imperative to consider climate change when planning and implementing corridor conservation. Fortunately, a diversity of tools exist that can facilitate corridor design across a wide variety of ecosystems.

Chapter 9

Ecological Connectivity in the Ocean

Mark H. Carr and Elliott L. Hazen

Recent years have seen rapid and substantial advances in our understanding of the environmental determinants that facilitate connectivity and the ecological corridors in the marine environment. Much of this advance stems from technological and genetic advances that have created new tools to better resolve the movement of organisms, the environmental processes associated with spatial and temporal patterns of movement, and the genetic and ecological consequences of connectivity. For example, attached satellite tags have revolutionized our ability to both discover and research the drivers of migration corridors in the open ocean, with new developments and sensors likely to continue to rapidly advance our knowledge (Block et al. 2011; Hays et al. 2016). But it is also motivated by growing recognition of the ecological and conservation significance of movement and connectivity of populations, and how that scales to communities and even more complex ecosystems. While movement and connectivity have become a focus of the design and application of spatial conservation approaches in the ocean for the past decade, there is now a sense of urgency to apply this knowledge to facilitate ecological and evolutionary adaptation of marine populations and mitigation to impacts of a changing global climate. In particular, how do we best design spatial conservation approaches to foster adaptation to and mitigate the detrimental ecological consequences of climate impacts in the ocean? The answer is wholly dependent on our understanding of movement and connectivity.

In order to manage species in the marine environment, we need to understand the timing and patterns of movement of marine organisms. Knowledge of the spatial connectivity of populations and communities in the marine environment has advanced a good deal in recent years. This recent research is largely focused on predicting how species migrations, larval dispersal, and distributions of marine species will adjust to a changing global climate. Moreover, we need to adapt conservation and management approaches, such as coastal and marine spatial planning and marine protected areas (MPAs) to accommodate a changing climate. Species movement patterns reflect the complex interactions between species' life-history traits and environmental factors, both of which differ markedly for species in marine versus terrestrial or freshwater environments (Carr et al. 2003). Though some large marine species that inhabit open ocean (pelagic) ecosystems for all or much of their lives (e.g., sharks, turtles, marine mammals) share similar key life-history traits with terrestrial species, differences between marine and terrestrial environments create marked differences in the configuration, dynamics, and predictability of movement corridors (Hays et al. 2016). Furthermore, the vast majority of marine species that inhabit coastal waters and seafloors produce young as larvae (animals) or spores (algae), capable of being dispersed vast distances from their parents by ocean currents. These differences in movement patterns and connectivity have profound implications not only for what constitutes corridors and the connectivity of populations and communities in the marine environment, but also for how we apply and design protected areas for conservation. In this chapter, we summarize the drivers of species movements, consider spatial connectivity and corridors, describe human threats to marine species and their corridors, and discuss implications for conservation approaches in an ever-changing ocean environment.

What Constitutes Pelagic Connectivity and Corridors?

In the marine environment and particularly in the open ocean, oceanographic features serve as both physical landmarks and habitat for many large pelagic animals. These features can have both vertical and horizontal structure that can serve as migration pathways. Vertically, water column structure varies from mixed at the surface to stratified at a depth where sharp gradients in temperature and salinity can serve to aggregate small plankton, beginning the cascade of foraging habitats for higher trophic level predators. These can be quite thin layers of water (1–10 meters), and yet still can serve important ecological functions (Benoit-Bird and McManus 2012). Below the photic zone (> 300 meters), a mix of fish, crustaceans, and cephalopods inhabit the deep scattering layer, providing an important prey resource for migrating predators in the open ocean (Klevjer et al. 2016). The avoidance of light by these prey species often results in diel vertical migration where species move from deep water to the surface at night to feed in more productive waters while reducing predation risk. These daily movements by organisms among multiple separate foraging niches—a deep prey resource for daytime diving animals and a shallower prey resource for nighttime feeding animals—can serve to connect deepwater nutrients to surface processes. Many ocean predators follow similar diel vertical migration behaviors when tracking resources, thereby contributing to this vertical intermixing of resources across ocean layers.

Horizontal features such as ocean currents for marine species and wind fields for seabirds can serve as direct drivers of species distribution by aggregating prey resources or serving as migratory corridors to reach new habitats (Weimerskirch et al. 2000; Luschi et al. 2003). Changes in ocean temperatures can result in direct physiological constraints to movement and indirect constraints by limiting prey availability. Temperature frontal boundaries in the ocean, where warm water meets cool and nutrient-rich water, are areas where prey species are often densely aggregated (Scales et al. 2014; Woodson and Litvin 2015). They allow cold-blooded predators to stay warm on one side, while foraging on the cooler more productive side of a front (Snyder et al. 2017). These features can be hard to measure in the marine environment, although remotely sensed satellite data have been particularly useful for this (Scales et al. 2018). By measuring where there is a high rate of change or a high standard deviation in temperature over the ocean surface these frontal boundaries can be identified (Hazen, Suryan et al. 2013). Temperature changes are not solely responsible for aggregation, so the ridges and valleys in the height of the ocean (sea surface height) can indicate areas where lower trophic level prey are passively aggregated, often creating a foraging opportunity for higher trophic levels (Kai et al. 2009). In addition to passive aggregation, areas of persistent and consistent phytoplankton blooms can serve as fuel for the bottom of the food web resulting in increased growth of zooplankton and in turn in foraging hotspots for top predators (Block et al. 2011; Suryan et al. 2012). Seasonal patterns in the marine realm can result in migration corridors where animals are moving to stay within a preferred temperature range and to maximize their foraging opportunities similar to terrestrial ungulates surfing the green wave (Block et al. 2011; Merkle et al. 2016). In this way, pelagic migratory pathways often connect productive foraging habitat to warm-water breeding habitat.

Highly migratory marine species also interact with their environment at multiple nested scales, such as annual migrations from breeding to foraging grounds (Kenney et al. 2001; fig. 9.1), seasonal migrations within foraging

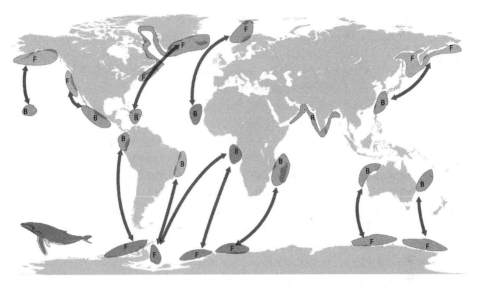

FIGURE 9.1 Global depiction of humpback whale migration corridors from warm water breeding and calving (B) to cool water, productive foraging (F) habitats. R identifies a potentially resident population. (Figure modified from: http://www .grida.no/resources/7650.)

grounds, and weekly or daily movements to take advantage of temporary food patches formed when prey aggregate and disperse. Some marine predators have been termed "central place foragers," such as seabirds, pinnipeds, and marine turtles that are tied to land to breed. They would have an increased requirement for pathways or corridors back and forth in the ocean to connect them back and forth to their terrestrial breeding locations. Breeding habitats are often surrounded by calm, warm water, in contrast to the cool, productive waters that underlie foraging habitat, further driving this need to migrate (fig. 9.1). Many species have been shown to have fidelity to exact migration routes with individuals retracing their tracks with extreme precision each year so that these routes can be considered important marine corridors (Costa et al. 2012; Abrahms et al. 2018).

Where Are the Major Pelagic Marine Corridors?

There are a number of physical features that have been identified as critical corridors for migration and foraging in the open ocean including but not limited to the Gulf Stream in the Atlantic (Musick and Limpus 1997); the

Antarctic Circumpolar Current in the Southern Ocean (Tynan 1998); the North Pacific Transition Zone (NPTZ) in the north Pacific (Polovina et al. 2017); and the equatorial upwelling zone in the central Pacific (Carlisle et al. 2017). These features are examples of important corridors because they can provide increased opportunity for adult dispersal or larval transport (see coastal section below for larval transport specifics) while also conferring greater foraging opportunities when prey are aggregated. Many of these corridors occur in areas beyond national jurisdiction and thus fall under international laws such as the United Nations Convention for the Law of the Seas. Therefore, conservation and management of these corridors require multinational collaboration and decision-making; one recent example includes efforts under way to identify ecologically/biologically significant areas (EBSAs). Many areas that support resident or migration hotspots in the world's oceans have been identified under the EBSA process including the North Pacific Transition Zone and the equatorial Pacific. Plans have been set in motion to encourage marine spatial planning as a way to further define EBSAs (Dunstan et al. 2016). Sadly, there exists no direct mechanism for conservation or management of identified EBSAs, highlighting the ongoing need for increased attention to the protection of the open ocean (Dunstan et al. 2016).

These open ocean EBSAs are vital for oceans to remain productive. For example, the NPTZ is often delineated by the Transition Zone Chlorophyll Front (TZCF). The TZCF is defined by the 0.2 mg chlorophyll $/1m^3$ border between the subarctic gyre and North Pacific gyre, and the NPTZ is defined by where warm tropical waters meet cooler more productive subarctic waters (Thorne et al. 2015; Polovina et al. 2017). These features harbor a rich fauna, serving both as critical foraging habitat for species such as neon flying squid (*Ommastrephes bartramii*), Hawaiian monk seals (*Neomonachus schauinslandi*), Laysan and Blackfoot albatross (*Phoebastria immutabilis* and *P. nigripes*), and Northern elephant seals (*Mirounga angustirostris*), all of which use the NPTZ to take advantage of increased foraging opportunities (Thorne et al. 2015; Polovina et al. 2017; Abrahms et al. 2018; fig. 9.2). In addition, it is an important corridor between the eastern and western Pacific for bluefin tuna (*Thunnus orientalis*) and loggerhead turtles (*Caretta caretta*), as both species migrate from western Pacific breeding grounds to eastern Pacific foraging grounds (Polovina et al. 2001; Block et al. 2011; Briscoe et al. 2016). The currents along the NPTZ likely aid migration and also provide increased foraging opportunities while transiting.

In the central Pacific, the convergence of currents (the westward equa-

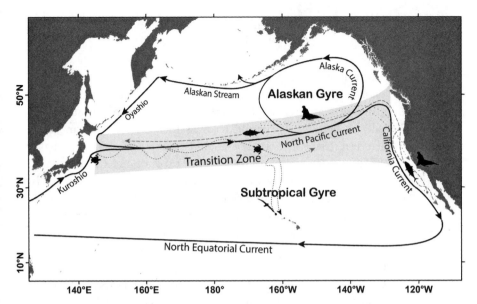

FIGURE 9.2 Examples of the use of the North Pacific Transition Zone (NPTZ). It serves as a migratory corridor for juvenile loggerhead turtles that disperse eastward (Briscoe et al. 2017), adult bluefin tuna that swim north and south along the California Current seasonally and then migrate westward along the NPTZ to reproduce (Block et al. 2011), Northern elephant seals that migrate from the California Current to use the NPTZ as foraging grounds (Costa et al. 2012), and Laysan albatross that use the NPTZ as a destination during brooding trips (Thorne et al. 2016). (Figure modified from Block et al. 2011, supplementary information.)

torial undercurrents with the eastward flowing equatorial counter currents) results in upwelling of nutrients and convergence of plankton and fish species (fig. 9.2). These features in turn attract migrating seabirds such as phalaropes and storm petrels that take advantage of the increased productivity to forage (Hunt Jr. 1990). However, El Niño Southern Oscillation (ENSO) variability can lead to dramatic changes in the strength of the undercurrents and in turn in upwelling productivity. During El Niño events, the north equatorial undercurrent is significantly weakened, and hence surface waters are warmer, resulting in mostly eastward flow. This is in contrast to La Niña events when the north equatorial undercurrent strengthens increasing westward flow and upwelling of nutrients to the photic zone (Wyrtki 1975; Johnson et al. 2000). This interannual variability in current patterns results in changes to both the location and importance of these marine corridors.

Threats to Pelagic Corridors and Potential
Conservation Approaches

Using ocean features as important migration cues, many pelagic animals make repeated migrations with high fidelity (Horton et al. 2017; fig. 9.2). However, marine features change in strength and location seasonally and interannually as broad-scale forcing changes the ocean ecosystem more rapidly than terrestrial corridors. For example, the NPTZ shifts seasonally over 1,000 kilometers as the subtropical gyre expands and contracts, and yet has also shown a steady northward shift over the past forty years that is predicted to continue over the next century (Polovina et al. 2008; Polovina et al. 2011). As the NPTZ moves farther north, it may be out of range for foraging pinnipeds and seabirds (Thorne et al. 2015; Polovina et al. 2017), and long-term warming may even negate its utility as a corridor (Hazen, Jorgensen et al. 2013).

Most nations have designated marine "roads" near ports as shipping lanes that can present a threat to large whales and other taxa (Redfern et al. 2013; Moore et al. 2018). Recent studies have examined locations of highly migratory predators and over twenty potential risks in the California Current and found areas of high overlap, particularly close to shore (Halpern et al. 2008; Halpern et al. 2009; Maxwell et al. 2013). It is worth noting that human uses of the ocean can change locations as demand for resources and shipping routes shift. Even when human uses remain fixed in location and time, the species distributions can change, ultimately modifying the overall risk and requiring additional flexibility in management as exemplified below (Maxwell et al. 2015).

A recent review described how oceanographic changes could result in new barriers to or pathways for animal migration, isolating or connecting populations (Briscoe et al. 2017). This could be caused by a change in thermal properties or current structure that aids migration, or it could be the appearance or disappearance of a physical barrier that may inhibit species from crossing. For example, equatorial upwelling and associated cool waters during La Niña conditions can act as a barrier to migration, because species tend to avoid crossing hemispheres (Briscoe et al. 2017). In contrast, anomalous warm ocean conditions created a pathway in the tropics off the coast of Brazil that resulted in increased numbers of Atlantic bluefin tuna (*Thunnus thynnus*) from 1960 to 1967, likely eschewing their usual migration patterns as a result of increased foraging opportunities (Fromentin et al. 2014). Marlin (*Makaira mazara*) in the Pacific often cross the equator during warm water periods resulting in increased connectivity between the north and south Pacific. An exception occurred in 2010 when La Niña conditions created a cold barrier that prevented these transequatorial migrations (Carlisle et al. 2017). In the northern Arctic, melting of sea ice is likely to open up new migration and foraging routes for Bowhead whales (*Balaena mysticetus*). New summer sea ice minima have already resulted in increased mixing of two genetically distinct populations (McKeon et al. 2016). Past barriers to population intermingling created by oceanographic features described above or the creation of isolated remnant populations from historic human harvesting, subsequently reinforced by strong behavioral affinity to close kin, have resulted in population divergence when breeding and feeding habitats are geographically separated (fig. 9.1). However, intermingling and metapopulation dynamics can result in genetic mixing when juveniles occasionally disperse from their natal colonies, for example, for seabirds, pinnipeds, and baleen whales (Inchausti and Weimerskirch 2002; Gonzales-Suarez et al. 2009; Baker et al. 2013). Given that many marine mammals are still recovering from hunting pressures of the past, they are likely recolonizing previous habitats and experiencing decreased overall genetic diversity (Gonzales-Suarez et al. 2009; Abadía-Cardoso et al. 2017). As new barriers to connectivity form or pathways open up, such as among foraging sites or between foraging sites and breeding grounds, the potential for genetic isolation or genetic mixing may shape future population structure.

There are not many fixed human-made features in the ocean that result in fragmentation that parallel those in terrestrial habitats. Shipping lanes are the closest corollary to terrestrial roads, and often overlap with baleen whale migration routes due to seasonally available prey aggregations (fig. 9.3). Ship strikes remain one of the greatest hindrances to population recovery for many large whale species, thus management approaches are being developed to reduce ship-strike risk and, in turn, aid the recovery of large whales (Redfern et al. 2013; Hazen et al. 2017). Historical data on whale presence in the Gulf of Maine was used to move shipping lanes slightly north to reduce ship-strike risk but this was an incredibly complicated process and may not reflect future distributions as climate changes foraging habitats (Wiley et al. 2011; Meyer-Gutbrod et al. 2018). Similar efforts have occurred on the west coast of the United States, where shipping lanes out of San Francisco have been moved to reduce ship-strike risk, and in Southern California air quality rules resulted in new, informal shipping routes that reduced strike risk (Redfern et al. 2013; Rockwood

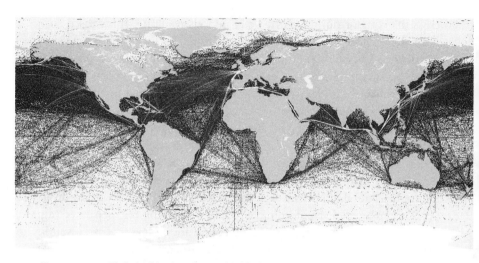

FIGURE 9.3 Global shipping lanes highlighting the potential for overlap between animal migration corridors and human uses (http://www.nceas.ucsb.edu/GlobalMarine/impacts).

et al. 2017). In New England, in areas where right whales were historically present, seasonal management areas are enacted in order to reduce mortality risk. Management actions include, for example, mandatory slowdown of ships to 10 knots or less. When three or more whales are present in an area, the National Marine Fisheries Service can also implement voluntary dynamic management areas that encourage mariners to slow to 10 knots. These efforts are similar to school crossing zones, but slow-down recommendations, whether voluntary or mandatory, have had limited success in actually reducing ship speeds and, in turn, reducing collision risk (Wiley et al. 2011; McKenna, Katz, et al. 2012; Van der Hoop et al. 2015). Citizen science tools such as Whale Alert allow people on the water to input whale sightings as they occur, and also can show the latest information on seasonal management areas to inform recreational boaters and the commercial shipping industry (Lewison et al. 2015).

In addition to directly causing mortality events, vessel noise is likely to have nonlethal but chronic effects on marine mammals, including increased stress levels and acoustic masking (McKenna, Ross et al. 2012; Shannon et al. 2016; Redfern et al. 2017). These chronic effects have been documented as increased stress hormones via fecal samples, which may have long-term effects on reproductive output and calf survival (Rolland et al. 2017). Already, melting sea ice in the Arctic is not only creating new habitat for Bowhead whales, but also the opportunity for new shipping lanes, which will likely result in additional threats to migration corridors (McKeon et al. 2016). But even with these stressors, many marine mammals still occupy these risky habitats (McKenna et al. 2015). Thus shipping lanes do not create clear barriers that can fragment habitat such as in terrestrial systems.

One of the greatest threats to pelagic migrants is marine bycatch, where interaction with fishing gear leads to acute stress or often mortality (Lewison et al. 2004). Numerous management approaches have been designed to minimize risk from fishing gear (lines, traps, and pots) that overlap with migratory corridors via changes in technology (sinking lines), acoustic deterrents (pingers), or changes in the fishing gear (circle hooks, net extenders; Lewison et al. 2004). However, there are still many areas of the ocean where regulations to prevent the problem are ineffective or do not apply. Variability in ocean conditions can lead to spatiotemporal changes in fishing patterns and/or migratory habitat that result in changes in overlap between migrating animals and fishing gear. For example, a marine heat wave in 2014–2015 resulted in a seasonal closure of crab fishing due to harmful algal blooms, and caused a high density of forage fish nearshore. When the crab fishery reopened late in the season with an unusually high harvest intensity, a record number of humpback whale entanglements resulted (Wilson et al. 2018). These cascading events highlight how changing climate and ocean conditions can lead to sudden impacts on a resource. Such anomalous conditions and ecosystem responses predicate the need for proactive management approaches that can identify and adapt when risk rates rise (Hobday et al. 2011; Hobday et al. 2013; Lewison et al. 2015; Maxwell et al. 2015).

On top of climate variability, extremes such as heat waves in the marine environment are occurring more frequently and with more severity (Bond et al. 2015; Hobday et al. 2016). These extremes have likely changed the location and timing of marine corridors (Anderson et al. 2013; Briscoe et al. 2017; Morley et al. 2018), and also potentially increased the exposure of migratory species to disturbance. In fact, coupled climate–species distribution models have predicted that warming in the central Pacific will reduce the efficacy of the NPTZ to serve as a migratory corridor by 2100 (Hazen, Jorgensen et al. 2013). As climate variability and change impact ocean properties new corridors may form or existing corridors may shift or be lost (Briscoe et al. 2017). When these changes are predictable (e.g., a poleward migration of a corridor), they are easier to address with current management structures, although changes in processes that span jurisdictional boundaries (e.g., US–Canada borders) can complicate management approaches (Mills et al. 2013). Nonlinear changes that often result from complex systems can present intractable problems that can be difficult to resolve such as when predators suddenly react to changes in prey distribution thus increasing their overlap with fishing gear, and leading to widespread entanglements (Wilson et al. 2018). To address the multiple scales at which pelagic animals interact with the environment requires a portfolio of management tools. Specifically, we can use tools that protect static locations such as MPAs in concert with tools that protect dynamic ocean features that regularly move to ensure protection is met. Dynamic tools (e.g., spatially dynamic marine protected areas) are designed to adjust (i.e., reconfigure or relocate) with changing conditions where new problems are likely to arise and require adaptive management action (Maxwell et al. 2015; Hazen et al. 2018). In contrast to static management areas (e.g., immovable marine protected areas, marine sanctuaries), dynamic management approaches that spatially track ocean features can in principle increase protection for migratory animals.

What Constitutes Connectivity and Corridors in the Coastal Ocean?

Connectivity is an important concept also in the coastal ocean (i.e., on continental shelves); it exists in different forms at multiple spatial scales. The vast majority of coastal-marine species have "bipartite" life histories in which adults are either sessile or demersal (i.e., attached to or closely associated with the seafloor, respectively), but produce offspring that are dispersed by ocean currents in the pelagic environment in the form of larvae (animals) or spores (algae). Movement of juveniles and adults of many of these species (plants, algae, invertebrates), like many terrestrial species, require connected seascapes during critical phases of their life cycles such as migrations between nursery and adult habitats for spawning. However, the range of these movements for bipartite organisms (e.g., coral reef species, estuarine species, rocky intertidal species) is limited (1 to 10s of kilometers; e.g., Kritzer and Sale 2010; Freiwald 2012) because of their strong habitat affinity and the discontinuous distribution of most coastal habitats (e.g., coral reefs, mangroves, seagrass beds, estuaries, kelp forests, rocky intertidal, deeper rocky reefs). Thus, in sharp contrast to pelagic species described in the previous sections, pelagic larval dispersal is responsible for most of the long-distance movements for species in the coastal ocean (10s to 100s of kilometers); it is determined by ocean currents that carry larvae from the location of birth to where they settle to spend the rest of their life (fig. 9.4). This means that much of the movement that occurs among local populations (also referred to as "subpopulations") and communities associated with discontinuous habitat patches separated by no more than a few kilometers is achieved by larval dispersal. As such, corridors for movement of coastal populations and communities are comprised of discontinuous habitat patches (see stepping-stones in chap. 5) and the ocean currents that transport offspring from one habitat patch to another. This form of corridor is fundamentally different from corridors of either pelagic marine species or terrestrial and freshwater species, but it needs to be considered in the design of protected areas in the face of climate change, habitat fragmentation, and habitat degradation.

Because the majority of offspring produced by adult fishes and invertebrates are carried away by ocean currents, the arrival of offspring produced elsewhere replenishes local populations and influences their size as well as the species composition of the community. These so-called "open" populations and the metapopulations and metacommunities they constitute make connectivity and corridors extremely important to coastal species and their habitats (Caley et al. 1996; Cowen and Sponaugle 2009; Kritzer and Sale 2010).

Species distributions and coastal ocean ecosystem compositions in general are determined by three key attributes: geomorphological features, oceanographic conditions, and water depth. A key geomorphological feature is the consolidation of the substratum: rocky versus sandy. These two substratum types support very different fishes, invertebrates, and macroalgae; the latter of which is largely confined to rocky substrata. Thus, the two substrata types not only support different species, but also create barriers to movement among like habitat patches. The communities associated with each of these substrata also differ markedly with water depth and temperature (e.g., Hamilton et al. 2010). For example, communities associated with rocky seafloors in the intertidal zone, shallow subtidal (0–30 meters depth), deeper subtidal (30–100 meters), and greater depths (> 100 meters) support dramatically different assemblages of fishes, invertebrates and macroalgae (fig. 9.5). Thus, connectivity needs to be maintained among these different substrata, which are largely defined by water temperature, substratum type, and depth within a region of similar oceanographic conditions.

Location, distance between patches, and patch size greatly influence connectivity in coastal oceans. Habitat patches that are located in areas more exposed to coastal currents or in current gyres and eddies that collect and concentrate larvae will receive more larvae than other patches. For a given

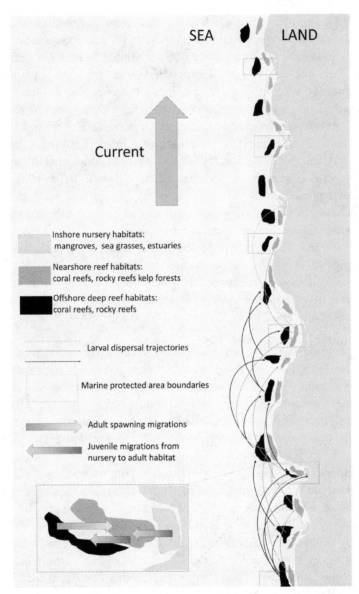

FIGURE 9.4 Patterns of population and community connectivity of coastal marine species created by larval dispersal between similar discontinuous habitats. Narrow arrows indicate larval dispersal between distinct habitat types or from offshore habitats to inshore nursery habitats. Larger arrows in inset indicate movement of adults from offshore habitat to inshore spawning habitat, and juvenile movement from inshore nursery habitats to offshore adult habitats. Marine protected areas are depicted by rectangles.

FIGURE 9.5 Distinct assemblages of demersal (i.e., bottom-associated) fishes associated with different combinations of substratum type (rocky reef versus soft-bottom)
and five bins of water depth along the coast of central California. The five depth
bins demarcated by white lines are intertidal, intertidal–30 meters, 31–100 meters,
101–200 meters, and greater than 200 meters. (Graphic courtesy of Emily Saarman
with fish images by Larry Allen.)

current velocity and pelagic larval duration (see below), the farther apart
patches are, the fewer larvae are transported between them and the lower is
their connectivity. Similarly, the smaller the habitat patches, the lower is the
likelihood of a larva encountering a particular patch. As described below,
size and quality of habitat patches also determine the number of larvae generated by adults in a patch.

The other key element of connectivity and corridors in coastal oceans are
the ocean currents that transport offspring from one habitat patch to another. Oceanographic processes that disperse larvae act at several spatial and
temporal scales (fig. 9.6) and vary in their predictability in space and time.
Large-scale features such as the Gulf Stream, the California Current, and
others (fig. 9.2), are generally predictable in the long term, multiple years
or more, but are altered by large-scale events, such as El Niño, that create very important episodes of dispersal. These large-scale current patterns

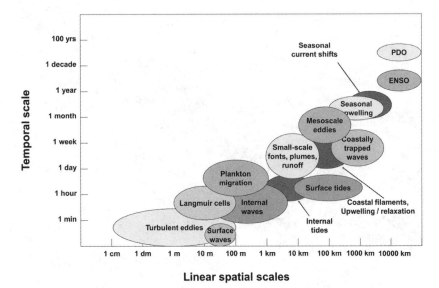

FIGURE 9.6 Stommel diagram of the wide variation in spatial and temporal scales of oceanographic processes that influence the direction and velocity of larval dispersal of coastal marine species. (After Carr and Syms 2006 and Dickey 2001.)

are nicely illustrated in in the Perpetual Ocean visualization (https://svs. gsfc.nasa.gov/10841). Co-occurring large- and smaller-scale processes provide opportunities for larvae to move in opposite direction depending on their depth and horizontal location. For example, along the west coast of North America, while the California Current moves from north to south, the deeper California Undercurrent moves south to north, as does a much smaller inshore manifestation of the Davidson Current. Through the timing of spawning and behavioral mechanisms described below, larvae can exploit those opposing processes to determine the direction and net distance of dispersal (e.g., Shanks and Eckert 2005).

Embedded within these larger-scale current patterns are smaller "mesoscale" processes of greater complexity and lower predictability. These include coastal upwelling, jets and gyres created by coastal headlands, and plumes of freshwater discharged from river mouths, among others. Even smaller "fine-scale" processes, such as coastal fronts (where water masses converge) can create barriers to dispersal (e.g., Galarza et al. 2009) or can collect larvae and shunt them to shore, creating hotspots of advection to settlement habitat to successfully complete the dispersal process and patterns of population and community connectivity (e.g., Woodson et al. 2012). The velocity and direction of these large-scale, mesoscale and fine-scale processes vary among years and seasonally, creating lower predictability than habitat-based corridors in terrestrial and freshwater environments.

The direction and distance that larvae are transported by these physical processes is determined by key life-history traits, heritable traits shared among individuals of a species, including when and where larvae are released, how long they remain viable in the pelagic environment, and larval behaviors that determine their exposure to particular water masses and in what habitat types they settle (e.g., Shanks and Eckert 2005). Together, these determine the influence of dispersal corridors on the patterns and magnitude of connectivity among suitable habitat patches.

Life-history traits of the pelagic larvae and bottom-associated juveniles and adults all influence how dispersal corridors determine patterns of population connectivity. Examples of adult life-history traits that interact with coastal corridors include the seasonality of reproduction, which determines when young are cast into the ocean currents; the seasonal current regimes they experience; and spawning locations that determine the exposure of larvae to different currents. Larval life-history traits include the duration that larvae remain in the pelagic environment, termed "pelagic larval duration," which influences how far they are carried by ocean currents (Kinlan and Gaines 2003; Shanks et al. 2003; Treml et al. 2012). Generally, species with longer pelagic larval durations disperse greater distances and this is reflected in greater gene flow and lower population heterogeneity (i.e., similar genetic composition among populations). The degree of genetic difference with increasing distance between populations is used to estimate dispersal distances ("isolation by distance"; Palumbi 2003). Behavioral traits of larvae include responses to a wide diversity of environmental cues, including light (phototaxis), temperature (thermotaxis), substratum type (geotaxis), and current direction and velocity (rheotaxis), that influence where individuals are located in water masses, which in turn determines the direction and velocity of their dispersal (e.g., Morgan and Fisher 2010; Morgan 2014; Leis 2018). The critical importance of these behaviors in modifying dispersal patterns is indicated by comparisons of oceanographic models that predict the dispersal of passive particles (i.e., no behavior) with the few examples of observed dispersal. Observed patterns of dispersal often indicate much shorter distances than predicted by the models and suggest that some populations are less "open" (i.e., more self-replenishing) than previously assumed (e.g., Shanks et al. 2003; Planes et al. 2009; Shanks 2009; D'Aloia et al. 2015). Unfortunately, larval behavior is the least understood aspect of the dispersal process and therefore of our understanding of population and community connectivity.

Another key life-history trait is ontogenetic use of habitats (i.e., how habitat use changes over an organism's lifetime). Some species are born from, settle into, and remain within a single habitat type throughout their life. Others use multiple habitats, including especially important coastal nursery habitats (e.g., estuaries, seagrass beds, mangroves) to which they settle and eventually migrate from to their adult habitats (Beck et al. 2001; Mumby et al. 2011; Igulu et al. 2014). Young juveniles may migrate from one critical habitat to another, but these are typically located relatively nearby, such that adult habitats in closer proximity to inshore nursery habitats support larger adult populations (Nagelkerken et al. 2002; Olds, Pitt et al. 2012; Olds, Connolly et al. 2012). This has critical implications for patterns of population and community connectivity and what constitutes the habitats that act as stepping-stones that facilitate connectivity for larval dispersal and post-settlement juvenile migration.

Demographic variables and their rates also influence how dispersal corridors determine patterns of connectivity. The number of larvae that disperse from one habitat patch to another is determined, in part, by the number, and size and age distribution of adults and their reproductive rates in the source population. This in turn determines the number of potential larvae transported between local populations. The number of adults is influenced by both the area and quality of habitat they inhabit. Small habitats of poor quality produce few larvae and can constitute "sink" populations in which the number of young produced is fewer than the number of young delivered to that local population. Thus, a minimum patch area to support a persistent local population is necessary for a local population to contribute to regional connectivity (López-Duarte et al. 2012; Cabral et al. 2016). Mortality rates of larvae in the pelagic environment and advection of larvae away from suitable adult habitat also determine the number of young transported between local populations. Finally, attributes of the recipient site will determine rates of post-settlement mortality and how numbers of larvae delivered to a patch translate to the size and reproductive potential of a local adult population (Cowen and Sponaugle 2009).

Together, patches of each coastal habitat/ecosystem along the coast and the ocean currents that transport larvae among them constitute dispersal corridors. Life-history traits and ecological processes determine the patterns and magnitude of connectivity across these coastal metapopulations and metacommunities.

Threats to Coastal Species, Ecosystems, and Their Connectivity

To date, threats to coastal species and their ecosystems have differed from terrestrial and freshwater species and ecosystems with respect to the relative importance of harvest versus habitat destruction. Historically, human impacts to marine species and ecosystems have been largely attributed to overfishing, the direct extraction of species from the ocean (McCauley et al. 2015), as compared to impacts to marine habitat. Thus, habitat fragmentation, especially with respect to connectivity, has been less a problem compared to terrestrial and freshwater environments. One key reason for this is because of the oceanographic component of corridors. Ocean currents that transport larvae from one population to another are rarely fragmented by human activities. An important exception to this has been the removal of vast quantities of coastal waters to cool the many power plants distributed along coastlines. Intakes of power plants entrain larvae and impinge larger juvenile stages, killing the vast majority of organisms that are entrained or impinged. These losses are significant enough to require costly means of mitigation (Strange 2012). Nonetheless, the rarity of fragmentation of coastal ocean currents is one important reason why corridors in marine systems are likely to contribute greatly to facilitating species range shifts in response to climate change.

However, it is the other component of coastal corridors, the discontinuous habitats that provide stepping-stones for the connectivity of species and communities along coastlines, which is vulnerable to habitat destruction and a source of fragmentation. Human activities impact habitat patches and the local populations they support in two ways. First, they can impair the quality of habitats and diminish their capacity to support natural populations and communities. One of the major impacts is associated with the various sources of coastal pollution. With the exception of occasional massive oil spills, the majority of coastal marine pollution comes from terrestrial runoff of nutrients and contaminants, especially through river inputs. The most obvious is the influx of high nutrient concentrations and pesticides associated with agricultural activities. The 16,000–22,000 km^2 "dead zone" of hypoxic waters in the Gulf of Mexico created by discharge from the Mississippi River is among the most notorious examples (Rabalais et al. 2002; Obenour et al. 2013; Van Meter et al. 2018). The other major source of habitat loss is associated with those habitats (mangroves, seagrasses, estuaries, wetlands) that have critical functional roles as nurseries for countless fishes and invertebrates. Mangrove deforestation, especially for shrimp aquaculture (20–50 percent globally), development, and agriculture has been a major concern of habitat loss and fragmentation (Valiela et al. 2001; McLeod and Salm 2006; Thomas et al. 2017). Another very important example is estuarine and wetland ecosystems, which have been lost or modified extensively (Gedan et al. 2009; Kirwan and Megonigal 2013). These and other sources of localized habitat loss remove or impair local habitats to support populations and ecosystems that facilitate regional connectivity. A second impact to corridors and connectivity is fishing mortality associated with local populations and ecosystems. Removal of adults diminishes the reproductive capacity of local populations and their contribution to larval production and connectivity.

The ecological consequences of climate change for coastal marine species and ecosystems are complex and variable, reflecting the wide variability in vulnerability of species, the habitats they inhabit, and the complex interactions among multiple stressors such as increased water temperature, hypoxia, ocean acidification, sea level rise, changes in currents and nutrient availability (Harley et al. 2006; Doney et al. 2011; Bruno et al. 2013). However, one of the most well-documented consequences is changes in species distributions across latitudes and depths as they track tolerable or favorable environmental conditions (Dulvy et al. 2008; Perry et al. 2005; Pinsky et al. 2013; Molinos et al. 2016). Shifts in depth distributions can be achieved by movement of adults over short distances (kilometers), but the redistribution of large numbers of individuals over long distances (100s of kilometers) is achieved by larval dispersal. Therefore, one fundamental conservation effort is to protect habitats and populations that facilitate the successful redistribution of species as their young disperse from less tolerable to more tolerable environmental conditions (McLeod et al. 2009; Carr et al. 2017). For example, it is estimated that 83 percent of current tidal wetlands along the US west coast will transition to unvegetated habitats by 2110 due to sea level rise (Thorne et al. 2018). Similarly, sea level rise is the predominant climate-related threat to mangroves (Mitra 2013). This would result in the entire loss of some estuaries that might otherwise act as stepping-stones for estuarine species to shift their distributions along coastlines.

Implications of Coastal Corridors for Species and Biodiversity Conservation

The significance of larval dispersal for determining the degree of population connectivity, including gene flow and the geographic range of species, has fundamental conservation implications. The goal is to ensure the integrity of corridors that facilitate population and community connectivity and enable species to redistribute in response to changing environmental conditions. As described above, with some exceptions, the ocean currents that transport larvae across coastal metapopulations are largely invulnerable to most human impacts other than those associated with climate change. Rather, it is ensuring the integrity (i.e., natural state) of local habitat patches and the populations, communities and ecosystems they support that is the key conservation objective. In the context of metapopulations, protection of patches of the same ecosystem type (e.g., rocky intertidal, coral reef, deep rocky reef) that provide required resources and conditions for individual species and communities create stepping-stones that enable species to replenish one another throughout their geographic ranges.

One conservation approach designed specifically for these purposes is the establishment of networks of marine protected areas (MPAs). MPA networks generally contain the same community type (e.g., rocky intertidal, coral reef, deep rocky reef) and are linked by larval dispersal (Carr et al. 2017; fig. 9.4). Thus, MPA networks encompass multiple instances of habitats and species that constitute those communities. The effectiveness of such networks requires considerations of habitat representation, size, spacing, and levels of protection.

Recognizing that species associate with particular habitats, MPA networks designed to protect biodiversity and facilitate population and community connectivity require that the variety of coastal ecosystems be represented throughout the network (Saarman et al. 2013). This habitat representation within and among MPAs is greatly facilitated by maps of habitat distribution (Young et al., forthcoming). Both the MPAs and the communities they contain need to be of sufficient size to ensure the persistence of populations large enough to contribute to connectivity across the network. The appropriate size of patches and MPAs is determined by the movement ranges of adult life stages and the variety of community types included within an MPA. To protect necessary larval sources, MPAs need to be of sufficient size to encompass the home ranges of mobile adults, especially fishes, to support and protect individuals throughout their entire lifetime. Because many species move among habitats (spawning, nursery, and adult habitats), MPAs that encompass multiple communities will protect the smaller scale corridors between them (Mumby 2006; Olds, Pitt et al. 2012; Olds, Connolly et al. 2012). MPAs that include multiple ecosystem types within them are also the most spatially efficient means (i.e., fewest number of MPAs) of achieving ecosystem representation across the network.

To ensure that populations, communities, and more inclusive ecosystems are connected by larval dispersal across the network, MPAs that include the same array of ecosystem types have to be spaced at distances that allow larvae to travel between them. Because of the great variation in dispersal distances among the many species in coastal ecosystems, spacing that supports intermediate dispersal distances is especially important. MPA sizes that encompass fish home ranges generally will encompass the short dispersal distances of spores of macroalgae and larvae of many invertebrates. Fishes and invertebrates with long dispersal distances (i.e., long larval durations) easily disperse between one or more MPAs. But spacing MPAs to ensure connectivity of intermediate distance disperses is important to facilitate connectivity of the complement of species within an ecosystem.

A key management element of any spatial protection is consideration of the levels and types of protection that ensure the persistence of species and ecosystems (Lester and Halpern 2008; Giakoumi et al. 2017; Sala et al. 2017). As described earlier, impacts of fishing on species and their ecosystems have been a primary concern and the focus of management in the form of no-take or partial take MPAs. However, other human impacts such as coastal pollution, habitat loss or degradation can also be restricted. Local and federal regulators now can restrict coastal discharges that might impair water quality within MPAs. Taken together, these design guidelines create MPA networks that enhance connectivity of species across biogeographic regions.

MPA networks might be an important mechanism to mitigate impacts of climate change, but are also vulnerable to the impacts of climate change. MPA networks connected by larval dispersal might mitigate climate impacts by facilitating the successful dispersal and recruitment to more favorable environments (Carr et al. 2017). One potential mechanism for this is by protecting genetic diversity of populations within MPAs. This strategy would help to ensure that those genotypes that are better adapted to changing environmental conditions in both the pelagic and the coastal environments would persist. The greater the genetic diversity of larvae, the higher the likelihood of individuals successfully founding new populations in novel conditions. Similarly, protected climate refugia, where populations experience slower rates of change, can then contribute to repopulating impacted populations via existing corridors of larval dispersal (McLeod et al. 2009; Mumby et al. 2011; Carr et al. 2017). Another management tool would be to eliminate the subsequent human take of successful recruits at their destination of dispersal, and thereby increase the likelihood of founding new populations.

There are, however, additional climate impacts that can diminish connectivity and the effectiveness of MPA networks either by altering larval behavior, reducing larval survival (Leis 2018), or by changing current patterns and creating spatial mismatches between patterns of dispersal and the location or spacing among MPAs (e.g., Fox et al. 2016). This could not only impair the role of MPA networks in facilitating the redistribution of species, but could diminish the conservation impact of MPAs altogether. This has two key implications for the design of MPA networks. First, by distributing the MPA network broadly, the likelihood of future mismatches between patterns of dispersal and successful delivery to an MPA is reduced. Second, by monitoring changes in current patterns and larval dispersal, the distribution of MPAs that constitute the network could be rearranged to maintain connectivity across the network. Thus, MPA networks will have to be adaptively managed to maintain the integrity of the network for mitigating climate impacts.

In sum, while we lack the ability to predict changes in marine connectivity with 100 percent accuracy, we have the ability to lay out adaptive management approaches that will continue to evolve as new data are collected and analyzed.

Additional research in marine connectivity is important to pursue because of the many complex biophysical interactions that determine connectivity and the ecological manifestations of climate change in the marine environment, which requires interdisciplinary studies involving both theoretical and empirical environmental (physical, chemical, and geological oceanographers) and biological (ecological, physiological, evolutionary) disciplines. Research in the next decade will rely on continued technological advances, management innovations, and funding opportunities to realize a deeper understanding necessary to inform critical and widespread ocean management and policy issues.

Chapter 10

Protecting and Restoring Corridors

In previous chapters, we discussed the importance of systematic conservation planning and described the tools used to identify priority sites for conservation of connectivity. Each site will have its own particular ecological, physical, economic, and social circumstances, which will dictate the steps required to secure the site as a functional corridor. Protecting and restoring corridors require some of the actions and considerations that we discussed in previous chapters, including identifying specific goals, mapping and modeling connectivity, incorporating climate change considerations, and collaborating with stakeholders. Additionally, ecological monitoring in and around the proposed corridor site should be performed to ensure the long-term desired outcomes.

In this chapter, we briefly describe general strategies that should be part of any implementation project and then examine the tools used to conserve corridors. In particular, we review the recent proliferation of incentive-based conservation tools, including purchasing a partial or full interest in land. Once land is identified or acquired, restoration is often necessary. Ecological restoration is too vast a subject to include in depth in this book; therefore, we cover general points pertinent to connectivity enhancements. The second half of this chapter provides corridor project case studies implemented at various spatial scales to illustrate principles discussed throughout the book.

It is challenging to apply science in the implementation of actual connectivity projects. Implementation is context-specific and depends on the level of development, fragmentation, land ownership pattern, socioeconomic factors, stakeholder interests along with their capacity to be supportive, available policy tools, and the natural communities to be connected

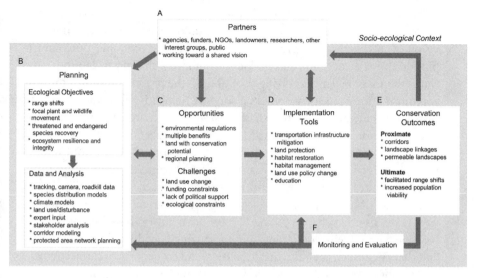

FIGURE 10.1 This framework for connectivity implementation includes (A) early partner engagement; (B) clear ecological objectives that drive data type and analysis; (C) opportunities and challenges that may advance or hinder implementation and should be addressed in the planning phase; (D) strategies to overcome challenges and ensure success; (E) resulting outcomes that increase connectivity and foster continued conservation by the partners; and finally (F) monitoring and project evaluation for adaptive management. (Modified from Keeley, Basson et al. 2018.)

(Worboys and Lockwood 2010; Fitzsimons et al. 2013; Brodie et al. 2016; fig. 10.1).

Opportunities and Challenges

In general, implementation requires developing a shared vision and involving locally appropriate partners (Worboys and Lockwood 2010). Oftentimes, communicating the vision and progress can be inspirational and garner support, such as for a proposed highway overpass in northern Los Angeles that received hundreds of thousands of dollars online (Popescu 2016; Save LA Cougars 2018).

Building partnerships is central to corridor implementation when land ownership is diverse. Partners can include public resource agencies, nonprofit conservation organizations, private landowner groups, and researchers. Together, a group of collaborators will ideally create a common vision through communication among partners. Establishing a common vision of success that integrates social, ecological, and economic outcomes proposed by partners and stakeholders can address challenges due to customs, values, or beliefs. Involving the public in the process through regular communications can increase public support, which is often vital to successful implementation. With a shared vision, priority areas for restoration or conservation can be determined (Beunen and Hagens 2009). The planning process should begin with clarifying the ecological objectives. Creating corridors to improve recovery prospects of threatened and endangered species requires different data and modeling approaches than facilitating range shifts of many species, or increasing ecosystem resilience and integrity. As discussed in chapters 7 and 8, there are many options for collecting baseline data as well as designing and prioritizing corridors. Opportunities and challenges such as land acquisition costs and co-benefits of connectivity conservation can also be considered in the planning process (Keeley, Basson et al. 2018).

Selecting places that might address not only connectivity but other interests to the community such as improved water quality, conservation of open space, carbon sequestration, and adaptation to climate change, can be an opportunity to increase support and funding for connectivity projects (e.g., Jongman 2008; Beunen and Hagens 2009). In the long term, clear policy and regulations at the local, regional, or national level specifically enabling connectivity conservation would be helpful, but many conservation tools can help advance connectivity, even without specific laws and policies on corridor conservation. Furthermore, opportunities for land acquisition may jump-start a corridor project and initiate a broader planning process.

Opportunities and challenges will also influence connectivity implementation strategies and tools. Outreach campaigns are an important strategy for building public support and increasing trust between nonprofit organizations, agencies, and local communities. Depending on the project, the targeted audience can be specific communities or private landowners in priority areas, or if considering longer-term horizons, children. The aim can be to share information about a specific connectivity project, if necessary with a formal public outreach strategy with in-depth and widespread media coverage. Moreover, one can generally educate communities about the issues associated with ongoing habitat fragmentation and the resulting need for landscape connectivity conservation action. Box 10.1 discusses an example of where signs and road closures were used to protect an amphibian population during seasonal movements. While many people think of large-scale connectivity, this example demonstrates that connectivity can be implemented to address opportunities at a variety of spatial scales. Charismatic flagship species, such as giant pandas (*Ailuropoda melanoleuca*), mountain lions (*Puma concolor*), or European bison (*Bison bonasus*), are good ambassadors for communicating the concept and need for connectivity conservation (Tiemann and Siebert 2008). Additionally, wildlife studies can be a good tool to engage with the public, because photos, videos, and movement paths of charismatic animals easily inspire people to action. Effective ways to distribute the information include social media, newspaper articles, newsletters, public presentations, workshops, school visits, field trips, volunteer days, and one-on-one communications with landowners (Fitzsimons et al. 2013). When communicating with the public, the use of stories and nontechnical, evocative language are most effective. We discuss other tools and strategies for corridor implementation in detail below.

Desired conservation outcomes can vary from different types of corridors (see chap. 4) to permeable landscapes, ultimately ensuring increased population viability and facilitating range shifts as adaptation to climate change. Monitoring and evaluation of implemented corridors are important for adaptive management, providing feedback to adjust to changes in the sociopolitical arena, new funding opportunities, unforeseen disturbances, new collaborative partners, and ensuring that project outcomes are reached. Therefore, we strongly recommend collecting baseline ecological and social information about a project area and then systematically quantifying changes at an appropriate temporal scale. Even when a corridor has successfully been established, constant vigilance is necessary to ensure that pressures from competing land uses do not compromise its functionality.

Increasingly, around the world, large-scale collaborative conservation efforts are forming (see www.globescapes.org). These opportunities represent a fundamental shift from protected-area conservation toward conserving a network of protected areas or seeking to achieve conservation at the scale that species and processes might need through time especially given climate change. One common aspect of collaborative projects is the inclusion of partners from local to subnational, and national and international groups and organizations. Most are very much bottom-up efforts. One new urban effort is green roofs, which, if implemented across enough roofs, can function as stepping-stone habitat (box 10.2). When top-down mechanisms are lacking, the challenge is that conservation implementation is often uneven across the focal region as resources and interest levels may vary across different jurisdictions. There are a number of efforts that focus on messaging and actions to incentivize targeted audiences, not just for

BOX 10.1

CAUTION: NEWTS CROSSING

In 1988 there was a meeting of the East Bay Regional Park District Board of Directors to receive public input into their developing land-use plan for Tilden Park (Alameda and Contra Costa Counties, California). One citizen asked if anything could be done about all of the newts (*Taricha* sp.) that were killed each winter as they crossed South Park Drive to reach their breeding stream, which followed much of the road. South Park Drive is a major road through the park and was increasingly being used by commuters. The board's response was to the effect that newts had been killed on that road for years and were doing just fine. The exchange initiated much discussion, however, and the issues raised were transmitted for comment to Dr. Robert C. Stebbins, herpetologist at the Museum of Vertebrate Zoology, University of California, Berkeley. He agreed that there was a problem and suggested that a study be performed to evaluate the level of newt carnage. This idea was accepted, and the board approved a plan to close the road (2 kilometers, or 1.25 miles) following heavy rains, effective in the fall of 1988 (fig. 10.2).

Road closure was met by a confused and sometimes angry public. Moreover, the policy of opening and closing the road all winter as rainstorms came and went added to the frustrations of motorists and the workload of the park staff. They did their best to educate drivers about the project, and in the autumn of 1989, Dr. Stebbins and a newly employed naturalist, Jessica Shepherd, began their research project. They marked off South Park Drive at 10-meter (33-foot) intervals, and regularly surveyed for newts, both alive and squashed. They found that newts were crossing the road all winter, rain or shine. In 1992, the park's supervising naturalist presented their results to the board, complete with a large jar of flattened newts. The board agreed to have the road closed for the entire rainy season, November 1 to April 1. Public reaction was mixed, but at least now drivers knew when the road was to be closed, and could plan accordingly.

This unusual management tactic to help newts move back and forth between their breeding stream and upland nonbreeding habitat continues to the present time. An unanticipated benefit of the plan is that when the road is closed, it becomes a favorite route for hikers, dog walkers, cyclists, and even equestrians. The public has accepted it as the normal course of the annual cycle in Tilden Park (Treadway 2017).

FIGURE 10.2 Sign indicating that road is closed for newt crossing, Tilden Regional Park, Alameda County, California. (Photo by William Z. Lidicker Jr.)

conservation of biodiversity, but on the benefits of conservation for humanity (Hilty et al. 2012). While such efforts have just emerged in the last few decades, increasing evidence suggests that shifting to a collaborative approach for large-scale conservation can result in stronger conservation outcomes.

Law and Policy Mechanisms

Few local or regional jurisdictions have so far established laws or funding streams pertaining to conservation of corridors at the time of writing. That said, proposals and establishment for legislation and regulation appears to be on the rise at the local and national level (Lausche et al. 2013). Increasingly, bodies in charge of transportation are encouraged or required to evaluate wildlife connectivity and to mitigate where connectivity is impaired or severed due to roads. For example, the state of Vermont has mapped out particular areas of concern that trigger their Department of Transportation to review any projects in those proposed regions with higher scrutiny (Austin et al. 2010). Likewise, California's Assembly Bill (No. 498) encourages the identification and conservation of wildlife corridors throughout the state. At the federal level, Tanzania just passed a wildlife corridor bill. Kenya and Brazil are advancing federal legislation on corridors. Similarly, a

BOX 10.2

GREEN ROOFS OFFER STEPPING-STONE CONNECTIVITY IN URBAN AREAS

To reduce the adverse impacts of urban development on biodiversity and improve environmental quality, promoting greening in the built environment (GRHC 2009) is an option (fig. 10.3). One approach to urban greening is the installation of "green roofs," that is, rooftops with a shallow soil cover supporting extensive vegetation. Originating in Germany, the inclusion of green roofs in building design has expanded rapidly in the United Kingdom and United States since the 1990s (Williams et al. 2014) for their environmental, aesthetic, and economic benefits (Oberndorfer et al. 2007). Green roofs reduce the volume and improve the quality of storm water runoff (e.g., Berndtsson et al. 2006; Villarreal and Bengtsson 2005); reduce energy needs by insulating individual buildings (Jaffal et al. 2012); mitigate the urban heat island effect (Peng and Jim 2013); reduce air and noise pollution (Yang et al. 2008; Coffman and Waite 2011), and contribute to carbon sequestration (Whittinghill et al. 2014; Rowe 2011). Beyond these ecosystem service benefits, green roofs benefit urban biodiversity by providing habitat for a greater abundance and diversity of both plants and animals than do conventional roofs (Oberndorfer et al. 2007). As a part of a system of wildlife habitat in urban and suburban areas, including parks and gardens, green roofs provide additional habitat for plants and animals and have the potential to function as stepping-stones for connectivity, enabling dispersal in greenspace networks (Coffman et al. 2014; Braaker et al. 2014); and migratory movements (Fernández-Cañero and González-Redondo 2010). For example, Chicago City Hall has been shown to facilitate the movement of many neotropical migrant bird species (Millett 2004), and the 2.5-hectare native grassland green roof on top of the Vancouver Convention Centre links bird and insect populations with nearby natural areas (Velasquez 2012). Many countries in the world are developing research, projects, and policies on using green roofs to enhance biodiversity (e.g., Coffman and Waite 2011). Green roofs have been designed to create habitat for migratory birds in many countries including England, Switzerland, and Canada (Fernández-Cañero and González-Redondo 2010), and have been proposed as sources of additional habitat to create viable metapopulations of endangered butterflies in business parks in the Netherlands (Snep et al. 2011).

federal wildlife corridor bill has repeatedly been put forward in the United States, but so far no action has been taken. Despite this, wildlife corridor language is increasingly seen in the United States, such as in transportation bills and land-management guidelines. Additionally, site-specific legislation can designate corridors. In South Korea, the protection of the Baekdu Daegan Mountay System Act (7038) created protected areas and buffers that

FIGURE 10.3 The 1-hectare (2.5-acre) green roof of the California Academy of Sciences in San Francisco is part of the green spaces in the densely populated city. (© 2008 California Academy of Sciences.)

ultimately protect a large mountain corridor along the peninsula (http://extwprlegs1.fao.org/docs/pdf/kor93916.pdf; Miller and Hyun 2011). Because of the increasing interest and focus on corridors, the IUCN Connectivity Specialist Group is writing standards for ecological connectivity. The rationale for such a document is that with so many countries considering new policies to increase connectivity and increasing numbers of corridor projects around the world, all would benefit from standards to guide and track such efforts.

Despite the rise in policy and management for connectivity, most efforts to conserve corridors rely on other types of existing policies to achieve connectivity. These policies may focus on land protection, species recovery, transportation, and natural resource conservation such as water quality protections. Other useful tools are economic incentives, including tax credits, payments for ecosystems services, and other market-driven tools such as conservation banking (Lausche et al. 2013).

Stewardship of Working Lands

Around the world, working landscapes make up a majority of what we referred to earlier as the matrix. Working landscapes support humanity through the production of natural resources for human consumption and also conserve components of biodiversity and ecosystem services necessary for continued production. As such, these landscapes often provide resources to support a diverse array of species, and can be permeable to allow for species movement as well as retain favorable abiotic conditions to promote ecosystem resilience over space and time. In the case of rangelands and forests, for example, working lands often provide critical resources for selected species and are essential for other species to disperse among habitat patches.

The stewardship of working lands represents an important mechanism for conserving connectivity. Voluntary stewardship programs encourage private landowners through education and community involvement to implement best management practices and conserve resources. These programs are often supported by government agencies and public universities that codevelop science-based information with land managers. Extending information on best management practices, including conserving biodiversity and maintaining permeability, can be a very cost-effective way to implement conservation across the matrix. Landowners participate in voluntary programs for a multitude of reasons that include increased access to information and local expertise, building connections with other local landowners, and public recognition for practicing sustainable methods.

Many countries have programs that encourage landowners to join community-based sustainable land stewardship groups. In Australia, for example, the Department of Agriculture, Fisheries and Forestry supports Landcare, a voluntary community group movement focused on improving natural resource management. Approximately four thousand groups operate mostly in rural Australia. From 2014 to 2018 the government invested $1 billion to conserve biodiversity and sustainable agriculture goals in phase I of the program and is planning a similar commitment for phase II (http://www.nrm.gov.au/national-landcare-program). A different effort in Western Australia, the Fitzgerald Biosphere Project, has "produced considerable benefits for corridors in the region" (Bradly 1991). This was accomplished through innovative farm plans that included planting native species in agricultural belts to serve as wildlife corridors. This example demonstrates that sustainable land use can be promoted by community organizations such as land care groups. These community organizations are often the best place to start working on habitat connectivity and climate resilience across working landscapes.

Getting people to change their behavior and adopt new practices can be a slow and daunting task. Children's education programs may provide long-term returns for conservation. Macedon Primary School in Victoria,

Australia, started the Slaty Creek Underpass Wildlife Project. The project was sponsored by the state roads management agency, and the children monitored wildlife use of a road underpass for an entire year. They worked with scientists and artists to help document the use of the underpass by wildlife (Abson 2004). The results yielded a better understanding of the species that use the underpass and an exhibition to communicate the findings of the project to the community. This illustrates a wonderful example of how conservation action can engage children and teach them about how to better manage human and natural systems.

Indigenous and local communities are the stewards of much of the world's remaining open space, and the success of many corridor projects will rely on them. In the past some conservation measures have contributed to further marginalization of indigenous groups. With indigenous cultures rapidly disappearing (Dalby 2003), a number of governments such as the United States, Australia, and Canada have begun exploring Indigenous Protected Areas and comanaged lands as a means to appropriately engage with and respect indigenous communities (Weaver 2015). Increasingly, stewardship of broader landscapes for connectivity in many countries will advance with indigenous and local empowerment.

Conflicts between protected areas and local people have a long history and can be attributed in part to the top-down approach generally used in the last century to establish protected areas (Peluso 1993). That approach had negative consequences for the people and for the protected areas and is an important context for working on conservation including corridors. Community-based conservation started in Africa as a backlash against regulating wildlife and land protection from the top down and is aimed at devolving natural resource management to local communities (Hulme and Murphree 1999). Similarly, the past oppression of indigenous people in Canada has led the current Minister of Environment and Climate Change, Catherine McKenna, to announce support for Indigenous Protected Areas as part of the country's conservation strategy (speech, 2017 Canadian Parks Conference, Banff). That objective goes beyond considering local communities as stakeholders in the decision-making process and engaging in shared visions as well as local empowerment to care for the land.

There is still a great deal that needs to be done for conservation interest groups and governments to truly integrate indigenous knowledge into land management and to empower local communities with control over land and wildlife (Sefa Dei 2000). One of the challenges is recognizing the diversity that exists within a single culture or local area and figuring out how to engage indigenous beliefs and histories in project development without simplifying them and weakening their integrity. We have a long way to go before we can define an effective approach to working with indigenous and local people to successfully conserve biodiversity, but we do know that empowering them in the process is an important and necessary first step.

Private Land Conservation

Increasingly, private organizations and governments are purchasing lands or easements on land to protect natural habitat, establish corridors, and improve connectivity. Such acquisitions compensate landowners for opportunity costs associated with necessary changes in land use. The most common mechanisms for doing this are (1) public or private acquisition of land and establishment of a protected area; (2) partial purchase of land rights that places limits on the use of the land; (3) land-management agreements that facilitate the protection of resources; and (4) purchasing concessions. These actions may occur sequentially on the same piece of land. For example, a landowner may enter a long-term management agreement to fence livestock out of creeks on a ranch and then sell or donate a conservation easement to prevent extensive development on the same property. These tools are most effective when combined with other tools such as education and regulations.

Multiple public and private partnerships and a combination of funding sources are used to finance incentive-based conservation actions. Despite the perception that incentive-based conservation is privately supported, much of the funding is public, including, for example, federal, state, and local tax breaks. Additional public funds come from federal land-management organizations, money paid for mitigation of environmental impacts, direct local taxes for open-space preservation, government general funds, and special bonds. Any transaction may combine funding from any of these sources, resulting in complex partnerships that make determining legal accountability difficult. The same issues for prioritizing land conservation that we discussed in chapter 7 should apply to the types of agreements described below.

Types of Agreements

A conservation agreement is a voluntary contract between a landowner and the holder of the agreement, which is often a public agency or land trust. This is generally a flexible, voluntary agreement negotiated with the landowner, and therefore a wide variety of agreements exist. Such contracts specify a conservation interest in the land and can impose land management requirements or use restrictions including for the protection and enhancement of natural resources. Land trusts typically conserve natural habitats and species, farmland, or recreational opportunities (Rissman and Merenlender 2008). Given this, it is important not to assume that lands in conservation easements necessarily contribute to biodiversity conservation or connectivity. A conservation agreement can be for a certain length of time or can be permanent and registered on the land title. Regardless, they require regular monitoring to prevent violations and are the subject of occasional legal battles by those seeking to change the land status.

One of the most prevalent types of conservation agreements used in the United States is the conservation easement, which can include temporary or permanent transfer of development rights. The conservation easement or covenant is one of the primary types of agreements that conservation land trusts use to protect environmental resources. Under conservation easements, land is retained in private ownership, and a land trust or government agency acquires nonpossessory interest in the property, restricting use for the preservation of natural resources, agriculture, or social and cultural amenities. The restrictions stay with the title of the property and are therefore transferred with the property if the land is sold. In return for donating or selling a conservation easement, a private landowner receives a reduction in taxes because the overall value of the property is presumed to be less, once some nonpossessory rights are removed. In Australia, permanent conservation agreements concerning management of aboriginal land are sometimes negotiated between indigenous communities and the government.

In most countries, the law dictates which type of institutions can enter a conservation agreement with landowners. These are generally public agencies and nonprofit conservation organizations, which include conservation land trusts. The Land Trust Alliance in the United States defines conservation land trusts as any organization that acts directly to conserve land, independent of the government. Conservation land trusts usually rely on tax relief from federal, state, and local governments and often seek public grants to support their activities. There is also a small but growing number of government agencies that function similarly to land trusts. Land trusts or similar institutions are often integral in implementing corridor protection and restoration projects. For a thorough examination of the history and practice of conservation trusts, read *Conservation Trusts* (Fairfax and Guenzler 2001) and *Buying Nature* (Fairfax et al. 2005).

While conservation agreements are generally seen as one of the most important tools for conservation of private lands, they are not without criticism. Unlike land-use regulations, decisions regarding private land incentives are usually made in agreements between private organizations or the government and individuals. Since these agreements are attached to property titles, the public is often not explicitly notified of the transactions, even if public funding is used to secure the protections. Conservation easements do not generally provide public access to the land for recreation and other uses, which can be disappointing for local residents. At the same time, this means that easement lands that restrict access often present the only refuge for species sensitive to recreation impacts (Reed and Merenlender 2008).

Management agreements are voluntary agreements between a landholder and another party, detailing the use and management of the land for a set period of time and are not binding if the land is sold. These can be very useful to enhance the biodiversity values of working lands. They are extremely flexible documents that can include land-management activities such as grazing regimes, fencing, riparian protection, tree planting, and restrictions on resource extraction. In the United States, the most widely known program that issues these types of agreements is the US Department of Agriculture's Conservation Reserve Program. It provides cost-share assistance to establish conservation practices on agricultural land by entering into ten- to fifteen-year contracts with landowners. In Europe, there are several agri-environment programs designed explicitly to conserve natural resources. Another good example is a program administered by a nonprofit conservation organization based in the United States called Ducks Unlimited, offering financial incentives to landowners willing to manage their land for waterfowl and other wetland wildlife under a ten-year agreement. Such payment programs are also becoming an increasingly popular tool for biodiversity conservation in tropical countries. The role of financial incentives and surrounding issues such as property rights, institutional arrangements, and monitoring are discussed in further depth by Ferraro (2001).

Conservation organizations have started to provide periodic payments as part of a fixed-term lease agreement in order to prevent the removal of natural resources in developing countries, which are referred to as conservation concessions. This mechanism has the potential to provide important protection from deforestation for the term of the lease, generally, fifteen to forty years, and possibly longer if renewed. There are potential disadvantages of these agreements for both parties. The investment in negotiating these agreements and the initial activities to improve the condition of the land can be lost if the land is sold. Very specific management guidelines may not be realistic in a rapidly changing environment. For example, restrictions on grazing regimes may be dependent on sufficient rainfall in any one year. Extreme weather in combination with management constraints could make livestock production at the site impossible in the future.

Restoring Land

It is generally desirable to conserve corridors that are relatively intact. However, some landscapes have been severely degraded by past land use such that insufficient networks of natural habitat remain, making it necessary to restore habitat. In some cases passive restoration, or what some refer to as "rest," may allow native plants to regenerate on their own; that may be the best way to establish habitat for corridors. In such cases, changes in land management are required. Landowners often will require compensation for activities such as fencing for riparian areas. This type of restoration also may require land conservation tools such as the conservation agreements discussed previously. In some areas, rest alone may never produce the desired results or may take too long, and more active restoration is required to provide habitat for the planned corridor project. Particular attention needs to be paid to the local habitat types and conditions, as well as desired outcomes. Again, a significant investment should be made in planning a restoration project prior to implementation.

In many cases, unwanted or invasive woody or herbaceous vegetation, may be present in important conservation areas. The removal of unwanted vegetation should always be done in small increments in order to avoid overly disturbing the site and exposing too much bare soil to the elements, which can result in the erosion of valuable topsoil and colonization of more weeds. Also, small disruptions are less likely to disturb wildlife living in these less desirable communities. In some cases, mature native trees need protection during restoration, especially for their root systems. This can be problematic in highly developed areas. Other important habitat features that should be maintained include naturally wet areas, stream banks and associated riparian areas, and rocky outcrops that provide important habitat for wildlife.

In the case of the country-wide restoration effort that is occurring in South Africa, scientists convinced forward-thinking decision makers that invasive woody plant species were using more water than native species, leading to reduced runoff and thereby reducing water levels (Le Maitre et al. 1996), an issue of concern due to a changing climate. Through the hard work of many organizations and individuals, the Working for Water program has grown into a widespread community-based effort to remove invasive plants that is benefiting water conservation, biodiversity, and resulting in much-needed employment (Turpie et al. 2008). Given a changing world, it is also important to consider the baseline for restoration. Is it appropriate or even realistic to aim for restoration of a past assemblage of species, or should efforts focus on species that are likely to persist into the future (Lawler et al. 2015)?

In some areas, more than just exotic plants need to be removed. Garbage such as cars, concrete, metal, and other undesirable materials, may need to be removed and perhaps replaced by more natural structural materials such as rocks and logs. In more contaminated areas, pollution remediation is a necessary first step. If land clearing is required, it is best to begin upslope, clear and replant small, noncontinuous sections of land incrementally over time so that slope stability and existing habitat structure can be maintained. It is always best to establish native vegetation prior to removing adjacent nonnative vegetation. If the corridor is being restored for a particular animal species, plantings may need to consider the species' dietary requirements. For example, the Peterson Creek corridor in Queensland, Australia, started in 1998 replanting trees across eight landowners' properties. This has allowed restoration of movement of the Lumbholtz tree kangaroo (*Dendrolagus lumholtzi*) and numerous other arboreal species including key food trees for wildlife. Not only has this project had significant wildlife effects, but it also has slowed runoff from rainfall, provided a windbreak, and reduced frost effects from mild frosts in close proximity to the corridor (Doug Burchill, pers. comm., 2018).

In addition to controlling weeds, it is important to discourage invasive vertebrate pests such as bird species that parasitize native species' nests, as well as nonnative rats, foxes, cats, and other exotic predators that consume native species (Burbidge and Manly 2002; Churcher and Lawton 1987). Reproductive success of yellow warblers in Montanan riparian corridors was heavily impacted by brown-headed cowbirds, parasitic birds that remove eggs from warbler nests and replace them with their own. Cowbird abundance was directly related to increased housing near corridors (Hansen et al. 2002).

Often a combination of passive and active restoration is required. For example, to increase connectivity between Morro do Diablo State Park and adjacent fragments of Atlantic Forest in the state of São Paulo, Brazil, local citizens worked with the Wildlife Trust and Instituto de Pesquisas

Ecológicas to protect natural habitat and plant trees (fig. 10.4). The goal is to provide additional habitat for several endemic species including the black lion tamarin (*Leon topithecus chrysopygus*) (http://www.ipe.org.br/en /news/1336-ipe-completes-the-largest-reforestation-corridor-in-brazil). In a similar effort in Australia, a Coxen's Fig Parrot Rainforest Restoration Project aims to restore the valuable subtropical lowland rainforest corridors throughout the Upper Pinbarren catchment. Efforts include increasing food trees for the endangered parrot, revegetating corridor linkages between parks, and more (Eco Logical Australia 2016).

Habitat restoration is a field unto itself and if pursued should be researched thoroughly. Useful places to learn more about the science of ecological restoration are two edited volumes: Palmer et al. (2016), and Clewell and Aronson (2013). For tips on how to do restoration see the *International Standards for the Practice of Ecological Restoration* (McDonald et al. 2016).

Finally, many people who work to restore ecosystems and recover species have an understanding of the importance of biodiversity and a personal connection with their local flora and fauna. This provides them with valuable knowledge about nature and in some cases connections with past generations who once may have subsisted primarily on natural resources from the immediate area. For example, in the book *Totem Salmon* (1999), the author, Freeman House, reveals the diverse reasons why members of the Matole River watershed group in California worked together to restore a river corridor that connects them to nature and to their heritage. We can gain a great deal of practical knowledge from people who are maintaining and enhancing connectivity. Therefore, below we include a variety of corridor projects to demonstrate important lessons.

Lessons from Corridor Projects

The number of corridor projects that seek to achieve conservation beyond protected areas worldwide is unknown. In 2001, the International Union for the Conservation of Nature (IUCN) identified over 150 ecological network projects focused on conserving biodiversity at the landscape or regional scale with an emphasis on ecological interconnectivity, restoring degraded ecosystems, and conserving buffer zones (Bennett and Witt 2001). The number of projects is certainly much larger today and would greatly increase if smaller-scale projects were included. An Internet search

FIGURE 10.4 Large-scale corridor restoration to connect protected areas in the state of São Paulo, Brazil, coordinated by Dr. Laury Cullen, research coordinator at the Instituto de Pesquisas Ecológicas (IPE; www.ipe.org.br). a. Before restoration, b. After restoration. (Photos by Laury Cullen Jr.)

reveals an astonishing number of wildlife corridor projects that seem to be under way, in a multitude of countries and at various scales. Clearly, the idea of reconnecting our landscapes has taken hold, and many public and private institutions are involved. Some of the most impressive efforts are being made by local volunteer groups and landowners with relatively few resources, for which it can be difficult to access information. The selected examples in this chapter include available information from a wide variety of countries and represent different spatial scales, from international networks to road underpasses.

Large-Scale Corridor Planning in Urban Areas: Lisbon's Master Plan

We pointed out in chapter 4 that landscape features such as riparian areas often serve as areas for de facto ecological connectivity. In urban areas, rivers and streams historically served as transportation corridors; more recently they are the focus of urban renewal projects, climate change adaptation measures, and biodiversity conservation. An effort is under way in the greater metropolitan area of Lisbon, Portugal, to rehabilitate green spaces along river corridors running north–south through deeply incised valleys to the coast. These spaces will improve quality of city life, offer alternatives to motorized travel, provide connected habitats for wildlife, and help adaptation to climate change by better managing more frequent and larger floods (Saraiva et al. 2002; Santos et al. 2015). This region supports the highest population density in Portugal (2.5 million people in nineteen municipalities); hence, the existing built environment greatly constrains restoration options. Rapid development in the area and a lack of environmental planning have led to problems with water quality from sewage, loss of stream function and habitat from concrete-lined channels, and other instream alterations resulting in increased flooding downstream.

The tremendous potential for restoration of these river corridors for both ecological and social values has been recognized (Silva et al. 2004; Kondolf et al. 2010; Saraiva 2018). The restoration of the Ribeira das Jardas was one of the first river rehabilitation projects in Portugal that was implemented under the concept of green infrastructure (fig. 10.5). Prior to the project, the Jardas stream was highly altered, degraded, and canalized within concrete walls. Consequently, the stream offered low habitat complexity as well as low amenity and recreational values (Kondolf et al. 2010).

A master plan was developed, establishing the goals of increasing identity, mobility, and sustainability, thereby improving urban quality of life. The master plan called for a public green area along the stream and the transformation of the concrete channel in the city center into a green corridor. It spelled out the goals of providing flood protection, regenerating aquatic and riparian habitat, providing space for leisure and recreation, and enhancing scenic and aesthetic values.

As part of the implementation, the sewage system treatment was upgraded and, consequently, water quality improved, creating conditions to restore some ecological functions (Saraiva 2018). To achieve continuity along the river corridor, some buildings were demolished that had been restricting the floodplain. In addition, banks and terraced floodplains were modified to increase permeability, and riparian vegetation was restored. Gathering places established on the creek-side terraces and a network of trails and cycling lanes are now attracting people from the surrounding densely populated neighborhoods to meet and recreate along the 'rediscovered' stream (NPK Consultants 2011).

While the plan addresses several improvements to the Ribeira das Jardas including transportation, social infrastructure, sewage systems, environmental protection, and nature conservation, one of the reasons for successful implementation of this project was the focus on public use and recreation. After the restoration, social uses of the greenway were the most visible benefits.

With this project, regional interest groups have developed a new approach to river management and flood protection involving an entire river catchment and the restoration of stream processes, rather than the more traditional engineered solutions used in the past. It is hoped that the stream restoration increases water infiltration and groundwater recharge, enhances hydrological connectivity, and restores riparian communities (fig. 10.5).

This effort, spearheaded by landscape architects, demonstrates the multiple benefits of urban stream restoration. The main goal of the Ribeira das Jardas project was to improve the aesthetic elements of the city landscape. Designs for individual streamside parks associated with urban renewal were the starting point; improved water quality, increased riparian vegetation, and wildlife connectivity are secondary benefits. While some native species may not be able to use the urban riparian corridor because of the intensive human use, the project nevertheless is an excellent example of how connectivity can be enhanced for human benefit and address urban ecology issues as well.

FIGURE 10.5 The condition of the Ribeira das Jardas near Lisbon, Portugal, before (above) and after (below) the rehabilitation project. (Photos by Graça Saraiva and António Manuel Silva.)

The Meadoway, Toronto, Ontario

The Meadoway is a Toronto initiative to revitalize a 16-kilometer electricity transmission corridor into 500 acres of greenspace that will extend from the city's downtown (fig. 10.6). Once completed, the corridor will be one of Canada's largest linear urban parks (https://themeadoway.ca/). Envisioned largely as a project for improving the urban living environment for people, it will also provide other ecological benefits. The project, announced in April 2018, is agreed to by the company that owns the power lines, and jointly headed by the Toronto and Region Conservation Authority (TRCA), the City of Toronto, and the W. Garfield Weston Foundation. The TRCA has assumed responsibility for maintaining the meadows including intensively monitoring for and controlling invasive species.

Toronto has already made important advances in conservation. In addition to an extensive system of urban parks, it is home to a unique network of ravines that covers 17 percent of its land, running from moraine communities north of the city south to Lake Ontario (Toronto Ravine Strategy 2017). The city has worked to conserve these north–south ravine systems. The Meadoway, however, will be unique in providing east–west connections, connecting four ravines and fifteen parks. In addition to improved habitat quality in the electricity transmission corridor, the Meadoway will be crucial in enhancing trail connectivity in Toronto because rapid population growth puts increasing pressure on the city's natural systems. While small, isolated patches of urban green spaces have proven valuable for conserving biodiversity in cities, reconnecting them through corridors such as the Meadoway more effectively conserves and increases biodiversity (Lepczyk et al. 2017).

Electricity transmission corridors are increasingly recognized as productive pollinator habitat. Vegetation in these corridors must be carefully managed to prevent tree growth, which encourages the growth of low-height and early successional plants ideal for pollinators (Wojcik and Buchmann 2012). The TRCA's commitment to plant native wildflowers and grasses in the Meadoway will support many bee and butterfly species native to Ontario that are in severe decline (Toronto Pollinator Protection Strategy 2017).

Further, the Meadoway project will be a model for future revitalization of over 500 kilometers and 4,200 acres of transmission corridors in use by the same company across Toronto (https://themeadoway.ca/). Transforming transmission corridors into conservation lands will mitigate loss of habitat for pollinators and other species and may increase the landscape's resilience to climate change.

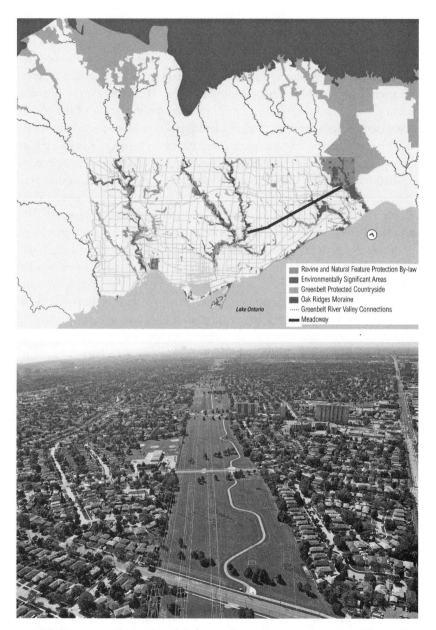

FIGURE 10.6 (above) Overview map of Toronto with The Meadoway highlighted; (below) Aerial view of The Meadoway looking west toward downtown Toronto. (Permission to use by Toronto and Region Conservation Authority [TRCA].)

Maintaining Connectivity within a Watershed: Sonoma Valley

The Sonoma Valley Wildlife Corridor (SVWC) is a regional-scale wildlife corridor encompassing approximately 10,000 acres between Sonoma Mountain and the Mayacamas Mountains (fig. 10.7). Now exemplified as a successful corridor implementation project, the inception of the SVWC was when a volunteer at the Sonoma Ecology Center recognized the unique value of Sonoma Valley and shared her vision to protect "the valley's last dark place at night" in the 1990s. What began as an effort to conserve watershed functionality and increase connectivity between Jack London State Park across the Sonoma Valley floor to Oak Hill Ranch and Bouverie Wildflower Preserve, the SVWC now includes dozens of protected parcels (60+ protected areas, 60+ conservation easements) and is part of a much larger multicounty network of linkages connecting habitats, providing a vital connection for wildlife movement within the northern San Francisco Bay Area.

Sonoma Valley is under intense pressure from increasing population and changing economies, which has resulted in the replacement of most of the lowland habitats, such as marshes, gently sloped streams, and oak savannas, with buildings, roads, and agriculture that present a barrier to animal movement (Merenlender 2000; Hilty and Merenlender 2004). Within this matrix of increasingly fragmented habitat, the SVWC is a swath of land that is highly suitable for conservation due to the presence of listed species and priority streams, as well as the critical connectivity offered between large protected lands in Sonoma Mountain and the Mayacamas Mountains.

The connectivity value of the corridor has been documented with estimates of parcel-scale landscape permeability (Sonoma County Agricultural Preservation and Open Space District 2015) that have been confirmed using wildlife abundance and movement data, including data from GPS collars, wildlife camera grids, underpass monitoring, and roadkill surveys (Sonoma Land Trust 2014a, b; Gray 2017). When the Nuns Fire burned over 57,000 acres throughout Sonoma Valley in October 2017, including 11,000 acres of protected lands in the northeast section of the corridor, existing cameras monitored wildlife movement during and after the fire. The data suggest that the lands in the corridor provided safe passage for animals to flee the fire. In addition to offering current habitat connectivity, the SVWC secures future climate benefits by linking habitat with a gradient of temperature, vegetation, and water availability that will allow plants and wildlife to adapt and shift in response to climate change. Climate benefit analyses show that by connecting the distinct coastal and interior climates, the SVWC is projected to provide net cooling between corridor termini by offering inland habitats access to coastal areas that are 1.02°C to 1.06°C cooler during summer months and offering access to higher elevation areas that are 0.12°C cooler during winter months (Gray and Merenlender 2015).

The large size of the corridor and the mix of both public and large private land ownership has required a tremendous amount of effort and coordination between multiple stakeholders to secure long-term commitments from the landowners and agencies to maintain permeability. Sonoma County residents have been essential in the implementation process, and ongoing outreach efforts aim to continue public engagement in order to work toward the shared vision of maintaining the scenic beauty and ecological integrity of Sonoma Valley. Diverse implementation strategies have been tailored to the needs of residents, including property acquisition, deed restriction, and new types of conservation easements as well as neighbor agreements that emphasize land management and restoration to maximize connectivity throughout the corridor. Targeted outreach materials have complemented these efforts, providing landowners with descriptions of best management practices for reducing impacts to wildlife movement by using minimal or wildlife-friendly fence designs, managing vegetation for fire safety, preventing pets from roaming freely in wildlands, minimizing outdoor night lighting, and avoiding pesticide use.

The wildlife and connectivity data have been aggregated into a geographic information system that has been essential for designing, planning, and communicating the project to the general public and decision makers. As a consequence, the SVWC has been designated a Habitat Connectivity Corridor in the Sonoma County 2020 General Plan, which recommends rezoning the land as a Biotic Habitat Area. These data have also been used to identify critical road crossing sites that have been shared with transportation agencies, resulting in partnerships to limit the construction of new roads and maintain wildlife crossing structures. A public hospital situated on 405 hectares (1,000 acres) of mostly undeveloped land at the narrowest part of the SVWC is scheduled to close in 2018, presenting an opportunity for repurposing the land to secure a critical connection at the center of the corridor (Sonoma County Agricultural Preservation and Open Space District 2015).

The adjustments needed to protect this corridor are expensive, and successful implementation depends on raising support and awareness within

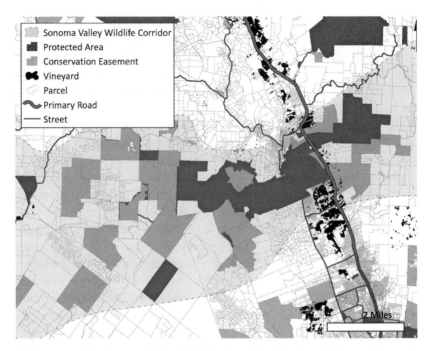

FIGURE 10.7 The Sonoma Valley Wildlife Corridor runs from Sonoma Mountain to the Mayacamas Mountains through the Sonoma Valley with extensive vineyard and residential land use on both sides that expands into neighboring wildlands. Conservation investments have been made to secure protected areas (dark gray) and conservation easements (light gray) throughout the area, although roads, vineyards (black), and residential development are prevalent at the narrow pinch point in the center of the corridor. (Permission to use by Morgan Gray.)

and beyond the local community. In 2017, the Gordon and Betty Moore Foundation awarded Sonoma Land Trust a $1.8 million grant to purchase land, support wildlife monitoring, improve habitat condition, and maintain connectivity within the SVWC. Press coverage and social media have shared the importance of the SVWC with a wide audience within the San Francisco Bay Area and beyond, from Instagram and Facebook posts of wildlife camera pictures to multiple episodes in a regional television program. The ability of the Sonoma Valley Wildlife Corridor to enhance connectivity has benefited from a collaborative approach to planning, securing, managing, and monitoring habitat. Beyond that, benefits came from integration of the natural and human systems within the region; and ongoing engagement with a wide audience.

Corridor Conservation in the Kilimanjaro Ecosystem, Southern Kenya

The Kilimanjaro Transboundary Landscape stretches from Amboseli National Park to the Chyulu Hills and Tsavo West National Parks in Kenya to Mt. Kilimanjaro National Park in Tanzania (fig. 10.8). Amboseli National Park, an area of 392 km² (151 square miles), forms the core of the protected area system on the Kenya side; community lands and group ranches surround the park. While Amboseli is world-renowned for its elephants, the park is too small to support viable populations of elephants, predators, and ungulates. Wildlife is dependent upon the areas outside the park for migration movements and resource use at different times of the year depending on rainfall patterns.

The Amboseli region has an elephant population of approximately 1,500 individuals. These elephants are a major driving force in the ecology of the ecosystem and the subject of one of the longest elephant studies in Africa by the Amboseli Trust for Elephants. Scientists have documented their movement patterns and that of other wildlife species. One of the main wildlife movement routes stretches from Amboseli to the Chyulu Mountains. In the wet season, mammals disperse out of Amboseli and move through Kimana Sanctuary to Chyulu West National Park.

The Amboseli ecosystem has been home to Maasai pastoralists for centuries. Pastoralism is the main economic activity and requires vast rangelands for grazing. Due to the increasing human population and changing lifestyles, pastoralists have started farming or leasing their land to farmers, especially near swamps. Traditionally, elephants and other wildlife have depended on these swamps for water and food; thus, the encroachment of farming has led to human–elephant conflict. Land use on these community and private lands vary greatly, yielding an inconsistent and unstable environment for elephants and other wildlife.

The greatest threat in the Kilimanjaro landscape is habitat fragmentation and loss. The group ranch land, a form of communal land, that surrounds the park has been subdivided into two-acre, ten-acre, and sixty-acre lots allocated to individual Maasai owners. The subdivision of the land is primarily due to a breakdown in communal systems, failure of the group ranch system to deliver equitable benefits and improve livelihoods to communities, and socioeconomic changes such as a more sedentary way of life. Landowners are selling their lots for development and agriculture, resulting in habitat loss, which is significantly fragmenting the landscape. This fragmentation puts the entire ecosystem at risk.

On top of it all, the rapid rate of change is alarming and is adding another threat of increasing severity. Dramatic increases in daily temperatures and variation in annual rainfall are contributing to prolonged droughts affecting pastoralists and wildlife alike (Amboseli Ecosystem Management Plan 2008; Altmann et al. 2002). Innovative solutions are required to maintain the rich biodiversity and resilience of the Amboseli ecosystem.

In 2008, the African Wildlife Foundation (AWF) launched a conservation lease program with individual Maasai landowners in an effort to curb habitat fragmentation within the spaces critical for wildlife movements between protected areas (Fitzgerald 2015). AWF's objectives were the following:

- Contribute to the survival of Amboseli National Park by protecting scientifically documented and strategic corridors.
- Provide direct incentives to landowners to keep their land open to wildlife.
- Prevent the conversion of land from open to agricultural, fenced, overgrazed or developed land.

AWF successfully established seven conservancies over 24,000 acres with 400 landowners:

- Kilitome, 90 landowners, 5,400 acres
- Nailepeau, 70 landowners, 4,200 acres
- Osupuko, 45 landowners, 2,700 acres
- Ole Polos Conservancy, 50 landowners, 3,000 acres
- Oltiyani Conservancy, 75 landowners, 4,500 acres
- Nalarami Conservancy, 70 landowners, 4,200 acres

The conservancies were created via a lease agreement between AWF and the landowners by which landowners receive a payment for ecosystem services. The lease outlines the purpose, duration, land-use restrictions, retained rights, payment requirements, violation procedures, and other relevant parameters (K. Fitzgerald, pers. comm., 2018). The lease prohibits activities that are not compatible with conservation. Grazing is permitted in compliance with a management plan. The community selected a Maasai attorney to meet with them in the absence of AWF and review the lease agreement before signing. By having this meeting without AWF, community members were free to voice concerns and changes were made as a result. The extensive community engagement process took approximately eight months and complied with the Free and Prior Informed Consent process where landowners were fully briefed on and exercised their voluntary rights.

In 2008, AWF started lease payments at 500 KSh/acre (Kenyan Shilling) with an annual increase of 2.5–3 percent, based on a valuation process. The rate in 2018 is 750 KSh/acre. Payments are made to each landowner through electronic transfer to individual bank accounts, as per request of the landowners. AWF helped landowners set up bank accounts. Payments are made twice a year. If there is a violation, AWF retains the right to withhold payment. Community scouts are recruited, trained, and managed by the Big Life Foundation (www.biglife.org).

Based on an average household of 7, the lease program is directly benefitting over 2,800 individuals, and this does not include employment benefits. Since the program started in 2008, more than US$1 million have been paid to the community members, diversifying their income and creating a more resilient community in addition to protecting a key wildlife corridor.

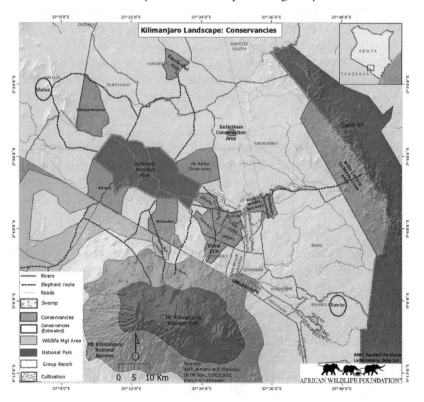

FIGURE 10.8 Protected areas in the Kilimanjaro landscape. (Permission to use by Kathleen Fitzgerald.)

While successful overall, a challenge of this program is that landowners who choose not to be part of the program, as is their right, may develop their land with uses incompatible with wildlife conservation, such as fencing, infrastructure, and farming. These developments may undermine the long-term viability of the conservancies.

Caribbean Challenge Initiative

The Caribbean has highly diverse marine and coastal habitats, featuring 10 percent of the world's coral reefs, 18 percent of the world's seagrass beds, and 12 percent of the world's mangrove forests (Knowles et al. 2015). These marine resources support tourism, the region's largest economic sector, and provide food security to several million people. However, marine ecosystems in the Caribbean are significantly degrading as a result of overfishing, pollution, and climate change. A 2014 IUCN report described coral reefs in the Caribbean as on the verge of collapse, with only an estimated 14.3 percent of reef area showing live coral cover (Jackson et al. 2014).

To address these challenges, the Caribbean Challenge Initiative (CCI) was founded by a group of Caribbean governments in 2008. Eleven of the thirty-eight countries and territories in the wider Caribbean region are currently committed to the CCI, in partnership with fifteen companies and a range of private foundations, multilateral agencies, and NGOs. The CCI's primary goal is to effectively conserve and manage 20 percent of the Caribbean's marine and coastal areas by 2020. This "20 by 20" goal essentially envisages the establishment of an ecological network of marine protected areas (MPAs, fig. 10.9). From 2008 to 2009 alone, the total area in the Caribbean covered by MPAs increased by approximately 32,000 km² (12,355 square miles), or around 35 percent (Knowles et al. 2015). Now, five of the eleven CCI countries and territories have met or exceeded the goal of 20 percent MPA coverage, and CCI participants continue to improve MPA enforcement.

Thus far, CCI efforts have mainly enhanced connectivity of Caribbean marine ecosystems by facilitating the establishment of larger MPAs. Although the intent is to design MPAs utilizing ecological network principals, this has been a challenge. Ecological connectivity in the oceans is an emerging field of study (chap. 9), and collecting connectivity information to guide decisions is difficult (Lagabrielle et al. 2014). The CCI has made important headway to integrate science and policy regarding marine

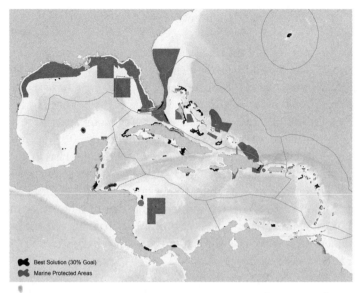

Best Solution (30% Goal)
Marine Protected Areas

FIGURE 10.9 High connectivity value reef units are selected by marine ecoregion and overlaid onto the current Marine Protected Area network. The map shows the results of the best solution for marine ecoregional coral connectivity (30 percent target set for local retention and by reef unit), overlaid onto the World Database on Protected Areas and The Nature Conservancy's Marine Protected Area Database of the Insular Caribbean.

ecosystem management. The initiative created a forum that helps countries incorporate connectivity knowledge, apply spatial planning tools, and use simulations for MPA planning on a regional scale, mainly through trainings and pilot projects.

Strong science supports the vision of an MPA network. One study integrated information on reef connectivity from coral larval dispersal patterns with economic and political information, and thus taking into account costs, ecological factors, and conservation targets in order to identify reefs that should be included in the expanding MPA network (Schill et al. 2015). Right now, less than 30 percent of reefs with high connectivity value are protected, but prospects for future protection are good. A secondary goal of the CCI was to establish reliable sources of funding to promote marine conservation over the long term, and this has already been highly successful. Future resources can be put toward efforts such as increasing area under protection, including the creation of more no-take zones, and strengthening MPA management.

Bhutan Conservation and Corridors

Bhutan might be one of the smallest countries in the world, but it hosts a diversity of habitats and wildlife including flagship species such as snow leopards (*Panthera uncia*), Bengal tigers (*Panthera tigris tigris*), and Asian elephants (*Elephas maximus*). It is listed as one of the biodiversity hotspots of the Eastern Himalayas, with habitats ranging from alpine to temperate and subtropical zones within a short span of about 150 kilometers from south to north. The geographical meeting of the Indo–Malayan and Pale-arctic realms has resulted in Bhutan hosting 200 species of mammals, 728 species of birds, and more than 5,600 species of vascular plants within the 72 percent forest cover of the country. Bhutan may also be one of the very few countries in the world where the promise to maintain 60 percent of its land under forest cover is enshrined in the country's constitution (Sonam Wangdi, pers. comm., 2018).

In the early 1970s, the King pronounced that nature conservation is a major national priority, with conservation being one of the four pillars of Bhutan's unique development philosophy of Gross National Happiness in which economic development should not take place at the cost of environmental health. In setting forth to honor this priority, Bhutan has achieved formal protection of more than 50 percent of its land. Not only does this honor the cultural priority of living life in harmony with nature, but this commitment also helps one of the country's top industries, ecotourism. One of the unique features of the protected areas is the interconnectivity provided by a biological corridors system that facilitates genetic intermixing of wildlife species (fig. 10.10).

Ever since the creation of this protected area system in the 1990s, sustainable financing for the management of these protected areas has been a challenge. Very recently the government adopted the idea of project finance for permanence, wherein it made a commitment to create a US$43 million fund. This will ensure the future management of these protected areas under the new funding program called the Bhutan for Life Initiative. Some of these funds come from the Green Climate Fund because these lands help the country with their commitment to carbon neutrality and also enhance water conservation since rivers that originate in Bhutan provide water for one-fifth of the world. Additionally, these protected areas are meant to help rural populations in particular to ensure that their surrounding wildlands are resilient during this time of climate change. "It is in this protected areas network, and the wildlife corridors that connect them, that most of the

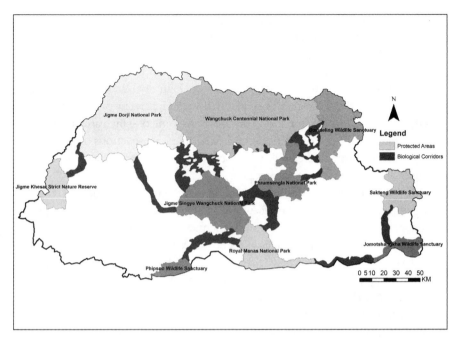

FIGURE 10.10 Corridors are connecting all protected areas in Bhutan. (Permission to use by Bhutan Nature Conservation Division.)

country's treasured natural resources can be found," Bhutan's prime minister Dasho Tshering Tobgay says (www.panda.org, accessed March 26, 2018).

Yellowstone to Yukon (Y2Y)

The Y2Y vision is to connect and protect 3,200 kilometers of habitat from south of Grand Teton National Park in Wyoming to the Arctic circle in the Yukon so that people and nature in that large region can thrive. This vision inherently incorporates many of the top recommendations for facilitating climate adaptation. It includes expansion and increased numbers of protected areas as well as maintaining and restoring connections between them, all the while limiting human-caused stressors in and adjacent to protected areas and corridors. Additionally, as a mountain ecosystem, it incorporates elevational and latitudinal connectivity throughout the region. Species are more likely to be able to find desirable microclimates because of the close proximity of different habitats resulting from the diversity of slope, aspect, and elevational characteristics inherent to mountain ecosystems (Graumlich and Francis 2011).

The vision of a connected landscape has helped support significant advances in conservation across the region (Hannibal 2013). In the first twenty years, the region advanced from 11 to 21 percent protected areas. Other types of improvements in conservation, some of which are not necessarily permanent, such as caribou critical habitat or restrictions on different types of recreation, increased from 1 percent to 31 percent (https://y2y.net/publications/20th-anniversary-report).

Improvements in connectivity for wildlife across roads has increased significantly as well. As we discussed earlier in the book, roads can present a serious barrier for animal movement. Some of the most visible work that is being done to reconnect our landscapes is mitigation of the impacts of roads on wildlife. An excellent book on roads and the problems posed for wildlife conservation, as well as ways to improve the situation, is the *Handbook of Road Ecology* (Van der Ree et al. 2015). Mitigating the impacts of roads is primarily done by public transportation departments, but local interest groups engage in advocating for and supporting them. In the Y2Y region, the prevalence of wildlife crossing structures across the region is increasingly expansive. More than forty crossing structures are already built in both Banff National Park and on the Flathead Reservation in Montana, as well as a number of other structures in other areas totaling more than 100 crossing structures across Y2Y, and more are planned (Francis et al., in prep.). The fencing and other structures on the Trans-Canada Highway in Banff National Park in Alberta (fig. 5.3) now represent one of the most well-studied mitigation systems in the world with more than seventeen years of research examining use by different species, which is often different for males and females, changes over time, and differs depending on the presence of various overpass and underpass structures (e.g., Clevenger and Waltho 2005; Sawaya et al. 2013; Ford et al. 2017). Normalizing safe passage for wildlife across busy highways in the Y2Y region will continue to be a priority.

Ultimately, conservation advances in Y2Y are achieved at local scales. One of the places where a collaboration among stakeholders has been working for more than a decade to advance core area conservation and connectivity is the transboundary region of the Cabinet Purcell Mountains in British Columbia, Montana, and Idaho (fig. 10.11). One aspect of the work has been to increase the protection of identified corridors. The three primary corridors identified by grizzly bear biologists are now almost entirely

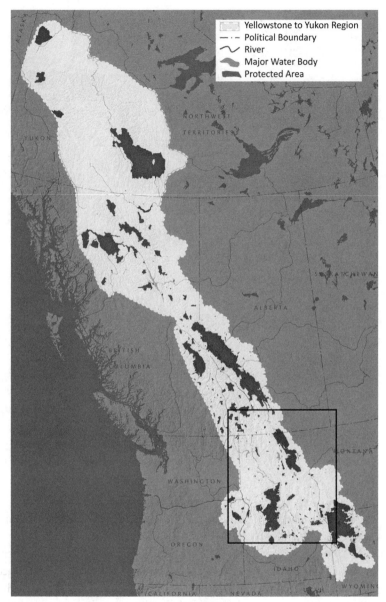

FIGURE 10.11 (a) Protected areas in the Yellowstone-to-Yukon (Y2Y) region. (b: on facing page) The grizzly bear distribution in the Y2Y transboundary region (inset in panel A). The three arrows point to three different linkages: The Duck Lake, Kidd Creek, and Yaak River linkage where private land acquisitions have secured movement corridors for grizzly bears. Map b by Michael Proctor; permission to use by the Yellowstone-to-Yukon Conservation Initiative.)

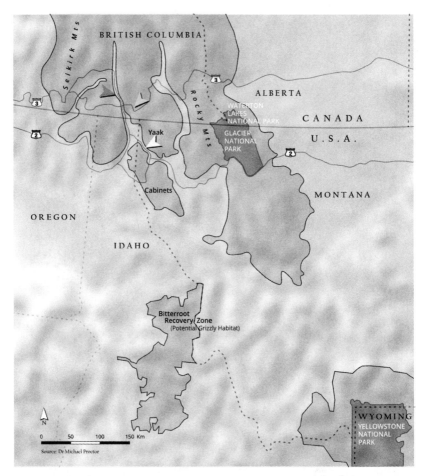

FIGURE 10.11 (b)

protected by easements or acquisition. At the time of writing, an additional land purchase is about to significantly increase the protection of one of these corridors. The conservation of core areas, corridor conservation, and progress in human–wildlife coexistence combined is helping to reconnect populations in the broader Y2Y region, ensuring that grizzly bear and other wildlife populations will continue to thrive. The grizzly bear population in the Cabinets-Yaak has increased from about ten bears in the early 1990s to sixty today, suggesting that this multifaceted partnership approach is working.

In summary, the above are a few examples of the many corridor projects that are going on around the world at multiple scales, from highway underpasses to comprehensive regional approaches. These efforts involve local communities, as well as national, and international organizations. We highly recommend that groups interested in enhancing and restoring connectivity visit and learn about other projects in order to take advantage of lessons learned. The most successful projects have strong collaborative relationships at all scales of governance and across interest groups. Building those relationships is essential to any planning process and will increase our understanding of human and natural systems.

Conclusion

The potential benefits as well as the challenges associated with maintaining and restoring connectivity depend a great deal on environmental, social, and historical context. Therefore, throughout this book, we included many different examples to illustrate (1) the consequences of habitat loss, fragmentation, and rapid climate change for biodiversity conservation; (2) how ecological principles inform our understanding of how species and communities respond to these disturbances; (3) how corridors are identified and designed for different purposes including to address climate change and challenges to marine connectivity; and (4) caveats and guidelines for project planning and implementation. By exploring the problem from a theoretical perspective and presenting practical examples, we aimed to make the information in this book useful to those studying and practicing conservation science in a wide variety of situations. Because context is so important, no one size fits all with respect to corridors. However, where possible, we provided ecological principles to consider and general steps that may improve the chances of successfully maintaining and restoring connectivity.

Understanding the historical context of a landscape is important. Geologic history plays a critical role in shaping natural ecosystems and, more recently, human history has determined the extent, type, and complexity of the matrix. Conservation biologists are increasingly recognizing the importance of the matrix on biodiversity both within and outside of protected areas. The matrix has particular influences on ecological networks because they often include narrow, marginal habitat surrounded by modified landscapes. The type and configuration of the matrix influences species invasions and impacts, and therefore influences which native species are likely to persist, and which exotic species are likely to invade.

Scale is the most defining aspect of a project and is influenced by the

274

nature of the problems addressed, the extent of existing barriers to connectivity, conservation objectives, and the institutions addressing the problem. The project scale, in turn, influences the processes that can be maintained or restored. For example, large-scale connectivity may be required to facilitate large-animal migration, whereas plant pollination may be accommodated within a smaller protected area network. We illustrated this range of needs by describing small-scale projects, from an urban river corridor in Lisbon, Portugal, to the efforts of defining ecological networks for all of Australia. Institutional participation greatly influences the scale at which a problem is addressed, since locally based environmental groups rarely take on large-scale planning efforts, and political entities find it difficult to work beyond their boundaries.

We know that animal and plant populations naturally vary from being completely isolated from one another, through intermittently connected metapopulations, to sharing a single widespread breeding population. Therefore, conserving and enhancing connectivity usually requires more than a single, minimum-size corridor. It is important not to rely on the minimum dimensions or on a limited number of landscape features to provide connectivity, because a variety of larger features will accommodate more species and be more resilient to change over time. Wider and redundant corridors can lessen problems sometimes associated with corridors such as genetic and demographic problems that can plague species in marginal and fragmented habitat.

No matter the scale of an existing project, it is important to at least consider the larger landscape context and long-term changes in climate in planning, design, and evaluation of ecological corridors. This is because large-scale processes often influence local-scale phenomena. Also, as time passes, new efforts may be started to address connectivity at larger scales that will need to build on existing corridors and reserves. This means a need to think about context across a larger area than the planned conservation area boundaries. In fact, we argue that protected area managers need to include working beyond their individual protected area in their mandate, with some focus on connectivity.

Corridors should be part of larger ecological networks consisting of blocks of natural landscapes, possibly buffered from the matrix, and connected by different types of corridors. These ecological networks can offer resilience to perturbation, even from large-scale stochastic events. They should be assessed for climate resilience and augmented where needed to be climate-wise and provide connections from current habitat to future suitable habitat, in addition to permitting daily, dispersal, or migratory movements. This will help ensure that they accommodate range shifts that are likely to occur.

Land-use and conservation planning are two sides of the same coin. We need to acknowledge that land-use and conservation planning must realistically be based on less than adequate information, reducing our ability to make accurate predictions regarding the response of species and natural communities to proposed changes across the landscape. Lack of funding may make it difficult to augment existing data and provide for monitoring the impacts of conservation strategies as they proceed. Even more challenging is the reality that planning, more often than not, proceeds in a social context that constrains conservation options, and biases the process toward minimal rather than optimal solutions. Given these constraints, land-use planning should adopt an adaptive management strategy to the extent possible, to allow for adjustments in the process as knowledge and experience are gained.

Mapping the history of land-use change is usually the first step to developing models used to forecast future land use. While there are many uncertainties surrounding forecasting models, it is important that we do our best to make predictions of where existing natural areas need protection to minimize future habitat loss and fragmentation. Protecting natural resources and connectivity that are threatened in the near term is generally more costly than protecting landscapes that are not under demand for development. However, critical environmental benefits may be lost due to land-use change in the immediate future; therefore, resources that are under immediate threat should be a priority for conservation. Land use continues to fragment and impact natural systems in many ways, as we discussed, but land-use change does not receive enough attention as a primary determinant of global change. Land use along the coastal ocean zone is having large impacts on marine ecosystems. Pollution from plastics, oil, nitrogen, and other waste products causes dead zones in the ocean. While human use of the oceans including overfishing and shipping is impacting marine life, what we do on land that spills over into the ocean as well as climate change caused from burning fossil fuels may be the primary drivers of fragmentation across the seascapes of the world. The problems associated with continued intense land use and its interaction with climate change that we do not address and manage today will be passed on to future generations, leaving them fewer options for conservation and imparting higher costs.

We caution that ultimately, no connected, protected area network will conserve biodiversity unless accompanied by measures to reduce human population and consumption, while increasing equal access to resources.

Failure to challenge our current growth and consumption paradigm means that development will continue to consume the open spaces we aim to protect. Several opportunities to stabilize population and consumption exist, and could be realized with the right policy interventions. Investing in health, education, and female empowerment is the best route toward reducing population growth rates, and is already occurring in many places undergoing urbanization and economic development (Sanderson et al. 2018). Globally, a large unmet need for family planning still exists (Bongaarts and Sinding 2009) and further investment could help to stabilize the global population at six billion people by 2100, instead of the nine to twelve billion projected without intervention (Bradshaw and Brook 2014). Reducing consumption by the wealthiest citizens needs to focus on reducing sprawling low-density development, vastly reducing food waste, and seeking efficiencies in energy and water use. These changes are essential but will take time. Until we achieve true sustainability, we must learn to live with nature and not at nature's expense. Only this way can we fulfill our moral and ethical responsibility to all life on Earth. Protecting the freedom for all life to roam needs to be a central course of action.

Future Research and Directions

We lack a great deal of information on ecological communities, landscape-level processes, and interactions between humans and the biosphere. Scientists often document changes in natural communities caused by human impacts, but in most cases we know very little about the underlying mechanisms driving the patterns of change we have documented, or the best ways to mitigate the negative effects. Places such as South America, Africa, and southern Asia deserve even more attention as many living systems remain relatively understudied in these parts of the world. Studying connectivity in marine systems is a very new direction and additional interdisciplinary research into the many complex biophysical interactions that determine connectivity and the ecological manifestations of climate change is important to pursue. Even in better-studied areas, more research in conservation biology, including experimental tests of conservation and climate adaptation strategies and treatments, as well as the social aspects of conservation, is needed.

Truly interdisciplinary research, designed from the outset by collaborations of physical, biological, and social scientists, and nonacademic partners, requires significant investments to build professional relationships and bridge intellectual domains. However, such collaborations hold great potential for addressing global change impacts. We need to understand the underlying issues, design and evaluate solutions, and identify the costs of, barriers to, and opportunities for implementation. The urgency to intervene, conserve threatened species, and improve degraded systems should not preclude capitalizing on opportunities to conduct long-term studies and controlled, experimental research on intervention effectiveness. By that same logic, developing global guidance and standards for corridors are challenging, but we should persist as they could be extremely helpful to managers implementing corridors and policymakers considering legislation to advance corridor conservation.

Learning more about how organisms move through landscapes, the cues they use to navigate, and the resources that are critical for their safe passage will improve our ability to design effective corridors. Equally important is a good understanding of how species respond to ecotones of various types. The interesting question of whether edge zones are stable over time or will gradually grow in width, further diminishing the effective size of habitat patches, needs to be addressed.

We have much to learn about the nature of interactions among species in communities, and we must strive to more accurately predict how those interactions will change if a species disappears or, conversely, an exotic or range-expanding species invades. In general, much more needs to be learned about the behavior of small populations and the generation of minimum threshold densities. In addition, the roles of parasites and diseases, both native and exotic, have been especially poorly investigated.

Moreover, attention needs to be paid to spatial ecosystem modeling that integrates biological, physical, and, if possible, social processes across large-scale ecosystems and entire regions to help inform conservation planners and decision makers. For example, environmental modeling needs to be combined with land-use change modeling to forecast the environmental consequences of likely future land-use change. These types of models and the increased scientific understanding and practical applications that they can provide will become even more critical in a prolonged era of rapid climate change.

In our efforts to reduce the severity of the largest and probably most rapid extinction episode in our planet's history, we will need all the tools in our toolbox and more. In addition to the obvious first steps, such as trying to maintain minimum viable populations of all native species, we need to protect the earth's remaining wilderness areas and the goods and services they provide, which means increased focus and allocation of resources for conservation programs. Identifying the location and extent of protected habitat that will best achieve multiple conservation benefits requires that we quantify the functional relationship between conserving landscapes and the desired results. This can be difficult when species occurrences and demographic data are limited, and require quantifying the combined value of multiple benefits, from species conservation to water resources and recreational amenities—all common objectives of conservation. Many of the benefits cannot be interchanged, and in some cases, one location may not be substitutable for another. Research is needed to examine the best methods for valuing combinations of conservation benefits to more effectively target conservation objectives and enhance connectivity for multiple purposes.

In reviewing the state of our understanding of habitat connectivity science and practice we believe additional research and actions listed in box C.1, many of which are discussed in context throughout the book, will provide new insights and even surprises going forward. A major challenge of all the suggested investigations listed is that all places are idiosyncratic; that is, planning must be informed by local information.

Conservation science and conservation practice are inextricably linked, and it is well recognized that ecosystem management is an adaptive process that requires continuous monitoring. Accordingly, we have emphasized throughout this book the importance of ecological monitoring both for effective conservation planning and project implementation. Implementing corridors as part of ecological networks can be achieved through conservation instruments including community-driven initiatives and public or privately funded incentives such as conservation easements, discussed in chapter 10. However, challenges remain: whether people are living with nature in the matrix or visiting a protected area to experience nature, there are inevitable conflicts between humans and some animals. Examples are crop-raiding elephants and aggressive bears, as well as sensitive plant communities, such as rocky cliffs and wetlands that we all should avoid trampling. A shift toward ecosystem stewardship may help as we expand the intermix between human communities and nature. This approach aims to sustain ecosystem and human well-being under conditions of uncertainty and change. Its principles include maintaining diverse options, supporting social learning and innovation, and creating flexible governance systems (Chapin et al. 2010). More social research on how to advance ecosystems management goals is a top priority if we are going to advance conservation in a highly populated world.

Moreover, for science to be useful to conservation, it must be continually translated and disseminated for applied purposes and become incorporated into education and outreach at all levels. A more concerted effort

BOX C.1

FUTURE RESEARCH AND DIRECTION TO IMPROVE OUR UNDERSTANDING OF CORRIDOR ECOLOGY

Organismal Ecology

- Experimental study of how organisms move and materials flow, adjusting to interventions such as stream restoration and flow recovery. Where possible, research should have controls, replication, and be long term.
- Empirical study of species dispersal distances in comparison with other types of movement, threshold distances, and the influence of the matrix on movement.
- Influence of sociality on dispersal capabilities across different matrixes and various plant and animal assemblages.
- Community ecology with a focus on the role of mutualisms and predator/ prey coactions affecting competitive interactions in patches with different community composition.
- Studies of the influence of habitat patch size and/or quality on the Allee effect for small populations resulting in potential changes to extinction rates and providing a greater understanding of extinction debt and the changes to connectivity that may result in increased persistence.

Corridor Design and Function

- Influence of corridor dimensions, especially width, on corridor function, especially in highly developed areas.
- Retrospective analysis of outcomes of past actions to enhance connectivity including corridor restoration and diversified farming across the matrix.
- Study of recreation impacts, including risk analyses for corridor function in high-visitation areas, changes in animal movement patterns to avoid people, and avoidance zones around human trails and paths.
- Comparison of different modeling approaches to determine the consistency in resulting connectivity maps.
- Improvement of corridor models by taking into account differential animal behavior states and individuality.
- Use of quickly improving technologies to design corridors based on empirical movement data of a suite of local species.
- Incorporation of climate-wise concepts and modeling approaches into landscape planning efforts to ensure that connectivity conservation action will result in more climate resilient landscapes.
- Conservation planning and tradeoff analyses of the implications of (a) improving habitat connectivity through increasing protected area size and number versus (b) corridor conservation, or (c) diversifying the matrix surrounding or between protected areas.

- Measures of the interaction between varying levels of habitat connectivity and disease spread or parasite loads and/or virulence levels.
- Concepts, planning, and policies of corridors in the air that may mitigate negative effects on wildlife such as wind farms, night-lights, and human-controlled flying objects.

Land-Use Dynamics
- Studies of the impact of exurban development on natural systems and the rate and pattern of low-density development; improved maps of this land-use type across countries are required.
- Assessment of the effectiveness of policies that are intended to promote infill and shift development away from areas that may impact the integrity of corridors and large ecological networks.
- Exploration of the relationship between landscape configuration, including the pattern and type of the matrix, and the spread of invasive species in various bioregions.
- Examination of interactions between land use, future water deficit under climate change, and consequences for stream flow to protect freshwater communities from extinction in many parts of the world.

Connectivity Implementation
- Studies of community engagement including how to foster social learning spaces. These are vital given the importance of a shared vision and community support for successful corridor implementation.
- Citizen science and use of technology to share information on local biodiversity with communities as a way to increase support for connectivity projects.
- Education on sustainable land management including ways to enhance riparian vegetation, maintain natural stream flows, control invasive species, manage pests, and minimize movement barriers such as fencing.
- Retrospective study of land conservation instruments and their implementation on working landscapes to improve matrix permeability and conservation outcomes to help improve and advance effectiveness of connectivity enhancing programs.
- Examine social values and ways to promote coexistence and even living with pride in or adjacent to wildlife corridors even when potential conflict may still exist.
- Identification of policies and conservation instruments to implement transboundary corridor conservation within and across countries and other jurisdictions.

needs to be made to connect young people to the natural world and deepen their understanding of our interdependence with nature.

Finally, we applaud the growing community of scientists, environmental activists, planners, land managers, and politicians who are working on behalf of the world's citizenry to prevent further fragmentation, to restore connectivity at all scales, and in the end to make the world healthier for all of its inhabitants. It is our hope that this book will prove to be useful to those students, researchers, practitioners, and policymakers engaged in the study and implementation of ecological networks.

REFERENCES

Abadía-Cardoso, A., N. B. Freimer, K. Deiner, and J. C. Garza. 2017. "Molecular population genetics of the northern elephant seal *Mirounga angustirostris*." *Journal of Heredity* 108 (6): 618–27.

Abbitt, R. J. F., J. M. Scott, and D. S. Wilcove. 2000. "The geography of vulnerability: Incorporating species geography and human development patterns into conservation planning." *Biological Conservation* 96 (2): 169–75. doi: 10.1016/s0006-3207(00)00064-1.

Åberg, J., G. Jansson, J. E. Swenson, and P. Angelstam. 1995. "The effect of matrix on the occurrence of hazel grouse (*Bonasa bonasia*) in isolated habitat fragments." *Oecologia* 103 (3): 265–69. doi: 10.1007/bf00328613.

Abrahms, B., E. L. Hazen, S. J. Bograd, J. S. Brashares, P. W. Robinson, K. L. Scales, D. E. Crocker, and D. P. Costa. 2018. "Climate mediates the success of migration strategies in a marine predator." *Ecology Letters* 21 (1): 63–71.

Abrahms, B., S. C. Sawyer, N. R. Jordan, J. W. McNutt, A. M. Wilson, and J. S. Brashares. 2017. "Does wildlife resource selection accurately inform corridor conservation?" *Journal of Applied Ecology* 54 (2): 412–22.

Abson, R. N. 2004. "The use by vertebrate fauna of Slaty Creek Wildlife Underpass, Calder Freeway, Black Forest, Macedon, Victoria." Master's thesis, University of Tasmania.

Ackerly, D. D. 2003. "Community assembly, niche conservatism, and adaptive evolution in changing environments." *International Journal of Plant Sciences* 164 (3): S165–S184. doi: 10.1086/368401.

Ackerly, D. D., S. R. Loarie, W. K. Cornwell, S. B. Weiss, H. Hamilton, R. Branciforte, and N. J. B. Kraft. 2010. "The geography of climate change: Implications for conservation biogeography." *Diversity and Distributions* 16 (3): 476– 87.

Ackerly, D. D., R. A. Ryals, W. K. Cornwell, S. R. Loarie, S. Veloz, K. D. Higgason, W. L. Silver, and T. E. Dawson. 2012. Potential impacts of climate change on biodiversity and ecosystem services in the San Francisco Bay Area. California Energy Commission.

Acreman, M. C., J. R. Blake, D. J. Booker, R. J. Harding, N. Reynard, J. O. Mountford, and C. J. Stratford. 2009. "A simple framework for evaluating regional wetland ecohydrological response to climate change with case studies from Great Britain." *Ecohydrology* 2 (1): 1–17.

Acton, E. J., J. W. Acton, and P. D. Acton. 2018. "Simulations of cheetah roaming demonstrate the effect of safety corridors on genetic diversity and human–cheetah conflict." *Journal of Emerging Investigators*. April: 1–8.

Adams, A. M. 2016. " 'Restoration economy' strives to protect pollinators, create jobs." *Scientific American* (21 November 2016). www.scientificamerican.com/article/ldquo-restoration-economy-rdquo-strives-to-protect-pollinators-create-jobs/.

Ahern, J. 1995. "Greenways as a planning strategy." *Landscape and Urban Planning* 33 (1–3): 13155. doi: 10.1016/0169-2046(95)02039-v.

Alagador, D., J. O. Cerdeira, and M. B. Araujo. 2014. "Shifting protected areas: Scheduling spatial priorities under climate change." *Journal of Applied Ecology* 51 (3): 703–13. doi: 10.1111/13652664.12230.

Alagador, D., J. O. Cerdeira, and M. B. Araujo. 2016. "Climate change, species range shifts and dispersal corridors: An evaluation of spatial conservation models." *Methods in Ecology and Evolution* 7 (7): 853–66. doi: 10.1111/2041-210x.12524.

Albrecht, T. 2004. "Edge effect in wetland-arable land boundary determines nesting success of Scarlet Rosefinches (*Carpodacus erythrinus*) in the Czech Republic." *The Auk* 121 (2): 361–71. doi: 10.2307/4090400.

Allen, C. H., L. Parrott, and C. Kyle. 2016. "An individual–based modelling approach to estimate landscape connectivity for bighorn sheep (*Ovis canadensis*)." *PeerJ* 4. doi: 10.7717/peerj.2001.

Allen, J. C., W. M. Schaffer, and D. Rosko. 1993. "Chaos reduces species extinction by amplifying local population noise." *Nature* 364 (6434): 229–32. doi: 10.1038/364229a0.

Altieri, M. A. 2010. "Agroecology: Environmentally Sound and Socially Just Alternatives to the Industrial Farming Model." In *Animal and Plant Productivity*, edited by R. J. Hudson. Paris: EOLSS Publications.

Altmann, J., S. C. Alberts, S. A. Altmann, and S. B. Roy. 2002. "Dramatic change in local climate patterns in the Amboseli basin, Kenya." *African Journal of Ecology* 40 (3): 248–51.

Amboseli Ecosystem Management Plan. 2008. Amboseli Ecosystem Stakeholders and KWS Biodiversity Planning, Assessment & Compliance Department. Geonode-rris.biopama.org/documents/867/download. Accessed 20 June 2018.

Ament, R., R. Callahan, M. McClure, M. Reuling, and G. Tabor. 2014. "Wildlife connectivity: Fundamentals for conservation action." *Center for Large Landscape Conservation*. Bozeman, MT.

Anderson, E. P., C. N. Jenkins, S. Heilpern, J. A. Maldonado-Ocampo, F. M. Carvajal-Vallejos, A. C. Encalada, J. F. Rivadeneira, M. Hidalgo, C. M. Cañas, and H. Ortega. 2018. "Fragmentation of Andes-to-Amazon connectivity by hydropower dams." *Science Advances* 4: eaao1642.

Anderson, J. J., E. Gurarie, C. Bracis, B. J. Burke, and K. L. Laidre. 2013. "Modeling climate change impacts on phenology and population dynamics of migratory marine species." *Ecological Modelling* 264: 83–97.

Anderson, M.G., A. Barnett, M. Clark, J. Prince, A. O. Sheldon, and B. Vickery. 2016. *Resilient and Connected Landscapes for Terrestrial Conservation*. The Nature Conservancy, Eastern Conservation Science, Eastern Regional Office. Boston.

Anderson, M. G., M. Clark, and A. O. Sheldon. 2014. "Estimating climate resilience for conservation across geophysical settings." *Conservation Biology* 28 (4): 959–70. doi: 10.1111/cobi.12272.

Anderson, P. K. 1970. "Ecological structure and gene flow in small mammals." *Mammal Review* 1 (2): 29–30. doi: 10.1111/j.1365-2907.1970.tb00294.x.

Andreassen, H. P., K. Hertzberg, and R. A. Ims. 1998. "Space-use responses to habitat fragmentation and connectivity in the root vole *Microtus oeconomus*." *Ecology* 79 (4): 1223. doi: 10.2307/176738.

Andreassen, H. P., and R. A. Ims. 2001. "Dispersal in patchy vole populations: Role of patch configuration, density dependence, and demography." *Ecology* 82 (10): 2911. doi: 10.2307/2679970.

Andreassen, H. P., R. A. Ims, and O. K. Steinset. 1996. "Discontinuous habitat corridors: Effects on male root vole movements." *Journal of Applied Ecology* 33 (3): 555. doi: 10.2307/2404984.

Andrén, H. 1994. "Effects of habitat fragmentation on birds and mammals in landscapes with different proportions of suitable habitat: A review." *Oikos* 71 (3): 355. doi: 10.2307/3545823.

Aparicio, E., M. J. Vargas, J. M. Olmo, and A. De Sostoa. 2000. "Decline of native freshwater fishes in a Mediterranean watershed on the Iberian Peninsula: A quantitative assessment." *Environmental Biology of Fishes* 59 (1): 11–19.

Aquilani, S. M., and J. S. Brewer. 2004. "Area and edge effects on forest songbirds in a non-agricultural upland landscape in northern Mississippi, USA." *Natural Areas Journal* 24 (4): 326–35.

Aragón, G., L. Abuja, R. Belinchón, and I. Martínez. 2015. "Edge type determines the intensity of forest edge effect on epiphytic communities." *European Journal of Forest Research* 134 (3): 443–51.

Arendt, R. 2004. "Linked landscapes: Creating greenway corridors through conservation subdivision design strategies in the northeastern and central United States." *Landscape and Urban Planning* 68 (2): 241–69.

Ascensão, F., L. Fahrig, A. P. Clevenger, R. T. Corlett, J. A. G. Jaeger, W. F. Laurance, and H. M. Pereira. 2018. "Environmental challenges for the Belt and Road Initiative." *Nature Sustainability* 1: 206–8.

Ashcroft, M. B. 2010. "Identifying refugia from climate change." *Journal of Biogeography* 37 (8): 1407–13.

Asher, N. 2016. Mountain highway underpass to help rare pygmy possums breed. *ABC News*. 16 Nov. 2016. www.abc.net.au/news/2016-11-16/highway-underpass-helping-pygmy-possums-mate/8029284. (Accessed 23 April 2018.)

Austin, J. M., C. Slesar, and F. M. Hammond. 2010. "Strategic wildlife conservation and transportation planning: The Vermont experience." In *Safe Passages: Highways, Wildlife, and Habitat Connectivity*, edited by J. P. Beckmann, A. P. Clevenger, M. P. Huijser and J. A. Hilty. Washington, DC: Island Press.

Ayres, D. R., K. Zaremba, C. M. Sloop, and D. R. Strong. 2008. "Sexual reproduction of cordgrass hybrids (*Spartina foliosa x alterniflora*) invading tidal marshes in San Francisco Bay." *Diversity and Distributions* 14 (2): 187–95.

Baguette, M., R. Michniewicz, and V. M. Stevens. 2017. "From genes to metapopulations." *Nature Ecology & Evolution* 1: 0130.

Baker, C. S., D. Steel, J. Calambokidis, E. Falcone, U. González-Peral, J. Barlow, A. M. Burdin, P. J. Clapham, J. K. B. Ford, and C. M. Gabriele. 2013. "Strong maternal fidelity and natal philopatry shape genetic structure in North Pacific humpback whales." *Marine Ecology Progress Series* 494: 291–306.

Banks, S. C., D. B. Lindenmayer, S. J. Ward, and A. C. Taylor. 2005. "The effects of habitat fragmentation via forestry plantation establishment on spatial genotypic structure in the small marsupial carnivore *Antechinus agilis*." *Molecular Ecology* 14 (6): 1667–80. doi: 10.1111/j.1365-294x.2005.02525.x.

Bastille-Rousseau, G., I. Douglas-Hamilton, S. Blake, J. M. Northrup, and G. Wittemyer. 2018. "Applying network theory to animal movements to identify properties of landscape space use." *Ecological Applications* 28 (3): 854–64.

Bastille-Rousseau, G., J. Wall, I. Douglas-Hamilton, and G. Wittemyer. 2018. "Optimizing the positioning of wildlife crossing structures using GPS telemetry." *Journal of Applied Ecology* 55: 2055–62.

Bates, P. D., and A. P. J. De Roo. 2000. "A simple raster-based model for flood inundation simulation." *Journal of Hydrology* 236 (1–2): 54–77.

Bates, P. D., M. D. Wilson, M. S. Horritt, D. C. Mason, N. Holden, and A. Currie. 2006. "Reach scale floodplain inundation dynamics observed using airborne synthetic aperture radar imagery: Data analysis and modelling." *Journal of Hydrology* 328 (1–2): 306–18.

Bawa, K. 1990. "Plant-pollinator interactions in tropical rain forests." *Annual Review of Ecology and Systematics* 21 (1): 399–422. doi: 10.1146/annurev.ecolsys.21.1.399.

Beck, M. W., K. L. Heck, K. W. Able, D. L. Childers, D. B. Eggleston, B. M. Gillanders, B. Halpern, C. G. Hays, K. Hoshino, and T. J. Minello. 2001. "The identification, conservation, and management of estuarine and marine nurseries for fish and invertebrates: A better understanding of the habitats that serve as nurseries for marine species and the factors that create site-specific variability in nursery quality will improve conservation and management of these areas." *AIBS Bulletin* 51 (8): 633–41.

Beckmann, J. P., A. P. Clevenger, M. P. Huijser, and J. A. Hilty, eds. 2010. *Safe Passages: Highways, Wildlife, and Habitat Connectivity*. Washington, DC: Island Press.

Beier, P. 1993. "Determining minimum habitat areas and habitat corridors for cougars." *Conservation Biology* 7 (1): 94–108. doi: 10.1046/j.1523-1739.1993.07010094.x.

———. 1995. "Dispersal of juvenile cougars in fragmented habitat." *Journal of Wildlife Management* 59 (2): 228. doi: 10.2307/3808935.

———. 2012. "Conceptualizing and designing corridors for climate change." *Ecological Restoration* 30 (4): 312–19. doi: 10.3368/er.30.4.312.

Beier, P., and B. Brost. 2010. "Use of land facets to plan for climate change: Conserving the arenas, not the actors." *Conservation Biology* 24 (3): 701–10. doi: 10.1111/j.1523-1739 .2009.01422.x.

Beier, P., M. L. Hunter, and M. Anderson. 2015. "Special section: Conserving nature's stage." *Conservation Biology* 29 (3): 613–17. doi: 10.1111/cobi.12511.

Beier, P., and S. Loe. 1992. "Checklist for evaluating impacts to wildlife movement corridors." *Wildlife Society Bulletin* 20: 434–40.

Beier, P., D. R. Majka, and S. L. Newell. 2009. "Uncertainty analysis of least-cost modeling for designing wildlife linkages." *Ecological Applications* 19 (8): 2067–77.

Beier, P., D. R. Majka, and W. D. Spencer. 2008. "Forks in the road: Choices in procedures for designing wildland linkages." *Conservation Biology* 22 (4): 836–51. doi: 10.1111 /j.1523-1739.2008.00942.x.

Beier, P., W. Spencer, R. F. Baldwin, and B. McRae. 2011. "Toward best practices for developing regional connectivity maps." *Conservation Biology* 25 (5): 879–92.

Beier, P., Maryann Van Drielen, and Bright O. Kankam. 2002. "Avifaunal collapse in West African forest fragments." *Conservation Biology* 16 (4): 1097–1111. doi: 10.1046 /j.1523-1739.2002.01003.x.

Beissinger, S. R., and D. R. McCullough. 2002. *Population Viability Analysis*. Chicago: University of Chicago Press.

Bélisle, M. 2005. "Measuring landscape connectivity: The challenge of behavioral landscape ecology." *Ecology* 86 (8): 1988–95.

Belote, R. T., M. S. Dietz, C. N. Jenkins, P. S. McKinley, G. H. Irwin, T. J. Fullman, J. C. Leppi, and G. H. Aplet. 2017. "Wild, connected, and diverse: Building a more resilient system of protected areas." *Ecological Applications* 27 (4): 1050–56. doi: 10.1002 /eap.1527.

Belote, R. T., M. S. Dietz, B. H. McRae, D. M. Theobald, M. L. McClure, G. H. Irwin, P. S. McKinley, J. A. Gage, and G. H. Aplet. 2016. "Identifying Corridors among

Large Protected Areas in the United States." *PLoS ONE* 11 (4). doi: 10.1371/journal
.pone.0154223.

Ben-David, M., T. A. Hanley, and D. M. Schell. 1998. "Fertilization of terrestrial vegetation
by spawning Pacific salmon: The role of flooding and predator activity." *Oikos* 83 (1): 47.
doi: 10.2307/3546545.

Bennett, A. F. 1990. "Habitat corridors and the conservation of small mammals in a fragmented
forest environment." *Landscape Ecology* 4 (2–3): 109–22. doi: 10.1007/bf00132855.

———. 1991. "Roads, roadsides and wildlife conservation: A review." In *Nature Conservation
2: The Role of Corridors*, edited by D. A. Saunders and R. J. Hobbs, 99–118. Chipping
Norton, New South Wales, Australia: Surrey Beatty.

———. 2003. *Linkages in the Landscape: The Role of Corridors and Connectivity in Wildlife Conservation*. Gland, Switzerland: IUCN.

Bennett, A. F., K. Henein, and G. Merriam. 1994. "Corridor use and the elements of corridor
quality: Chipmunks and fencerows in a farmland mosaic." *Biological Conservation* 68 (2):
155–65. doi: 10.1016/0006-3207(94)90347-6.

Bennett, G. 2004. *Integrating Biodiversity Conservation and Sustainable Use: Lessons Learned
from Ecological Networks*. Gland, Switzerland, and Cambridge, England: IUCN.

Bennett, G., and P. Witt. 2001. "The development and application of ecological networks: A
review of proposals, plans and programmes." In *IUCN Report B1142*. Gland, Switzerland: World Conservation Union and AIDEnvironment.

Benoit-Bird, K. J., and M. A. McManus. 2012. "Bottom-up regulation of a pelagic community
through spatial aggregations." *Biology Letters* 8 (5): 813–16.

Bentley, J. M., C. P. Catterall, and G. C. Smith. 2000. "Effects of fragmentation of Araucarian
vine forest on small mammal communities." *Conservation Biology* 14 (4): 1075–87. doi:
10.1046/j.1523-1739.2000.98531.x.

Bentrup, G. 2008. *Conservation Buffers: Design Guidelines for Buffers, Corridors, and Greenways*:
US Department of Agriculture, Forest Service, Southern Research Station, Asheville, NC.

Berger, J. 2004. "The last mile: How to sustain long-distance migration in mammals." *Conservation Biology* 18 (2): 320–31. doi: 10.1111/j.1523-1739.2004.00548.x.

Berger, J., and S. L. Cain. 2014. "Moving beyond science to protect a mammalian migration
corridor." *Conservation Biology* 28 (5): 1142–50.

Berndtsson, J. Czemiel, T. Emilsson, and L. Bengtsson. 2006. "The influence of extensive vegetated roofs on runoff water quality." *Science of the Total Environment* 355 (1–3): 48–63.

Beunen, R., and J. E. Hagens. 2009. "The use of the concept of Ecological Networks in nature
conservation policies and planning practices." *Landscape Research* 34 (5): 563–80. doi:
10.1080/01426390903184280.

Bird, B. L., L. C. Branch, and D. L. Miller. 2004. "Effects of coastal lighting on foraging behavior of beach mice." *Conservation Biology* 18 (5): 1435–39. doi: 10.1111/j.1523-1739
.2004.00349.x.

Bissonette, J. A., and S. Broekhuizen. 1995. "Martes populations as indicators of habitat spatial patterns: The need for a multiscale approach." In *Landscape Approaches in Mammalian
Ecology and Conservation*, edited by W. Z. Lidicker Jr., 95–121. Minneapolis: University
of Minnesota Press.

Blazquez-Cabrera, S., A. Gastón, P. Beier, G. Garrote, M. Á. Simón, and S. Saura. 2016. "Influence of separating home range and dispersal movements on characterizing corridors
and effective distances." *Landscape Ecology* 31 (10): 2355–66.

Block, B. A., Ian. D. Jonsen, S. J. Jorgensen, A. J. Winship, S. A. Shaffer, S. J. Bograd, E. L. Hazen, D. G. Foley, G. A. Breed, and A.L. Harrison. 2011. "Tracking apex marine predator movements in a dynamic ocean." *Nature* 475 (7354): 86.

Boesing, A. L., E. Nichols, and J. Paul Metzger. 2018. "Biodiversity extinction thresholds are modulated by matrix type." *Ecography* 41: 1520–33.

Bolger, D. T., T. A. Scott, and J. T. Rotenberry. 1997. "Breeding bird abundance in an urbanizing landscape in coastal southern California." *Conservation Biology* 11 (2): 406–21. doi: 10.1046/j.1523-1739.1997.96307.x.

——. 2001. "Use of corridor-like landscape structures by bird and small mammal species." *Biological Conservation* 102 (2): 213–24. doi: 10.1016/s0006-3207(01)00028-3.

Bollinger, E. K., and T. A. Gavin. 2004. "Responses of nesting bobolinks (*Dolichonyx oryzivorus*) to habitat edges." *The Auk* 121 (3): 767. doi: 10.1642/0004-8038(2004)121[0767:ronbdo]2.0.co;2.

Bond, N. A., M. F. Cronin, H. Freeland, and N. Mantua. 2015. "Causes and impacts of the 2014 warm anomaly in the NE Pacific." *Geophysical Research Letters* 42 (9): 3414–20.

Bongaarts, J., and S. W. Sinding. 2009. "A response to critics of family planning programs." *International Perspectives on Sexual and Reproductive Health* 35 (1): 39–44.

Boose, A. B., and J. S. Holt. 1999. "Environmental effects on asexual reproduction in *Arundo donax*." *Weed Research* 39 (2): 117–27. doi: 10.1046/j.1365-3180.1999.00129.x.

Bow Corridor Ecosystem Advisory Group. 2012. Wildlife corridor and habitat patch guidelines for the Bow Valley. Town of Canmore, Town of Banff, Municipal District of Bighorn, Banff National Park, Government of Alberta.

Bowers, M. A., and J. L. Dooley. 1999. "A controlled, hierarchical study of habitat fragmentation: Responses at the individual, patch, and landscape scale." *Landscape Ecology* 14 (4): 381–89.

Braaker, S., J. Ghazoul, M. K. Obrist, and M. Moretti. 2014. "Habitat connectivity shapes urban arthropod communities: The key role of green roofs." *Ecology* 95 (4): 1010–21.

Bradly, K. 1991. "A data bank is never enough: The local approach to landcare." In *Nature Conservation 2: The Role of Corridors*, edited by D. A. Saunders and R. J. Hobbs, 377–86. Chipping Norton, New South Wales, Australia: Surrey.

Bradshaw, C. J. A., and B. W. Brook. 2014. "Human population reduction is not a quick fix for environmental problems." *Proceedings of the National Academy of Sciences* 111 (46): 16610–15.

Brambilla, M., E. Caprio, G. Assandri, D. Scridel, E. Bassi, R. Bionda, C. Celada et al. 2017. "A spatially explicit definition of conservation priorities according to population resistance and resilience, species importance and level of threat in a changing climate." *Diversity and Distributions* 23 (7): 727–38. doi: 10.1111/ddi.12572.

Brearley, G., J. Rhodes, A. Bradley, G. Baxter, L. Seabrook, D. Lunney, Y. Liu, and C. McAlpine. 2013. "Wildlife disease prevalence in human-modified landscapes." *Biological Reviews* 88 (2): 427–42.

Breckheimer, I., N. M. Haddad, W. F. Morris, A. M. Trainor, W. R. Fields, R. Jobe, B. R. Hudgens, A. Moody, and J. R. Walters. 2014. "Defining and evaluating the umbrella species concept for conserving and restoring landscape connectivity." *Conservation Biology* 28 (6):1584–93.

Briscoe, D. K., A. J. Hobday, A. Carlisle, K. Scales, J. P. Eveson, H. Arrizabalaga, J. N. Druon, and J. M. Fromentin. 2017. "Ecological bridges and barriers in pelagic ecosystems." *Deep-Sea Research Part II–Topical Studies in Oceanography* 140: 182–92.

Briscoe, D. K., D. M. Parker, S. Bograd, E. Hazen, K. Scales, G. H. Balazs, M. Kurita et al. 2016. "Multi–year tracking reveals extensive pelagic phase of juvenile loggerhead sea turtles in the North Pacific." *Movement Ecology* 4 (1): 23. doi: 10.1186/s40462-016-0087-4.

Brodie, J. F., M. Paxton, K. Nagulendran, G. Balamurugan, G. R. Clements, G. Reynolds, A. Jain, and J. Hon. 2016. "Connecting science, policy, and implementation for landscape-scale habitat connectivity." *Conservation Biology* 30 (5): 950–61. doi: 10.1111/cobi.12667.

Brooker, L., M. Brooker, and P. Cale. 1999. "Animal dispersal in fragmented habitat: Measuring habitat connectivity, corridor use, and dispersal mortality." *Conservation Ecology* 3 (1). doi: 10.5751/es-00109-030104.

Brosset, A., P. Charles-Dominique, A. Cockle, J. Cosson, and D. Masson. 1996. "Bat communities and deforestation in French Guiana." *Canadian Journal of Zoology* 74 (11): 1974–82. doi: 10.1139/z96-224.

Brost, B. M., and P. Beier. 2012. "Use of land facets to design linkages for climate change." *Ecological Applications* 22 (1): 87–103.

Brown, A., and D. Bell. 2016. Banff Park hopes to save more animals with $26M boost to highway fence. https://www.cbc.ca/news/canada/calgary/banff-fence-bear-death-1.3668135: *CBC News*.

Brown, J. H. 1978. "The theory of insular biogeography and the distribution of boreal birds and mammals." *Great Basin Naturalist Memoirs*: 209–27.

Brown, J. H., and M. V. Lomolino. 1998. *Biogeography*. 2nd ed. Sunderland, MA: Sinauer.

Bruno, J. F., C. D. G. Harley, and M. T. Burrows. 2013. "Climate change and marine communities." In *Marine Community Ecology and Conservation*, edited by M. D. Bertness, B. R. Silliman, J. F. Bruno, and J. J. Stachowicz. Sunderland, MA.: Sinauer.

Buma, B., and C. A. Wessman. 2011. "Disturbance interactions can impact resilience mechanisms of forests." *Ecosphere* 2 (5): 1–13.

Bunn, S. E., and A. H. Arthington. 2002. "Basic principles and ecological consequences of altered flow regimes for aquatic biodiversity." *Environmental Management* 30 (4): 492–507.

Bunnell, D. B., J. V. Adams, O. T. Gorman, C. P. Madenjian, S. C. Riley, E. F. Roseman, and J. S. Schaeffer. 2010. "Population synchrony of a native fish across three Laurentian Great Lakes: Evaluating the effects of dispersal and climate." *Oecologia* 162 (3): 641–51.

Burbidge, A. A., and B. F. J. Manly. 2002. "Mammal extinctions on Australian islands: Causes and conservation implications." *Journal of Biogeography* 29 (4): 465–73. doi: 10.1046/j.1365-2699.2002.00699.x.

Burke, D. M., and E. Nol. 1998. "Influence of food abundance, nest-site habitat, and forest fragmentation on breeding ovenbirds." *The Auk* 115 (1): 96–104. doi: 10.2307/4089115.

Burkey, T. V. 1997. "Metapopulation extinction in fragmented landscapes: Using bacteria and protozoa communities as model ecosystems." *American Naturalist* 150 (5): 568–91. doi: 10.1086/286082.

Burrows, M. T., D. S. Schoeman, A. J. Richardson, J. G. Molinos, A. Hoffmann, L. B. Buckley, P. J. Moore et al. 2014. "Geographical limits to species-range shifts are suggested by climate velocity." *Nature* 507 (7493): 492. doi: 10.1038/nature12976.

Buza, L., A. Young, and P. Thrall. 2000. "Genetic erosion, inbreeding and reduced fitness in fragmented populations of the endangered tetraploid pea *Swainsona recta*." *Biological Conservation* 93 (2): 177–86. doi: 10.1016/s0006-3207(99)00150-0.

Cabral, R. B., S. D. Gaines, M. T. Lim, M. P. Atrigenio, S. S. Mamauag, G. C. Pedemonte, and P. M. Aliño. 2016. "Siting marine protected areas based on habitat quality and extent provides the greatest benefit to spatially structured metapopulations." *Ecosphere* 7 (11).

Caley, M. J., M. H. Carr, M. A. Hixon, T. P. Hughes, G. P. Jones, and B. A. Menge. 1996. "Recruitment and the local dynamics of open marine populations." *Annual Review of Ecology and Systematics* 27 (1): 477–500.

Cameron, D. R., J. Marty, and R. F. Holland. 2014. "Whither the rangeland?: Protection and conversion in California's rangeland ecosystems." *PLoS ONE* 9 (8): e103468.

Cantrell, R. S., C. Cosner, and W. F. Fagan. 1998. "Competitive reversals inside ecological reserves: The role of external habitat degradation." *Journal of Mathematical Biology* 37 (6): 491–533. doi: 10.1007/s002850050139.

Capinha, C., D. Rödder, H. M. Pereira, and H. Kappes. 2014. "Response of non-native European terrestrial gastropods to novel climates correlates with biogeographical and biological traits." *Global Ecology and Biogeography* 23 (8): 857–66.

Carlisle, A. B., R. E. Kochevar, M. C. Arostegui, J. E. Ganong, M. Castleton, J. Schratwieser, and B. A. Block. 2017. "Influence of temperature and oxygen on the distribution of blue marlin (*Makaira nigricans*) in the Central Pacific." *Fisheries Oceanography* 26 (1): 34–48.

Carr, M. H., J. E. Neigel, J. A. Estes, S. Andelman, R. R. Warner, and J. L. Largier. 2003. "Comparing marine and terrestrial ecosystems: Implications for the design of coastal marine reserves." *Ecological Applications* 13 (sp1): 90–107.

Carr, M. H., S. P. Robinson, C. Wahle, G. Davis, S. Kroll, S. Murray, E. J. Schumacker, and M. Williams. 2017. "The central importance of ecological spatial connectivity to effective coastal marine protected areas and to meeting the challenges of climate change in the marine environment." *Aquatic Conservation: Marine and Freshwater Ecosystems* 27: 6–29.

Carr, M. H., and C. Syms. 2006. "Chapter 15: Recruitment." In *The Ecology of Marine Fishes: California and Adjacent Waters*, edited by L. G. Allen, D. J. Pondella II, and M. H. Horn, 411–27. Berkeley: University of California Press.

Carroll, C., J. J. Lawler, D. R. Roberts, and A. Hamann. 2015. "Biotic and climatic velocity identify contrasting areas of vulnerability to climate change." *PLoS ONE* 10 (10): e0140486.

Carvalho, K. S., and H. L. Vasconcelos. 1999. "Forest fragmentation in central Amazonia and its effects on litter-dwelling ants." *Biological Conservation* 91 (2–3): 151–57. doi: 10.1016/s0006-3207(99)00079-8.

Chapin, F. S., S. R. Carpenter, G. P. Kofinas, C. Folke, N. Abel, W. C. Clark, P. Olsson, D. M. S. Smith, B. Walker, and O. R. Young. 2010. "Ecosystem stewardship: Sustainability strategies for a rapidly changing planet." *Trends in Ecology & Evolution* 25 (4): 241–49.

Chester, C. C., J. A. Hilty, and S. C. Trombulak. 2012. "Climate change science, impacts, and opportunities." In *Climate and Conservation: Landscape and Seascape Science, Planning, and Action*, edited by J. A. Hilty, C. C. Chester, and M. S. Cross, 3–15. Washington, DC: Island Press/Center for Resource Economics.

Chetkiewicz, C. L. B., and M. S. Boyce. 2009. "Use of resource selection functions to identify conservation corridors." *Journal of Applied Ecology* 46 (5): 1036–47.

Chetkiewicz, C. L. B., C. C. S. Clair, and M. S. Boyce. 2006. "Corridors for conservation: Integrating pattern and process." *Annual Review of Ecology Evolution and Systematics* 37: 317–42. doi: 10.1146/annurev.ecolsys.37.091305.110050.

Cho, J., Y. W. Lee, P. J. F. Yeh, K. S. Han, and S. Kanae. 2014. "Satellite-based assessment of large-scale land cover change in Asian arid regions in the period of 2001–2009." *Environmental Earth Sciences* 71 (9): 3935–44. doi: 10.1007/s12665-013-2778-0.

Christie, M. R., and L. L. Knowles. 2015. "Habitat corridors facilitate genetic resilience irrespective of species dispersal abilities or population sizes." *Evolutionary Applications* 8 (5): 454–63. doi: 10.1111/eva.12255.

Churcher, P. B., and J. H. Lawton. 1987. "Predation by domestic cats in an English village." *Journal of Zoology* 212 (3): 439–55. doi: 10.1111/j.1469-7998.1987.tb02915.x.

Clarke, D. J., K. A. Pearce, and J. G. White. 2006. "Powerline corridors: Degraded ecosystems or wildlife havens?" *Wildlife Research* 33 (8): 615–26. doi: 10.1071/WR05085.

Clements, F. E. 1897. "Peculiar zonal formations of the Great Plains." *The American Naturalist* 31 (371): 968–70. doi: 10.1086/276738.

Clevenger, A. P., B. Chruszcz, and K. E. Gunson. 2001. "Highway mitigation fencing reduces wildlife-vehicle collisions." *Wildlife Society Bulletin* 29 (2): 646–53.

Clevenger, A. P., and N. Waltho. 2005. "Performance indices to identify attributes of highway crossing structures facilitating movement of large mammals." *Biological Conservation* 121 (3): 453–64. doi: 10.1016/j.biocon.2004.04.025.

Clewell, A. F., and J. Aronson. 2013. *Ecological Restoration: Principles, Values, and Structure of an Emerging Profession.* Washington, DC: Island Press.

Cockburn, A. 1992. "Habitat heterogeneity and dispersal: Environmental and genetic patchiness." In *Animal Dispersal*, edited by N. C. Stenseth and W. Z. Lidicker Jr. Dordrecht: Springer.

———. 2003. "Cooperative breeding in oscine passerines: Does sociality inhibit speciation?" *Proceedings of the Royal Society B: Biological Sciences* 270 (1530): 2207–14. doi: 10.1098/rspb.2003.2503.

Coffman, R. R., S. Cech, and R. Gettig. 2014. "Living architecture and biological dispersal." *Journal of Living Architecture* 1 (5): 1–9.

Coffman, R. R., and T. Waite. 2011. "Vegetated roofs as reconciled habitats: Rapid assays beyond mere species counts." *Urban Habitats* 6 (1): 1–10.

Collingham, Y. C., and B. Huntley. 2000. "Impacts of habitat fragmentation and patch size upon migration rates." *Ecological Applications* 10 (1): 131–44.

Compton, B. W., K. McGarigal, S. A. Cushman, and L. R. Gamble. 2007. "A resistant-kernel model of connectivity for amphibians that breed in vernal pools." *Conservation Biology* 21 (3): 788–99.

Cook, J. E. 1996. "Implications of modern successional theory for habitat typing: A review." *Forest Science* 42 (1): 67–75.

Coristine, L. E., P. Soroye, R. N. Soares, C. Robillard, and J. T. Kerr. 2016. "Dispersal limitation, climate change, and practical tools for butterfly conservation in intensively used landscapes." *Natural Areas Journal* 36 (4): 440–52.

Corporate Ecoforum and The Nature Conservancy. 2012. *The New Business Imperative: Valuing Natural Capital.*

Correa Ayram, C. A., M. E. Mendoza, A. Etter, and D. R. P. Salicrup. 2016. "Habitat connectivity in biodiversity conservation: A review of recent studies and applications." *Progress in Physical Geography* 40 (1): 7–37. doi: 10.1177/0309133315598713.

Costa, D. P., G. A. Breed, and P. W. Robinson. 2012. "New insights into pelagic migrations: Implications for ecology and conservation." *Annual Review of Ecology, Evolution, and Systematics* 43: 73–96.

Costanza, R., R. d'Arge, R. de Groot, S. Farber, M. Grasso, B. Hannon, K. Limburg et al. 1997. "The value of the world's ecosystem services and natural capital." *Nature* 387 (6630): 253–60. doi: 10.1038/387253a0.

Cove, M. V., R. M. Spínola, V. L. Jackson, J. C. Sáenz, and O. Chassot. 2013. "Integrating occupancy modeling and camera-trap data to estimate medium and large mammal detection and richness in a Central American biological corridor." *Tropical Conservation Science* 6 (6): 781–95.

Cowen, R. K., and S. Sponaugle. 2009. "Larval dispersal and marine population connectivity." *Annual Review of Marine Science* 1: 443–66.

Cox, P. A., and T. Elmqvist. 2000. "Pollinator extinction in the Pacific Islands." *Conservation Biology* 14 (5): 1237–39. doi: 10.1046/j.1523-1739.2000.00017.x.

Cox, P. A., T. Elmqvist, E. D. Pierson, and W. E. Rainey. 1991. "Flying foxes as strong interactors in South Pacific Island ecosystems: A conservation hypothesis." *Conservation Biology* 5 (4): 448–54. doi: 10.1111/j.1523-1739.1991.tb00351.x.

Cramer, P. C., and K. M. Portier. 2001. "Modeling Florida panther movements in response to human attributes of the landscape and ecological settings." *Ecological Modelling* 140 (1–2): 51–80. doi: 10.1016/s0304-3800(01)00268-x.

Creech, T. G., C. W. Epps, R. J. Monello, and J. D. Wehausen. 2014. "Using network theory to prioritize management in a desert bighorn sheep metapopulation." *Landscape Ecology* 29 (4): 605–19.

Crispo, E., J. Moore, J. A. Lee-Yaw, S. M. Gray, and B. C. Haller. 2011. "Broken barriers: Human-induced changes to gene flow and introgression in animals." *BioEssays* 33 (7): 508–18.

Crome, F., J. Isaacs, and L. Moore. 1994. "The utility to birds and mammals of remnant riparian vegetation and associated windbreaks in the tropical Queensland uplands." *Pacific Conservation Biology* 1 (4): 328. doi: 10.1071/pc940328.

Crooks, K. R., and M. E. Soulé. 1999. "Mesopredator release and avifaunal extinctions in a fragmented system." *Nature* 400 (6744): 563–66. doi: 10.1038/23028.

Crooks, K. R., and M. Sanjayan. 2006. *Connectivity Conservation*, Vol. 14. Cambridge, England: Cambridge University Press.

Crooks, K. R., A. V. Suarez, D. T. Bolger, and M. E. Soulé. 2001. "Extinction and colonization of birds on habitat islands." *Conservation Biology* 15 (1): 159–72. doi: 10.1111/j.1523-1739.2001.99379.x.

Croteau, J. 2016. Alberta bear experts warn of conflicts with cyclists as woman recovers from attack. https://globalnews.ca/news/2840129/alberta-bear–experts-warn-of-conflicts-with-cyclists-as-woman-recovers-from-attack/. Accessed 25 June 2018: *Global News*.

Cushman, S. A., and E. L. Landguth. 2012. "Multi-taxa population connectivity in the Northern Rocky Mountains." *Ecological Modelling* 231: 101–12. doi: 10.1016/j.ecolmodel.2012.02.011.

Cushman, S. A., K. S. McKelvey, J. Hayden, and M. K. Schwartz. 2006. "Gene flow in complex landscapes: Testing multiple hypotheses with causal modeling." *American Naturalist* 168 (4): 486–99.

Dai, L., D. Vorselen, K. S. Korolev, and J. Gore. 2012. "Generic indicators for loss of resilience before a tipping point leading to population collapse." *Science* 336 (6085): 1175–77.

Daily, G. C., ed. 1997. *Nature's Services: Societal Dependence on Natural Ecosystems*. Washington, DC: Island Press.

Daily, G. C., and K. Ellison. 2002. *The New Economy of Nature: The Quest to Make Conservation Profitable*. Washington, DC: Island Press.

Dalby, A. 2003. *Language in Danger: The Loss of Linguistic Diversity and the Threat to Our Future*. New York: Columbia University Press.

Dale, S. 2001. "Female-biased dispersal, low female recruitment, unpaired males, and the extinction of small and isolated bird populations." *Oikos* 92 (2): 344–56. doi: 10.1034 /j.1600-0706.2001.920217.x.

Dale, S., K. Mork, R. Solvang, and A. J. Plumptre. 2000. "Edge effects on the understory bird community in a logged forest in Uganda." *Conservation Biology* 14 (1): 265–76. doi: 10.1046/j.1523-1739.2000.98340.x.

Aloia, D., Cassidy C, S. M. Bogdanowicz, R. K. Francis, J. E. Majoris, R. G. Harrison, and P. M. Buston. 2015. "Patterns, causes, and consequences of marine larval dispersal." *Proceedings of the National Academy of Sciences* 112 (45): 13940–45.

Danielson, B. J. 1991. "Communities in a landscape: The influence of habitat heterogeneity on the interactions between species." *American Naturalist* 138 (5): 1105–20. doi: 10 .1086/285272.

Darling, F. F. 1938. *Bird Flocks and the Breeding Cycle: A Contribution to the Study of Avian Sociality.* Cambridge, England: Cambridge University Press.

Date, E. M., H. A. Ford, and H. F. Recher. 1991. "Frugivorous pigeons, stepping stones, and weeds in northern New South Wales." In *Nature Conservation 2: The Role of Corridors*, edited by D. A. Saunders and R. J. Hobbs, 141–45. Chipping Norton, New South Wales, Australia: Surrey Beatty.

Davis, M. B., and R. G. Shaw. 2001. "Range shifts and adaptive responses to Quaternary climate change." *Science* 292 (5517): 673–79.

DeChaine, E. G., and A. P. Martin. 2004. "Historic cycles of fragmentation and expansion in *Parnassius smintheus* (Papilionidae) inferred using mitochondrial DNA." *Evolution* 58 (1): 113–27. doi: 10.1111/j.0014-3820.2004.tb01578.x.

Delattre, P., B. De Sousa, E. Fichet-Calvet, J. P. Quéré, and P. Giraudoux. 1999. "Vole outbreaks in a landscape context: Evidence from a six-year study of *Microtus arvalis*." *Landscape Ecology* 14 (4): 401–12. doi: 10.1023/a:1008022727025.

Derworiz, C. 2016. How do the animals cross the road in Banff National Park? http://calgary herald.com/news/local-news/how-do–the-animals-cross-the–road-in–banff-national -park: *Calgary Herald.*

Diamond, J. M. 1973. "Distributional ecology of New Guinea birds: Recent ecological and biogeographical theories can be tested on the bird communities of New Guinea." *Science* 179 (4075): 759–69. doi: 10.1126/science.179.4075.759.

———. 1976. "Island biogeography and conservation: Strategy and limitations." *Science* 193 (4257):1027–29. doi: 10.1126/science.193.4257.1027.

Dickey, T. D. 2001. "The role of new technology in advancing ocean biogeochemical research." *Oceanography* 14: 108–20.

Dickson, B. G., C. M. Albano, B. H. McRae, J. J. Anderson, D. M. Theobald, L. J. Zachmann, T. D. Sisk, and M. P. Dombeck. 2016. "Informing strategic efforts to expand and connect protected areas using a model of ecological flow, with application to the western United States." *Conservation Letters* 10 (5): 564–71.

Dijak, W. D., and F. R. Thompson. 2000. "Landscape and edge effects on the distribution of mammalian predators in Missouri." *Journal of Wildlife Management* 64 (1): 209. doi: 10.2307/3802992.

Dilkina, B., R. Houtman, C. P. Gomes, C. A. Montgomery, K. S. McKelvey, K. Kendall, T. A. Graves, R. Bernstein, and M. K. Schwartz. 2017. "Trade-offs and efficiencies in optimal budget-constrained multispecies corridor networks." *Conservation Biology* 31 (1): 192–202. doi: 10.1111/cobi.12814.

Dilts, T. E., P. J. Weisberg, P. Leitner, M. D. Matocq, R. D. Inman, K. E. Nussear, and

T. C. Esque. 2016. "Multiscale connectivity and graph theory highlight critical areas for conservation under climate change." *Ecological Applications* 26 (4): 1223–37. doi: 10.1890/15-0925.

Dixo, M., J. P. Metzger, J. S. Morgante, and K. R. Zamudio. 2009. "Habitat fragmentation reduces genetic diversity and connectivity among toad populations in the Brazilian Atlantic Coastal Forest." *Biological Conservation* 142 (8):1560–69.

Dobrowski, S. Z., and S. A. Parks. 2016. "Climate change velocity underestimates climate change exposure in mountainous regions." *Nature Communications* 7.

Dobson, A. 2003. "Metalife!" *Science* 301 (5639): 1488–90. doi: 10.1126/science.1090481.

Doerr, V. A. J., T. Barrett, and E. D. Doerr. 2011. "Connectivity, dispersal behaviour and conservation under climate change: A response to Hodgson et al." *Journal of Applied Ecology* 48 (1): 143–47. doi: 10.1111/j.1365–2664.2010.01899.x.

Doherty, T. S., and D. A. Driscoll. 2018. "Coupling movement and landscape ecology for animal conservation in production landscapes." *Proceedings of the Royal Society B* 285 (1870): 20172272.

Doney, S. C. 2010. "The growing human footprint on coastal and open-ocean biogeochemistry." *Science* 328 (5985): 1512–16.

Doney, S., M. Ruckelshaus, J. E. Duffy, J. P. Barry, F. Chan, C. A. English, H. M. Galindo, J. M. Grebmeier, A. B. Hollowed, and N. Knowlton. 2011. "Climate change impacts on marine ecosystems." *Annual Review of Marine Science* 4: 1–27.

Donovan, T. M., and C. H. Flather. 2002. "Relationships among North American songbird trends, habitat fragmentation, and landscape occupancy." *Ecological Applications* 12 (2): 364. doi: 10.2307/3060948.

Drapeau, P., A. Leduc, J. Giroux, J. L. Savard, Y. Bergeron, and W. L. Vickery. 2000. "Landscape-scale disturbances and changes in bird communities of boreal mixed-wood forests." *Ecological Monographs* 70 (3): 423. doi: 10.2307/2657210.

Driscoll, M. J. L., and T. M. Donovan. 2004. "Landscape context moderates edge effects: Nesting success of wood thrushes in central New York." *Conservation Biology* 18 (5): 1330–38. doi: 10.1111/j.1523-1739.2004.00254.x.

Dudley, N., ed. 2008. *Guidelines for Applying Protected Area Management Categories.* Gland, Switzerland: IUCN.

Dulvy, N. K., S. I. Rogers, S. Jennings, V. Stelzenmüller, S. R. Dye, and H. R. Skjoldal. 2008. "Climate change and deepening of the North Sea fish assemblage: A biotic indicator of warming seas." *Journal of Applied Ecology* 45 (4): 1029–39.

Dunn, R. R. 2000. "Isolated trees as foci of diversity in active and fallow fields." *Biological Conservation* 95 (3): 317–21. doi: 10.1016/s0006-3207(00)00025-2.

Dunning, J. B., R. Borgella, K. Clements, and G. K. Meffe. 1995. "Patch isolation, corridor effects, and colonization by a resident sparrow in a managed pine woodland." *Conservation Biology* 9 (3): 542–50. doi: 10.1046/j.1523-1739.1995.09030542.x.

Dunstan, C., and B. Fox. 1996. "The effects of fragmentation and disturbance of rainforest on ground-dwelling small mammals on the Robertson Plateau, New South Wales, Australia." *Journal of Biogeography* 23 (2): 187–201. doi: 10.1046/j.1365-2699.1996.d01-220.x.

Dunstan, P. K., N. J. Bax, J. M. Dambacher, K. R. Hayes, P. T. Hedge, D. C. Smith, and A. D. M. Smith. 2016. "Using ecologically or biologically significant marine areas (EBSAs) to implement marine spatial planning." *Ocean & Coastal Management* 121: 116–27.

Dybas, C. L. 2004. California wine country clashes with ecosystem. *Washington Post*, June 21.

Eco Logical Australia. 2016. Noosa biodiversity plan: Biodiversity assessment report. https://www.noosa.qld.gov.au/documents/40217326/40856521/2016-12-12%20General%20Committee%20-%20Attachment%201%20to%20Item%206%20-%20Biodiversity%20Assessment%20Report.pdf. Accessed 6 June 2018. Prepared for Noosa Shire Council.

Edmands, S. 1999. "Heterosis and outbreeding depression in interpopulation crosses spanning a wide range of divergence." *Evolution* 53 (6): 1757. doi: 10.2307/2640438.

Ehrlich, P., and A. Ehrlich. 2004. *One with Nineveh: Politics, Consumption, and the Human Future.* Washington, DC: Island Press.

Elliot, N. B., S. A. Cushman, D. W. Macdonald, and A. J. Loveridge. 2014. "The devil is in the dispersers: Predictions of landscape connectivity change with demography." *Journal of Applied Ecology* 51 (5): 1169–78. doi: 10.1111/1365-2664.12282.

Epps, C. W., D. R. McCullough, J. D. Wehausen, V. C. Bleich, and J. L. Rechel. 2004. "Effects of climate change on population persistence of desert-dwelling mountain sheep in California." *Conservation Biology* 18 (1): 102–13. doi: 10.1111/j.1523-1739.2004.00023.x.

Epps, C. W., B. M. Mutayoba, L. Gwin, and J. S. Brashares. 2011. "An empirical evaluation of the African elephant as a focal species for connectivity planning in East Africa." *Diversity and Distributions* 17 (4): 603–12.

Epps, C. W., J. D. Wehausen, V. C. Bleich, S. G. Torres, and J. S. Brashares. 2007. "Optimizing dispersal and corridor models using landscape genetics." *Journal of Applied Ecology* 44 (4): 714–24.

Estes, J. A., and D. O. Duggins. 1995. "Sea otters and kelp forests in Alaska: Generality and variation in a community ecological paradigm." *Ecological Monographs* 65 (1): 75–100. doi: 10.2307/2937159.

Estes, J. A., J. Terborgh, J. S. Brashares, M. E. Power, J. Berger, W. J. Bond, S. R. Carpenter, T. E. Essington, R. D. Holt, and J. B. C. Jackson. 2011. "Trophic downgrading of planet Earth." *Science* 333 (6040): 301–6.

Evans, F. C. 1942. "Studies of a small mammal population in Bagley Wood, Berkshire." *The Journal of Animal Ecology* 11 (2): 182. doi: 10.2307/1355.

Eycott, A. E., G. B. Stewart, L. M. Buyung-Ali, D. E. Bowler, K. Watts, and A. S. Pullin. 2012. "A meta-analysis on the impact of different matrix structures on species movement rates." *Landscape Ecology* 27 (9): 1263–78.

Fábos, J. G. 2004. "Greenway planning in the United States: Its origins and recent case studies." *Landscape and Urban Planning* 68 (2–3): 321–42. doi: 10.1016/j.landurbplan.2003.07.003.

Fagan, M. E., R. S. DeFries, S. E. Sesnie, J. P. Arroyo-Mora, and R. L. Chazdon. 2016. "Targeted reforestation could reverse declines in connectivity for understory birds in a tropical habitat corridor." *Ecological Applications* 26 (5): 1456–74. doi: 10.1890/14-2188.

Fahrig, L. 2003. "Effects of habitat fragmentation on biodiversity." *Annual Review of Ecology, Evolution, and Systematics* 34 (1): 487–515. doi: 10.1146/annurev.ecolsys.34.011802.132419.

———. 2017. "Ecological responses to habitat fragmentation per se." *Annual Review of Ecology, Evolution, and Systematics* 48: 1–23.

Fairfax, S. K., and D. Guenzler. 2001. *Conservation Trusts.* Lawrence: University Press of Kansas.

Fairfax, S. K., L. Reymond, L. Gwin, M. A. King, and L. A. Watt. 2005. *Buying Nature: The Limits of Land Acquisition as a Conservation Strategy.* Boston: MIT Press.

Faith, D. P., and P. A. Walker. 2002. "The role of trade-offs in biodiversity conservation planning: Linking local management, regional planning and global conservation efforts." *Journal of Biosciences* 27 (4): 393–407. doi: 10.1007/bf02704968.

Fan, D. M., Z. X. Sun, B. Li, Y. X. Kou, R. G. J. Hodel, Z. N. Jin, and Z. Y. Zhang. 2017. "Dispersal corridors for plant species in the Poyang Lake Basin of southeast China identified by integration of phylogeographic and geospatial data." *Ecology and Evolution* 7 (14): 5140–48. doi: 10.1002/ece3.2999.

Farhadinia, M. S., M. Ahmadi, E. Sharbafi, S. Khosravi, H. Alinezhad, and D. W. Macdonald. 2015. "Leveraging trans-boundary conservation partnerships: Persistence of Persian leopard (*Panthera pardus saxicolor*) in the Iranian Caucasus." *Biological Conservation* 191: 770–78.

Farris, K. L., M. J. Huss, and S. Zack. 2004. "The role of foraging woodpeckers in the decomposition of ponderosa pine snags." *Condor* 106 (1): 50. doi: 10.1650/7484.

Fausch, K. D. 2015. *For the Love of Rivers: A Scientist's Journey*. Corvallis, Oregon: Oregon State University Press.

Fernández-Cañero, R., and P. González-Redondo. 2010. "Green roofs as a habitat for birds: A review." *Journal of Animal and Veterinary Advances* 9: 2041–52.

Ferraro, P. J. 2001. "Global habitat protection: Limitations of development interventions and a role for conservation performance payments." *Conservation Biology* 15 (4): 990–1000. doi: 10.1046/j.1523-1739.2001.015004990.x.

Ferraz, G., G. J. Russell, P. C. Stouffer, R. O. Bierregaard, S. L. Pimm, and T. E. Lovejoy. 2003. "Rates of species loss from Amazonian forest fragments." *Proceedings of the National Academy of Sciences* 100 (24): 14069–73. doi: 10.1073/pnas.2336195100.

Fitzgerald, K. H. 2015. "The silent killer: Habitat loss and the role of African protected areas to conserve biodiversity." In *Protecting the Wild*, edited by G. Wuerthner, E. Crist, and T. Butler. Washington, DC: Island Press.

Fitzsimons, J., I. Pulsford, and G. Wescott. 2013. "Lessons from large-scale conservation networks in Australia." *Parks* 19 (1): 115–25.

Fleishman, E., J. R. Thomson, E. L. Kalies, B. G. Dickson, D. S. Dobkin, and M. Leu. 2014. "Projecting current and future location, quality, and connectivity of habitat for breeding birds in the Great Basin." *Ecosphere* 5 (7). doi: 10.1890/es13-00387.1.

Folke, C., S. Carpenter, B. Walker, M. Scheffer, T. Elmqvist, L. Gunderson, and C. S. Holling. 2004. "Regime shifts, resilience, and biodiversity in ecosystem management." *Annual Review of Ecology, Evolution, and Systematics* 35 (1): 557–81. doi: 10.1146/annurev.ecolsys.35.021103.105711.

Fontaine, C., and A. Gonzalez. 2005. "Population synchrony induced by resource fluctuations and dispersal in an aquatic microcosm." *Ecology* 86 (6): 1463–71. doi: 10.1890/04-1400.

Ford, A. T., M. Barrueto, and A. P. Clevenger. 2017. "Road mitigation is a demographic filter for grizzly bears." *Wildlife Society Bulletin* 41 (4): 712–19.

Forman, R. T. T. 1991. "Landscape corridors: From theoretical foundations to public policy." In *Nature Conservation 2: The Role of Corridors*, edited by D. A. Saunders and R. J. Hobbs, 71–84. Chipping Norton, New South Wales, Australia: Surrey Beatty.

———. 1995. *Land Mosaics: The Ecology of Landscapes and Regions*. Cambridge, England: University of Cambridge.

Forman, R. T. T., and M. Gordon. 1986. *Landscape Ecology*. New York: John Wiley.

Forman, R. T. T., D. Sperling, J. A. Bissonette, A. P. Clevenger, C. D. Cutshall, V. H. Dale, L. Fahrig et al. 2003. *Road Ecology*. Washington, DC: Island Press.

Forys, E. A., and S. R. Humphrey. 1996. "Home range and movements of the Lower Keys marsh rabbit in a highly fragmented habitat." *Journal of Mammalogy* 77 (4): 1042–48. doi: 10.2307/1382784.

Foster, M. L., and S. R. Humphrey. 1995. "Use of highway underpasses by Florida panthers and other wildlife." *Wildlife Society Bulletin (1973–2006)* 23 (1): 95–100.

Fox, A. D., L. A. Henry, D. W. Corne, and J. M. Roberts. 2016. "Sensitivity of marine protected area network connectivity to atmospheric variability." *Royal Society Open Science* 3 (11): 160494.

Frakes, R. A., R. C. Belden, B. E. Wood, and F. E. James. 2015. "Landscape analysis of adult Florida panther habitat." *PLoS ONE* 10 (7): e0133044.

Frankham, R., J. D. Ballou, M. D. B. Eldridge, R. C. Lacy, K. Ralls, M. R. Dudash, and C. B. Fenster. 2011. "Predicting the probability of outbreeding depression." *Conservation Biology* 25 (3): 465–75.

Frankham, R., J. D. Ballou, K. Ralls, M. Eldridge, M. R. Dudash, C. B. Fenster, R. C. Lacy, and P. Sunnucks. 2017. *Genetic Management of Fragmented Animal and Plant Populations.* Oxford: Oxford University Press.

Freeman, M. T., P. I. Olivier, and R. J. van Aarde. 2018. "Matrix transformation alters species-area relationships in fragmented coastal forests." *Landscape Ecology* 33: 307–22.

Freeman, M. C., C. M. Pringle, and C. R. Jackson. 2007. "Hydrologic connectivity and the contribution of stream headwaters to ecological integrity at regional scales." *Journal of the American Water Resources Association* 43 (1): 5–14.

Freiwald, J. 2012. "Movement of adult temperate reef fishes off the west coast of North America." *Canadian Journal of Fisheries and Aquatic Sciences* 69 (8): 1362–74.

Fremier, A. K., M. Kiparsky, S. Gmur, J. Aycrigg, R. K. Craig, L. K. Svancara, D. D. Goble, B. Cosens, F. W. Davis, and J. M. Scott. 2015. "A riparian conservation network for ecological resilience." *Biological Conservation* 191: 29–37. doi: 10.1016/j.biocon.2015.06.029.

Friedman, D. S. 1997. "Walking on the wild side." *Landscape Architecture* (September): 52–57.

Fritz, R., and G. Merriam. 1993. "Fencerow habitats for plants moving between farmland forests." *Biological Conservation* 64 (2): 141–48. doi: 10.1016/0006-3207(93)90650-p.

Fromentin, J.-M., G. Reygondeau, S. Bonhommeau, and G. Beaugrand. 2014. "Oceanographic changes and exploitation drive the spatio-temporal dynamics of Atlantic bluefin tuna (*Thunnus thynnus*)." *Fisheries Oceanography* 23 (2): 147–56.

Fry, G. L. A. 1994. "The role of field margins in the landscape." *Monographs–British Crop Protection Council*: 31–39.

Furrer, R. D., and G. Pasinelli. 2016. "Empirical evidence for source–sink populations: A review on occurrence, assessments and implications." *Biological Reviews* 91 (3): 782–95.

Gagnon, J. W., N. L. Dodd, S. Sprague, C. Loberger, R. Nelson, S. Boe, and R. E. Schweinsburg. 2014. Evaluation of measures to promote desert bighorn sheep highway permeability: U.S. Route 93. Final Report 677. Phoenix: Arizona Department of Transportation.

Galarza, J. A., J. Carreras-Carbonell, E. Macpherson, M. Pascual, R. Severine, G. F. Turner, and C. Rico. 2009. "The influence of oceanographic fronts and early-life-history traits on connectivity among littoral fish species." *Proceedings of the National Academy of Sciences* 5: 1473–8.

Gallo, J. A., and R. Greene. 2018. "Connectivity analysis software for estimating linkage priority." Available at: https://doi.org/10.6084/m9.figshare.5673715. Accessed 4 May 2018.

Game, E. T., G. Lipsett-Moore, E. Saxon, N. Peterson, and S. Sheppard. 2011. "Incorporating climate change adaptation into national conservation assessments." *Global Change Biology* 17: 3150–60.

Gangodagamage, C., E. Barnes, and E. Foufoula-Georgiou. 2007. "Scaling in river corridor widths depicts organization in valley morphology." *Geomorphology* 91 (3–4): 198–215.

Garmendia, A., V. Arroyo-Rodríguez, A. Estrada, E. J. Naranjo, and K. E. Stoner. 2013. "Landscape and patch attributes impacting medium- and large-sized terrestrial mammals in a fragmented rain forest." *Journal of Tropical Ecology* 29 (4): 331–44.

Gascon, C., T. E. Lovejoy, R. O. Bierregaard Jr., J. R. Malcolm, P. C. Stouffer, H. L. Vasconcelos, W. F. Laurance, B. Zimmerman, M. Tocher, and S. Borges. 1999. "Matrix habitat and species richness in tropical forest remnants." *Biological Conservation* 91 (2–3): 223–29. doi: 10.1016/s0006-3207(99)00080-4.

Gedan, K. B., B. R. Silliman, and M. D. Bertness. 2009. "Centuries of human-driven change in salt marsh ecosystems." *Annual Review of Marine Science* 1: 117–41.

Gehlbach, F. R., and R. S. Baldridge. 1987. "Live blind snakes (*Leptotyphlops dulcis*) in eastern screech owl (*Otus asio*) nests: A novel commensalism." *Oecologia* 71 (4): 560–63. doi: 10.1007/bf00379297.

George, N. 2017. "Human–wildlife conflict in India: 1 human killed every day." *Phys.org— News and Articles on Science and Technology*. phys.org/news/2017-08–human-wildlife-conflict-india-human-day.html. Accessed 23 April 2018.

Gerow, K., N. C. Kline, D. E. Swann, and M. Pokorny. 2010. "Estimating annual vertebrate mortality on roads at Saguaro National Park, Arizona." *Human–Wildlife Interactions* 4 (2): 283–92.

Gharrett, A. J., and W. W. Smoker. 1991. "Two generations of hybrids between even- and odd-year pink salmon (*Oncorhynchus gorbuscha*): A test for outbreeding depression?" *Canadian Journal of Fisheries and Aquatic Sciences* 48 (9): 1744–49. doi: 10.1139/f91-206.

Giakoumi, S., C. Scianna, J. Plass-Johnson, F. Micheli, K. Grorud-Colvert, P. Thiriet, J. Claudet, G. Di Carlo, A. Di Franco, and S. D. Gaines. 2017. "Ecological effects of full and partial protection in the crowded Mediterranean Sea: A regional meta-analysis." *Scientific Reports* 7 (1): 8940.

Gibeau, M. L., A. P. Clevenger, S. Herrero, and J. Wierzchowski. 2002. "Grizzly bear response to human development and activities in the Bow River Watershed, Alberta, Canada." *Biological Conservation* 103 (2): 227–36. doi: 10.1016/s0006-3207(01)00131-8.

Gilbert-Norton, L., R. Wilson, J. R. Stevens, and K. H. Beard. 2010. "A meta-analytic review of corridor effectiveness." *Conservation Biology* 24: 660–68.

Gilchrist, A., A. Barker, and J. F. Handley. 2016. "Pathways through the landscape in a changing climate: The role of landscape structure in facilitating species range expansion through an urbanised region." *Landscape Research* 41 (1): 26–44. doi: 10.1080/01426397.2015 .1045466.

Gill, J. L., J. L. Blois, B. Benito, S. Dobrowski, M. L. Hunter Jr., and J. L. McGuire. 2015. "A 2.5-million-year perspective on coarse-filter strategies for conserving nature's stage." *Conservation Biology* 29 (3): 640–48. doi: 10.1111/cobi.12504.

Gilman, S. E., M. C. Urban, J. Tewksbury, G. W. Gilchrist, and R. D. Holt. 2010. "A framework for community interactions under climate change." *Trends in Ecology & Evolution* 25 (6): 325–31.

Gimona, A., L. Poggio, J. G. Polhill, and M. Castellazzi. 2015. "Habitat networks and food security: Promoting species range shift under climate change depends on life history and the dynamics of land use choices." *Landscape Ecology* 30 (5): 771–89. doi: 10.1007 /s10980-015-0158-8.

Glista, D. J., T. L. DeVault, and J. A. DeWoody. 2009. "A review of mitigation measures for reducing wildlife mortality on roadways." *Landscape and Urban Planning* 91 (1): 1–7.

Gonzalez, A. 2000. "Community relaxation in fragmented landscapes: The relation between species richness, area and age." *Ecology Letters* 3 (5): 441–48. doi: 10.1046/j.1461-02 48.2000.00171.x.

Gonzales-Suarez, M., R. Flatz, D. Aurioles-Gomboa, P. W. Hedrick, and L. R. Gerber. 2009. "Isolation by distance among California sea lion populations in Mexico: Redefining management stocks." *Molecular Ecology* 18 (6): 1088–99.

Graumlich, L., and W. L. Francis. 2011. *Moving Toward Climate Change Adaptation: The Promise of the Yellowstone to Yukon Conservation Initiative for Addressing the Region's Vulnerability to Climate Disruption.* Yellowstone to Yukon Conservation Initiative.

Gray, M. 2009. "Advances in wildlife crossing technologies." *Public Roads* 73 (2): 14–21.

———. 2017. Analysis of the Sonoma Valley wildlife corridor for maintaining connectivity: Using camera trap array data to evaluate habitat use by wildlife. Prepared for Sonoma Land Trust.

Gray, M., and A. M. Merenlender. 2015. Analysis of the Sonoma Developmental Center property for maintaining connectivity along the Sonoma Valley corridor: Implications for wildlife movement and climate change adaptation. Prepared for Sonoma Land Trust.

Grenfell, B. T., B. M. Bolker, and A. Kleczkowski. 1995. "Seasonality and extinction in chaotic metapopulations." *Proceedings of the Royal Society B: Biological Sciences* 259 (1354): 97–103. doi: 10.1098/rspb.1995.0015.

Grenfell, B., and J. Harwood. 1997. "(Meta)population dynamics of infectious diseases." *Trends in Ecology & Evolution* 12 (10): 395–99.

Green Roofs for Healthy Cities (GRHC). 2009. Biodiversity research on green roofs: Draft final research protocol. Toronto, Canada. Available from http://www.greenroofs.org. Accessed 5 July 2018.

Groom, M. J. 1998. "Allee effects limit population viability of an annual plant." *American Naturalist* 151 (6): 487. doi: 10.2307/2463323.

Groves, C. R., E. T. Game, M. G. Anderson, M. Cross, C. Enquist, Z. Ferdana, E. Girvetz et al. 2012. "Incorporating climate change into systematic conservation planning." *Biodiversity and Conservation* 21 (7): 1651–71. doi: 10.1007/s10531-012-0269-3.

Guarnizo, C. E., and D. C. Cannatella. 2013. "Geographic determinants of gene flow in two sister species of tropical Andean frogs." *Journal of Heredity* 105 (2): 216–25.

Gude, J. A., M. S. Mitchell, R. E. Russell, C. A. Sime, E. E. Bangs, L. D. Mech, and R. R. Ream. 2012. "Wolf population dynamics in the US Northern Rocky Mountains are affected by recruitment and human-caused mortality." *Journal of Wildlife Management* 76 (1): 108–18.

Guichard, F., S. A. Levin, A. Hastings, and D. Siegel. 2004. "Toward a dynamic metacommunity approach to marine reserve theory." *BioScience* 54 (11): 1003. doi: 10.1641/0006 -3568 (2004)054[1003:tadmat]2.0.co;2.

Guiot, J., and W. Cramer. 2016. "Climate change: The 2015 Paris Agreement thresholds and Mediterranean basin ecosystems." *Science* 354 (6311): 465–68.

Haas, C. A. 1995. "Dispersal and use of corridors by birds in wooded patches on an agricultural landscape." *Conservation Biology* 9 (4): 845–54. doi: 10.1046/j.1523-1739.1995.09 040845.x.

Haddad, N. M. 1999. "Corridor use predicted from behaviors at habitat boundaries." *American Naturalist* 153 (2): 215. doi: 10.2307/2463582.

Haddad, N. M., L. A. Brudvig, J. Clobert, K. F. Davies, A. Gonzalez, R. D. Holt, T. E. Lovejoy, J. O. Sexton, M. P. Austin, and C. D. Collins. 2015. "Habitat fragmentation and its lasting impact on Earth's ecosystems." *Science Advances* 1 (2): e1500052.

Haddad, N. M., L. A. Brudvig, E. I. Damschen, D. M. Evans, B. L. Johnson, D. J. Levey, J. L. Orrock et al. 2014. "Potential negative ecological effects of corridors." *Conservation Biology* 28 (5): 1178–87. doi: 10.1111/cobi.12323.

Haddad, N. M., and J. J. Tewksbury. 2005. "Low-quality habitat corridors as movement conduits for two butterfly species." *Ecological Applications* 15 (1): 250–57. doi: 10.1890 /03-5327.

Haight, R. G., S. A. Snyder, and C. S. Revelle. 2005. "Metropolitan open-space protection with uncertain site availability." *Conservation Biology* 19 (2): 327–37. doi: 10.1111/ j.1523-1739.2005 .00151.x.

Hailey, T. L., and R. DeArment. 1972. "Droughts and fences restrict pronghorns." *Proceedings of the Biennial Pronghorn Antelope Workshop* 5: 22–24.

Haines, J. 2004. FWP moves bears out of town. *Bozeman* (Montana) *Daily Chronicle*, October 8, 2004.

Hale, J. D., A. J. Fairbrass, T. J. Matthews, G. Davies, and J. P. Sadler. 2015. "The ecological impact of city lighting scenarios: Exploring gap crossing thresholds for urban bats." *Global Change Biology* 21 (7): 2467–78.

Halpern, B. S., C. V. Kappel, K. A. Selkoe, F. Micheli, C. M. Ebert, C. Kontgis, C. M. Crain, R. G. Martone, C. Shearer, and S. J. Teck. 2009. "Mapping cumulative human impacts to California Current marine ecosystems." *Conservation Letters* 2 (3): 138–48.

Halpern, B. S., S. Walbridge, K. A. Selkoe, C. V. Kappel, F. Micheli, C. D'agrosa, J. F. Bruno, K. S. Casey, C. Ebert, and H. E. Fox. 2008. "A global map of human impact on marine ecosystems." *Science* 319 (5865): 948–52.

Hamann, A., D. R. Roberts, Q. E. Barber, C. Carroll, and S. E. Nielsen. 2015. "Velocity of climate change algorithms for guiding conservation and management." *Global Change Biology* 21 (2): 997–1004. doi: 10.1111/gcb.12736.

Hamilton, S. L., J. E. Caselle, D. P. Malone, and M. H. Carr. 2010. "Incorporating biogeography into evaluations of the Channel Islands marine reserve network." *Proceedings of the National Academy of Sciences* 107 (43): 18272–77.

Hampe, A. 2005. "Fecundity limits in *Frangula alnus* (Rhamnaceae) relict populations at the species' southern range margin." *Oecologia* 143 (3): 377–86.

Hampe, A., and R. J. Petit. 2005. "Conserving biodiversity under climate change: The rear edge matters." *Ecology Letters* 8 (5): 461–67. doi: 10.1111/j.1461-0248.2005.00 739.x.

Hannah, L., L. Flint, A. D. Syphard, M. A. Moritz, L. B. Buckley, and I. M. McCullough. 2014. "Fine-grain modeling of species' response to climate change: Holdouts, stepping-stones, and microrefugia." *Trends in Ecology & Evolution* 29 (7): 390–97. doi: 10.1016/j .tree.2014.04.006.

Hannah, L., D. Panitz, and G. Midgley. 2012. "Cape floristic region, South Africa." In *Climate and Conservation: Landscape and Seascape Science, Planning, and Action*, edited by J. A. Hilty, C. C. Chester, and M. S. Cross, 80–91. Washington, DC: Island Press.

Hannibal, M. E. 2013. *Spine of the Continent: The Race to Save America's Last, Best Wilderness*. Lanham, MD: Rowman & Littlefield.

Hannon, S. J., C. A. Paszkowski, S. Boutin, J. DeGroot, S. E. Macdonald, M. Wheatley, and B. R. Eaton. 2002. "Abundance and species composition of amphibians, small mammals, and songbirds in riparian forest buffer strips of varying widths in the boreal mixedwood of Alberta." *Canadian Journal of Forest Research* 32 (10): 1784–1800. doi: 10.1139/x02-092.

Hanophy, W. 2009. *Fencing with Wildlife in Mind*. Denver: Colorado Parks and Wildlife.

Hansen, A. J., R. Rasker, B. Maxwell, J. J. Rotella, J. D. Johnson, A. W. Parmenter, U. Langner, W. B. Cohen, R. L. Lawrence, and M. P. V. Kraska. 2002. "Ecological causes and consequences of demographic change in the New West: As natural amenities attract people and commerce to the rural west, the resulting land-use changes threaten biodiversity, even in protected areas, and challenge efforts to sustain local communities and ecosystems." *BioScience* 52 (2): 151–62.

Hanski, I. 1998. "Metapopulation dynamics." *Nature* 396 (6706): 41.

———. 1999. *Metapopulation Ecology*. Oxford: Oxford University Press.

———. 2001. "Spatially realistic theory of metapopulation ecology." *Naturwissenschaften* 88 (9): 372–381.

Hanski, I., and M. E. Gilpin. 1997. *Metapopulation Biology: Ecology, Genetics and Evolution*. San Diego, CA: Academic Press.

Hanski, I., J. Alho, and A. Moilanen. 2000. "Estimating the parameters of survival and migration of individuals in metapopulations." *Ecology* 81 (1): 239. doi: 10.2307/177147.

Hanson, J. S., G. P. Malanson, and M. P. Armstrong. 1990. "Landscape fragmentation and dispersal in a model of riparian forest dynamics." *Ecological Modelling* 49 (3–4): 277–96. doi: 10.1016/0304-3800(90)90031-b.

Hansson, L. 1995. "Development and application of landscape approaches in mammalian ecology." In *Landscape Approaches in Mammalian Ecology and Conservation*, edited by W. Z. Lidicker Jr., 20–39. Minneapolis: University of Minnesota Press.

Harley, C. D. G., A. R. Hughes, K. M. Hultgren, B. G. Miner, C. J. B. Sorte, C. S. Thornber, L. F. Rodriguez, L. Tomanek, and S. L. Williams. 2006. "The impacts of climate change in coastal marine systems." *Ecology Letters* 9 (2): 228–41.

Haroldson, M. A., and K. Frey. 2002. "Grizzly bear mortalities." In *Yellowstone Grizzly Bear Investigations: Annual Report of the Interagency Study Team, 2001*, edited by C. C. Schwartz and M. A. Haroldson, 23–28. Bozeman, MT: United States Geological Survey.

Harris, L. D., T. Hoctor, D. Maehr, and J. Sanderson. 1996. "The role of networks and corridors in enhancing the value and protection of parks and equivalent areas." In *National Parks and Protected Areas*, edited by R. G. Wright, 173–97. Cambridge, England: Blackwell Science.

Harris, L. D., and J. Scheck. 1991. "From implications to applications: The dispersal corridor principle applied to the conservation of biological diversity." In *Nature Conservation 2: The Role of Corridors*, edited by D. A. Saunders and R. J. Hobbs, 189–20. Chipping Norton, New South Wales, Australia: Surrey Beatty.

Harrison, S. 1991. "Local extinction in a metapopulation context: An empirical evaluation." *Biological Journal of the Linnean Society* 42:73–88.

———. 1999. "Local and regional diversity in a patchy landscape: Native, alien, and endemic herbs on Serpentine." *Ecology* 80 (1): 70. doi: 10.2307/176980.

Hartley, M. J., and M. L. Hunter. 1998. "A meta-analysis of forest cover, edge effects, and artificial nest predation rates." *Conservation Biology* 12 (2): 465–69. doi: 10.1046/j.1523 -1739.1998.96373.x.

Hastings, A. 2003. "Metapopulation persistence with age-dependent disturbance or succession." *Science* 301 (5639): 1525–26. doi: 10.1126/science.1087570.

Hauer, F. R., H. Locke, V. J. Dreitz, M. Hebblewhite, W. H. Lowe, C. C. Muhlfeld, C. R. Nelson, M. F. Proctor, and S. B. Rood. 2016. "Gravel-bed river floodplains are the ecological nexus of glaciated mountain landscapes." *Science Advances* 2 (6). doi: 10.1126 /sciadv.1600026.

Havens, K., P. Vitt, S. Still, A. T. Kramer, J. B. Fant, and K. Schatz. 2015. "Seed sourcing for restoration in an era of climate change." *Natural Areas Journal* 35 (1): 122–33.

Havlick, D. 2004. "Roadkill." *Conservation in Practice* (Winter): 30–34.

Haydon, D., and H. Steen. 1997. "The effects of large- and small-scale random events on the synchrony of metapopulation dynamics: A theoretical analysis." *Proceedings of the Royal Society of London B: Biological Sciences* 264 (1386): 1375–81.

Hays, G. C., L. C. Ferreira, A. M. M. Sequeira, M. G. Meekan, C. M. Duarte, H. Bailey, F. Bailleul, W. D. Bowen, M. J. Caley, and D. P. Costa. 2016. "Key questions in marine megafauna movement ecology." *Trends in Ecology & Evolution* 31 (6): 463–75.

Hazen, E. L., S. Jorgensen, R. R. Rykaczewski, S. J. Bograd, D. G. Foley, I. D. Jonsen, S. A. Shaffer, J. P. Dunne, D. P. Costa, and L. B. Crowder. 2013. "Predicted habitat shifts of Pacific top predators in a changing climate." *Nature Climate Change* 3 (3): 234.

Hazen, E. L., D. M. Palacios, K. A. Forney, E. A. Howell, E. Becker, A. L. Hoover, L. Irvine, M. DeAngelis, S. J. Bograd, and B. R. Mate. 2017. "WhaleWatch: A dynamic management tool for predicting blue whale density in the California Current." *Journal of Applied Ecology* 54 (5): 1415–28.

Hazen, E. L., K. L. Scales, S. M. Maxwell, D. K. Briscoe, S. J. Bograd, S. Kohin, S. Benson, T. Eguchi, L. B. Crowder, and R. L. Lewison. 2018. "Supporting sustainable fisheries: An eco-informatic solution to ocean bycatch." *Nature*. Forthcoming.

Hazen, E. L., R. M. Suryan, J. A. Santora, S. J. Bograd, Y. Watanuki, and R. P. Wilson. 2013. "Scales and mechanisms of marine hotspot formation." *Marine Ecology Progress Series* 487: 177–83.

Hedrick, P. W. 1996. "Genetics of metapopulations: Aspects of a comprehensive perspective." In *Metapopulations and Wildlife Conservation*, edited by D. R. McCullough, 29–51. Washington, DC: Island Press.

Heller, N. E., J. Kreitler, D. D. Ackerly, S. B. Weiss, A. Recinos, R. Branciforte, L. E. Flint, A. L. Flint, and E. Micheli. 2015. "Targeting climate diversity in conservation planning to build resilience to climate change." *Ecosphere* 6 (4). doi: 10.1890/es14-00313.1.

Heller, N. E., and E. S. Zavaleta. 2009. "Biodiversity management in the face of climate change: A review of 22 years of recommendations." *Biological Conservation* 142 (1): 14–32. doi: 10.1016/j.biocon.2008.10.006.

Hellmund, P. C., and D. Smith. 2013. *Designing Greenways: Sustainable Landscapes for Nature and People*. Washington, DC: Island Press.

Helm, A., I. Hanski, and M. Pärtel. 2006. "Slow response of plant species richness to habitat loss and fragmentation." *Ecology Letters* 9 (1): 72–77.

Henein, K., and G. Merriam. 1990. "The elements of connectivity where corridor quality is variable." *Landscape Ecology* 4 (2–3): 157–70. doi: 10.1007/bf00132858.

Henle, K., D. B. Lindenmayer, C. R. Margules, D. A. Saunders, and C. Wissel. 2004. "Species survival in fragmented landscapes: Where are we now?" *Biodiversity and Conservation* 13 (1): 1–8. doi: 10.1023/b:bioc.0000004311.04226.29.

Heske, E. J. 1995. "Mammalian abundances on forest-farm edges versus forest interiors in southern Illinois: Is there an edge effect?" *Journal of Mammalogy* 76 (2): 562–68. doi: 10.2307/1382364.

Hewitt, G. 2000. "The genetic legacy of the Quaternary ice ages." *Nature* 405 (6789): 907–13.

Hewitt, G. M. 1993. "After the ice: *Parallelus* meets *Erythropus* in the Pyrenees." In *Hybrid Zones and the Evolutionary Process*, edited by R. G. Harrison, 140–64. New York: Oxford University Press.

Hickler, T., K. Vohland, J. Feehan, P. A. Miller, B. Smith, L. Costa, T. Giesecke, S. Fronzek, T. R. Carter, and W. Cramer. 2012. "Projecting the future distribution of European potential natural vegetation zones with a generalized, tree species-based dynamic vegetation model." *Global Ecology and Biogeography* 21 (1): 50–63.

Hilty, J. A. 2001. "Use of riparian corridors by wildlife in the oak woodland vineyard landscape." Berkeley: University of California.

Hilty, J. A., C. Brooks, E. Heaton, and A. M. Merenlender. 2006. "Forecasting the effect of land-use change on native and non-native mammalian predator distributions." *Biodiversity and Conservation* 15 (9): 2853–71. doi: 10.1007/s10531-005-1534-5.

Hilty, J. A., C. C. Chester, and M. S. Cross. 2012. *Climate and Conservation: Landscape and Seascape Science, Planning, and Action*. Washington, DC: Island Press.

Hilty, J. A., and A. M. Merenlender. 2004. "Use of riparian corridors and vineyards by mammalian predators in Northern California." *Conservation Biology* 18 (1): 126–35. doi: 10.1111/j.1523-1739.2004.00225.x.

Hilty, J. A., A. M. Merenlender, and W. Z. Lidicker Jr. 2006. *Corridor Ecology: The Science and Practice of Linking Landscapes for Biodiversity Conservation*. Washington, DC: Island Press.

Hobbs, R. J. 1992. "The role of corridors in conservation: Solution or bandwagon?" *Trends in Ecology & Evolution* 7 (11): 389–92. doi: 10.1016/0169-5347(92)90010-9.

Hobday, A. J., L. V. Alexander, S. E. Perkins, D. A. Smale, S. C. Straub, E. C. J. Oliver, J. A. Benthuysen, M. T. Burrows, M. G. Donat, and M. Feng. 2016. "A hierarchical approach to defining marine heatwaves." *Progress in Oceanography* 141: 227–38.

Hobday, A. J., J. R. Hartog, C. M. Spillman, and O. Alves. 2011. "Seasonal forecasting of tuna habitat for dynamic spatial management." *Canadian Journal of Fisheries and Aquatic Sciences* 68 (5): 898–911.

Hobday, A. J., S. M. Maxwell, J. Forgie, and J. McDonald. 2013. "Dynamic ocean management: Integrating scientific and technological capacity with law, policy, and management." *Stanford Environmental Law Journal* 33: 125.

Hodges, M. F., Jr., and D. G. Krementz. 1996. "Neotropical migratory breeding bird communities in riparian forests of different widths along the Altamaha River, Georgia." *Wilson Bulletin*: 496–506.

Hodgson, J. A., C. D. Thomas, S. Cinderby, H. Cambridge, P. Evans, and J. K. Hill. 2011. "Habitat re-creation strategies for promoting adaptation of species to climate change." *Conservation Letters* 4 (4): 289–97. doi: 10.1111/j.1755-263X.2011.00177.x.

Hodgson, J. A., C. D. Thomas, C. Dytham, J. M. J. Travis, and S. J. Cornell. 2012. "The speed of range shifts in fragmented landscapes." *PLoS ONE* 7 (10). doi: 10.1371/journal.pone.0047141.

Hodgson, J. A., C. D. Thomas, B. A. Wintle, and A. Moilanen. 2009. "Climate change, connectivity and conservation decision making: Back to basics." *Journal of Applied Ecology* 46 (5): 964–69. doi: 10.1111/j.1365-2664.2009.01695.x.

Hodgson, J. A., D. W. Wallis, R. Krishna, and S. J. Cornell. 2016. "How to manipulate landscapes to improve the potential for range expansion." *Methods in Ecology and Evolution* 7 (12): 1558–66. doi: 10.1111/2041-210X.12614.

Hoffmann, A. A. 2003. "Low potential for climatic stress adaptation in a rainforest *Drosophila* species." *Science* 301 (5629): 100–102. doi: 10.1126/science.1084296.

Holland, M. B. 2012. "Mesoamerican biological corridor." In *Climate and Conservation: Landscape and Seascape Science, Planning, and Action*, edited by J. A. Hilty, C. C. Chester, and M. S. Cross. Washington, DC: Island Press.

Holling, C. S., and L. H. Gunderson. 2002. "Resilience and adaptive cycles." In *Panarchy: Understanding Transformations in Human and Natural Systems*, edited by L. H. Gunderson and C. S. Holling, 25–62. Washington, DC: Island Press.

Horne, J. S., E. O. Garton, S. M. Krone, and J. S. Lewis. 2007. "Analyzing animal movements using Brownian bridges." *Ecology* 88 (9): 2354–63.

Horton, T. W., N. Hauser, A. N. Zerbini, M. P. Francis, M. L. Domeier, A. Andriolo, D. P. Costa, P. W. Robinson, C. A. J. Duffy, and N. Nasby-Lucas. 2017. "Route fidelity during marine megafauna migration." *Frontiers in Marine Science* 4: 422.

House, F. 1999. *Totem Salmon*. Boston: Beacon Press.

Huijser, M. 2004. "Evaluation of wildlife crossing structures on US Highway 93 in Montana." *Task Force on Ecology and Transportation Newsletter*. Washington, DC: Transportation Research Board of the National Academies.

Huijser, M. P., J. W. Duffield, A. P. Clevenger, R. J. Ament, and P. T. McGowen. 2009. "Cost–benefit analyses of mitigation measures aimed at reducing collisions with large ungulates in the United States and Canada: A decision support tool." *Ecology and Society* 14 (2).

Huijser, M. P., A. McGowen Kociolek, P. Hardy, A. Clevenger, and R. Ament. 2007. Wildlife-vehicle collision and crossing mitigation measures: A toolbox for the Montana Department of Transportation. Prepared for the State of Montana Department of Transportation in cooperation with the U.S. Department of Transportation Federal Highway Administration. FHWA/MT-07-002/8117-34.

Hulme, D., and M. Murphree. 1999. "Communities, wildlife and the 'new conservation' in Africa." *Journal of International Development* 11 (2): 277–85. doi: 10.1002/(sici)1099-1328(199903/04)11:2<277::aid-jid582>3.0.co;2-t.

Humphrey, J. W., K. Watts, E. Fuentes-Montemayor, N. A. Macgregor, A. J. Peace, and K. J. Park. 2015. "What can studies of woodland fragmentation and creation tell us about ecological networks? A literature review and synthesis." *Landscape Ecology* 30 (1): 21–50.

Hunt, G. L., Jr. 1990. "The pelagic distribution of marine birds in a heterogeneous environment." *Polar Research* 8 (1): 43–54. doi: 10.3402/polar.v8i1.6802.

Hunter, M. L. 2005. "A mesofilter conservation strategy to complement fine and coarse filters." *Conservation Biology* 19 (4): 1025–29.

Hylander, K., C. Nilsson, and T. Gothner. 2004. "Effects of buffer-strip retention and clearcutting on land snails in boreal riparian forests." *Conservation Biology* 18 (4): 1052–62. doi: 10.1111/j.1523-1739.2004.00199.x.

Igulu, M. M., I. Nagelkerken, M. Dorenbosch, M. G. G. Grol, A. R. Harborne, I. A. Kimirei, P. J. Mumby, A. D. Olds, and Y. D. Mgaya. 2014. "Mangrove habitat use by juvenile reef fish: Meta-analysis reveals that tidal regime matters more than biogeographic region." *PLoS ONE* 9 (12): e114715.

Imbach, P. A., B. Locatelli, L. G. Molina, P. Ciais, and P. W. Leadley. 2013. "Climate change and plant dispersal along corridors in fragmented landscapes of Mesoamerica." *Ecology and Evolution* 3 (9): 2917–32. doi: 10.1002/ece3.672.

Inchausti, P., and H. Weimerskirch. 2002. "Dispersal and metapopulation dynamics of an oceanic seabird, the wandering albatross, and its consequences for its response to long-line fisheries." *Journal of Animal Ecology* 71 (5): 765–70.

Inman, R. M., R. R. Wigglesworth, K. H. Inman, M. K. Schwartz, B. L. Brock, and J. D. Rieck. 2004. "Wolverine makes extensive movements in Greater Yellowstone Ecosystem." *Northwest Science* 78: 261–66.

IPCC. 2014. Climate change 2014: Synthesis report. Contribution of Working Groups 1, 2, and 3 to the Fifth Assessment Report of the Intergovernmental Panel on Climate Change. Core Writing Team: R. K. Pachauri and L. A. Meyer, eds. Geneva, Switzerland: IPCC.

Ismail, S. A., J. Ghazoul, G. Ravikanth, C. G. Kushalappa, R. U. Shaanker, and C. J. Kettle. 2017. "Evaluating realized seed dispersal across fragmented tropical landscapes: A two-fold approach using parentage analysis and the neighbourhood model." *New Phytologist* 214 (3): 1307–16.

IUCN Red List of Threatened Species. 2017. "*Panther leo* ssp." *Persica.* http://dx.doi.org/10.2305 /IUCN.UK.2008.RLTS.T15952A5327221.en.

Jackson, J., M. Donovan, K. Cramer, and V. Lam. 2014. "Status and trends of Caribbean coral reefs: 1970–2012." IUCN Report. Retrieved from Global Coral Reef Monitoring Network website: https://www.iucn.org/content/status-and–trends-caribbean-coral -reefs-1970-2012-0.

Jackson, S. T., and J. T. Overpeck. 2000. "Responses of plant populations and communities to environmental changes of the late Quaternary." *Paleobiology* 26 (sp4): 194–220.

Jacobs, S., N. Denddoncker, and H. Keune. 2013. *Ecosystem Services*, 1st ed. Amsterdam: Elsevier.

Jaeger, J. A. G., A. Spanowicz, J. Bowman, and A. P. Clevenger. 2017. "Monitoring the use and effectiveness of wildlife passages for small and medium-sized mammals along Highway 175: Main results and recommendations." *Concordia University News Bulletin* No. 8.

Jaffal, I., S.-E. Ouldboukhitine, and R. Belarbi. 2012. "A comprehensive study of the impact of green roofs on building energy performance." *Renewable Energy* 43: 157–64.

Jantz, P., S. Goetz, and N. Laporte. 2014. "Carbon stock corridors to mitigate climate change and promote biodiversity in the tropics." *Nature Climate Change* 4 (2): 138–42. doi: 10 .1038/nclimate2105.

Jantz, S. M., B. Barker, T. M. Brooks, L. P. Chini, Q. Huang, R. M. Moore, J. Noel, and G. C. Hurtt. 2015. "Future habitat loss and extinctions driven by land-use change in biodiversity hotspots under four scenarios of climate-change mitigation." *Conservation Biology* 29 (4): 1122–31.

Jarvie, M. 2017. "Highway crossings credited for decline in national park wildlife deaths." *Calgary Herald.* Accessed 14 January 2019.

Jenkins, C. N., K. S. Van Houtan, S. L. Pimm, and J. O. Sexton. 2015. "US protected lands mismatch biodiversity priorities." *Proceedings of the National Academy of Sciences* 112 (16): 5081–86.

Jenness, J., D. Majka, and P. Beier. 2011. Corridor designer evaluation tools. Environmental Research, Development and Education for the New Economy (ERDENE). Flagstaff: Northern Arizona University.

Jewitt, D., P. S. Goodman, B. F. N. Erasmus, T. G. O'Connor, and E. T. F. Witkowski. 2017. "Planning for the mmaintenance of floristic diversity in the face of land cover and climate change." *Environmental Management* 59 (5): 792–806. doi: 10.1007/s00267 -017-0829-0.

Jiang, X., S. A. Rauscher, T. D. Ringler, D. M. Lawrence, A. P. Williams, C. D. Allen, A. L. Steiner, D. M. Cai, and N. G. McDowell. 2013. "Projected future changes in vegetation in western North America in the twenty-first century." *Journal of Climate* 26 (11): 3671–87.

Johnson, G. C., M. J. McPhaden, G. D. Rowe, and K. E. McTaggart. 2000. "Upper equatorial Pacific Ocean current and salinity variability during the 1996–1998 El Niño–La Niña cycle." *Journal of Geophysical Research: Oceans* 105 (C1): 1037–53.

Johnson, L. B., and G. E. Host. 2010. "Recent developments in landscape approaches for the study of aquatic ecosystems." *Journal of the North American Benthological Society* 29 (1): 41–66.

Johnson, T. L., and J. F. Cully. 2004. "Drainages as potential corridors for the spread of sylvatic plague in black-tailed prairie dogs." Arcata, CA: American Society of Mammalogists 84th Annual Meeting Abstracts.

Jones, J. C. 2004. "Honey bee nest thermoregulation: Diversity promotes stability." *Science* 305 (5682): 402–4. doi: 10.1126/science.1096340.

Jongman, R. 2004. "The context and concept of ecological networks." In *Ecological Networks and Greenways Cconcept, Design, Implementation*, edited by R. Jongman and G. Pungetti, 7–32. Cambridge, England: Cambridge University Press.

——. 2008. "Ecological networks are an issue for all of us." *Journal of Landscape Ecology* 1 (1): 7–13.

Jordán, F. 2000. "A reliability-theory approach to corridor design." *Ecological Modelling* 128 (2–3): 211–20. doi: 10.1016/s0304-3800(00)00197-6.

Kai, E. T., V. Rossi, J. Sudre, H. Weimerskirch, C. Lopez, E. Hernandez-Garcia, F. Marsac, and V. Garçon. 2009. "Top marine predators track Lagrangian coherent structures." *Proceedings of the National Academy of Sciences* 106 (20): 8245–50.

Kaiser, J. 2001. "Bold corridor project confronts political reality." *Science* 293 (5538): 2196–99. doi: 10.1126/science.293.5538.2196.

Kalela, O., L. Kilpeläienen, T. Koponen, and J. Tast. 1971. "Seasonal differences in habitats of the Norwegian lemming, *Lemmus lemmus* (L.), in 1959 and 1960 at Kilpisjärvi, Finnish Lapland." *Annals of the Academy of Science Fennicae, 4A: Biology* 178:1–22.

Karanth, K. U., and J. D. Nichols. 1998. "Estimation of tiger densities in India using photographic captures and recaptures." *Ecology* 79 (8): 2852. doi: 10.2307/176521.

Kasten, K., C. Stenoien, W. Caldwell, and K. S. Oberhauser. 2016. "Can roadside habitat lead monarchs on a route to recovery?" *Journal of Insect Conservation* 20 (6): 1047–57.

Katz, D. 2017. "Canmore's wildlife corridors are overrun with people and dogs: Researchers." http://calgaryherald.com/news/local-news/canmores-wildlife-corridors-are–overrun -with-people-and–dogs-researchers. Accessed 25 June 2018: *Calgary Herald*.

Kays, R., M. C. Crofoot, W. Jetz, and M. Wikelski. 2015. "Terrestrial animal tracking as an eye on life and planet." *Science* 348 (6240): aaa2478.

Keeley, A. T. H., D. D. Ackerly, D. R. Cameron, N. E. Heller, P. R. Huber, C. A. Schloss,

J. H. Thorne, and A. M. Merenlender. 2018. "New concepts, models, and assessments of climate-wise connectivity." *Environmental Research Letters* 13: 073002.

Keeley A. T. H., G. Basson, D. R. Cameron, N. E. Heller, P. R. Huber, C. A. Schloss, J. H. Thorne, and A. M. Merenlender. 2018. "Making habitat connectivity a reality." *Conservation Biology*. Forthcoming.

Keeley, A. T. H., P. Beier, and J. W. Gagnon. 2016. "Estimating landscape resistance from habitat suitability: Effects of data source and nonlinearities." *Landscape Ecology*: 1–12. doi: 10.1007/s10980-016-0387-5.

Keeley, A. T. H., P. Beier, B. W. Keeley, and M. E. Fagan. 2017. "Habitat suitability is a poor proxy for landscape connectivity during dispersal and mating movements." *Landscape and Urban Planning* 161: 90–102.

Keeling, M. J., and B. T. Grenfell. 1997. "Disease extinction and community size: Modeling the persistence of measles." *Science* 275 (5296): 65–67.

Keller, L. F., P. Arcese, J. N. M. Smith, W. M. Hochachka, and S. C. Stearns. 1994. "Selection against inbred song sparrows during a natural population bottleneck." *Nature* 372 (6504): 356–57. doi: 10.1038/372356a0.

Kemper, J., R. M. Cowling, and D. M. Richardson. 1999. "Fragmentation of South African renosterveld shrublands: Effects on plant community structure and conservation implications." *Biological Conservation* 90 (2): 103–11. doi: 10.1016/s0006-3207(99)00021-x.

Kendall, B. E., O. N. Bjørnstad, J. Bascompte, T. H. Keitt, and W. F. Fagan. 2000. "Dispersal, environmental correlation, and spatial synchrony in population dynamics." *American Naturalist* 155 (5): 628–36.

Kenney, R. D., C. A. Mayo, and H. E. Winn. 2001. "Migration and foraging strategies at varying spatial scales in western North Atlantic right whales: A review of hypotheses." *Journal of Cetacean Research and Management* 2 (Special Issue): 251–60.

Keppel, G., and G. Wardell-Johnson. 2015. "Refugial capacity defines holdouts, microrefugia and stepping-stones: A response to Hannah et al." *Trends in Ecology & Evolution* 30 (5): 233–34.

Keppel, G., K. Mokany, G. W. Wardell-Johnson, B. L. Phillips, J. A. Welbergen, and A. E. Reside. 2015. "The capacity of refugia for conservation planning under climate change." *Frontiers in Ecology and the Environment* 13 (2): 106–12. doi: 10.1890/140055.

Keppel, G., K. P. Van Niel, G. W. Wardell-Johnson, C. J. Yates, M. Byrne, L. Mucina, A. G. T. Schut, S. D. Hopper, and S. E. Franklin. 2012. "Refugia: Identifying and understanding safe havens for biodiversity under climate change." *Global Ecology and Biogeography* 21 (4): 393–404.

Khan, J. A. 1995. "Conservation and management of Gir Lion Sanctuary and National Park, Gujarat, India." *Biological Conservation* 73 (3): 183–88. doi: 10.1016/0006-3207(94) 00107-2.

King, D. I., C. R. Griffin, and R. M. Degraff. 1996. "Effects of clearcutting on habitat use and reproductive success of the ovenbird in forested landscapes." *Conservation Biology* 10 (5): 1380–86. doi: 10.1046/j.1523-1739.1996.10051380.x.

Kinlan, B. P., and S. D. Gaines. 2003. "Propagule dispersal in marine and terrestrial environments: A community perspective." *Ecology* 84 (8): 2007–20.

Kinley, T. A., and C. D. Apps. 2001. "Mortality patterns in a subpopulation of endangered mountain caribou." *Wildlife Society Bulletin* 29 (1): 158–64.

Kirwan, M. L., and J. P. Megonigal. 2013. "Tidal wetland stability in the face of human impacts and sea-level rise." *Nature* 504 (7478): 53.

Kitzes, J., and A. M. Merenlender. 2013. "Extinction risk and tradeoffs in reserve site selection for species of different body sizes." *Conservation Letters* 6 (5): 341–49.

Klein, C., K. Wilson, M. Watts, J. Stein, S. Berry, J. Carwardine, M. S. Smith, B. Mackey, and H. Possingham. 2009. "Incorporating ecological and evolutionary processes into continental-scale conservation planning." *Ecological Applications* 19 (1): 206–17. doi: 10 .1890/07-1684.1.

Klemens, M. W. 2000. "From information to action: Developing more effective strategies to conserve turtles." In *Turtle Conservation*, edited by M. W. Klemens, 239–58. Washington, DC: Smithsonian Institution Press.

Klevjer, T. A., X. Irigoien, A. Røstad, E. Fraile-Nuez, V. M. Benítez-Barrios, and S. Kaartvedt. 2016. "Large scale patterns in vertical distribution and behaviour of mesopelagic scattering layers." *Scientific Reports* 6: 19873.

Kneitel, J. M., and T. E. Miller. 2003. "Dispersal rates affect species composition in metacommunities of *Sarracenia purpurea* Inquilines." *American Naturalist* 162 (2): 165–71. doi: 10.1086/376585.

Knowles, J. E., E. Doyle, S. R. Schill, L. M. Roth, A. Milam, and G. T. Raber. 2015. "Establishing a marine conservation baseline for the insular Caribbean." *Marine Policy* 60: 84–97.

Kondolf, G. M., K. Podolak, and A. Gaffney. 2010. "From high rise to coast: Revitalizing Ribeira da Barcarena." Berkeley, California. Retrieved from https://escholarship.org/uc /item/3q77s4ss: University of California Water Resources Center.

Kormann, U., C. Scherber, T. Tscharntke, N. Klein, M. Larbig, J. J. Valente, A. S. Hadley, and M. G. Betts. 2016. "Corridors restore animal-mediated pollination in fragmented tropical forest landscapes." *Proceedings of the Royal Society B* 283 (1823): 20152347.

Kozakiewicz, M., and J. Szacki. 1995. "Movements of small mammals in a landscape: Patch restriction or nomadism?" In *Landscape Approaches in Mammalian Ecology and Conservation*, edited by W. Z. Lidicker Jr., 78–94. Minneapolis: University of Minnesota Press.

Kremen, C., V. Razafimahatratra, R. P. Guillery, J. Rakotomalala, A. Weiss, and J.-S. Ratsisompatrarivo. 1999. "Designing the Masoala National Park in Madagascar based on biological and socioeconomic data." *Conservation Biology* 13 (5): 1055–68. doi: 10.1046 /j.1523-1739.1999.98374.x.

Kremen, C., and T. Ricketts. 2000. "Global perspectives on pollination disruptions." *Conservation Biology* 14 (5): 1226–28. doi: 10.1046/j.1523-1739.2000.00013.x.

Kritzer, J. P., and P. F. Sale. 2010. *Marine Metapopulations*. Amsterdam: Elsevier.

Krosby, M., R. Norheim, D. Theobald, and B. McRae. 2014. "Riparian climate-corridors: Identifying priority areas for conservation in a changing climate." Climate Impacts Group, University of Washington, Seattle.

Kruchek, B. L. 2004. "Use of tidal marsh and upland habitats by the marsh rice rat (*Oryzomys palustris*)." *Journal of Mammalogy* 85 (3): 569–75. doi: 10.1644/beh-016.

Kubeš, J. 1996. "Biocentres and corridors in a cultural landscape. A critical assessment of the 'territorial system of ecological stability.'" *Landscape and Urban Planning* 35 (4): 231–40. doi: 10.1016/s0169-2046(96)00321-0.

Kubisch, A., T. Degen, T. Hovestadt, and H. J. Poethke. 2013. "Predicting range shifts under global change: The balance between local adaptation and dispersal." *Ecography* 36 (8): 873–82. doi: 10.1111/j.1600-0587.2012.00062.x.

Lagabrielle, E., E. Crochelet, M. Andrello, S. R. Schill, S. Arnaud-Haond, N. Alloncle, and B. Ponge. 2014. "Connecting MPAs–eight challenges for science and management." *Aquatic Conservation: Marine and Freshwater Ecosystems* 24 (S2): 94–110.

Lamont, B. B., P. G. L. Klinkhamer, and E. T. F. Witkowski. 1993. "Population fragmenta-tion may reduce fertility to zero in *Banksia goodii*? A demonstration of the Allee effect." *Oecologia* 94 (3): 446–50. doi: 10.1007/bf00317122.

Lane, S. N., K. F. Bradbrook, K. S. Richards, P. A. Biron, and A. G. Roy. 1999. "The applica-tion of computational fluid dynamics to natural river channels: Three-dimensional versus two-dimensional approaches." *Geomorphology* 29 (1–2): 1–20.

LaPoint, S., P. Gallery, M. Wikelski, and R. Kays. 2013. "Animal behavior, cost-based cor-ridor models, and real corridors." *Landscape Ecology* 28 (8): 1615–30. doi: 10.1007/s10 980-013-9910-0.

Larkin, J. L., D. S. Maehr, T. S. Hoctor, M. A. Orlando, and K. Whitney. 2004. "Landscape linkages and conservation planning for the black bear in west-central Florida." *Animal Conservation* 7 (1): 23–34.

Larson, C. L., S. E. Reed, A. M. Merenlender, and K. R. Crooks. 2016. "Effects of recreation on animals revealed as widespread through a global systematic review." *PLoS ONE* 11 (12): e0167259.

Larsen, L. G., J. Choi, M. K. Nungesser, and J. W. Harvey. 2012. "Directional connectivity in hydrology and ecology." *Ecological Applications* 22 (8): 2204–20.

Laurance, S. G. W. 2004. "Responses of understory rain forest birds to road edges in central Amazonia." *Ecological Applications* 14: 1344–57.

Laurance, S. G., and W. F. Laurance. 1999. "Tropical wildlife corridors: Use of linear rain-forest remnants by arboreal mammals." *Biological Conservation* 91 (2–3): 231–39. doi: 10.1016/s0006-3207(99)00077-4.

Laurance, W. F. 1990. "Comparative responses of five arboreal marsupials to tropical forest fragmentation." *Journal of Mammalogy* 71 (4): 641–53. doi: 10.2307/1381805.

———. 1991. "Ecological correlates of extinction proneness in Australian tropical rain for-est mammals." *Conservation Biology* 5 (1): 79–89. doi: 10.1111/j.1523-1739.1991.tb 00390.x.

———. 1995. "Extinction and survival of rainforest mammals in a fragmented tropical land-scape." In *Landscape Approaches in Mammalian Ecology and Conservation*, edited by W. Z. Lidicker Jr., 46–63. Minneapolis: University of Minnesota Press.

———. 1997. "Hyper-disturbed parks: Edge effects and the ecology of isolated rain-forest reserves in tropical Australia." In *Tropical Forest Remnants: Ecology, Management, and Con-servation of Fragmented Communities*, edited by W. F. Laurance, and R. O. G. Bierregaard Jr., 71–83. Chicago: University of Chicago Press.

Laurance, W. F., R. O. Bierregaard Jr., C. Gascon, R. K. Didham, A. P. Smith, A. J. Lynam, V. M. Viana, et al. 1997. "Tropical forest fragmentation: Synthesis of a diverse and dy-namic discipline." In *Tropical Forest Remnants: Ecology, Management, and Conservation of Fragmented Communities*, edited by W. F. Laurance and R. O. G. Bierregaard Jr., 502–14. Chicago: University of Chicago Press.

Laurance, W. F., T. E. Lovejoy, H. L. Vasconcelos, E. M. Bruna, R. K. Didham, P. C. Stouffer, C. Gascon, R. O. Bierregaard, S. G. Laurance, and E. Sampaio. 2002. "Ecosystem de-cay of Amazonian forest fragments: A 22-year investigation." *Conservation Biology* 16 (3): 605–18.

Lausche, B., M. Farrier, J. Verschuuren, A. G. M. La Viña, and A. Trouwborst. 2013. "The legal aspects of connectivity conservation: A concept paper." Vol. 1. Gland, Switzerland. Available at: https://portals.iucn.org/library/sites/library/files/documents/EPLP-085 -vol001.pdf: IUCN.

Lawler, J. J., D. D. Ackerly, C. M. Albano, M. G. Anderson, S. Z. Dobrowski, J. L. Gill, N. E. Heller, R. L. Pressey, E. W. Sanderson, and S. B. Weiss. 2015. "The theory behind, and the challenges of, conserving nature's stage in a time of rapid change." *Conservation Biology* 29 (3): 618–29. doi: 10.1111/cobi.12505.

Lawler, J. J., A. S. Ruesch, J. D. Olden, and B. H. McRae. 2013. "Projected climate-driven faunal movement routes." *Ecology Letters* 16 (8): 1014–22. doi: 10.1111/ele.12132.

Lawler, J., J. Watson, and E. Game. 2015. "Conservation in the face of climate change: Recent developments." *F1000Research* 4. doi: 10.12688/f1000research.6490.1.

Le Maitre, D. C., B. W. Van Wilgen, R. A. Chapman, and D. H. McKelly. 1996. "Invasive plants and water resources in the Western Cape Province, South Africa: Modelling the consequences of a lack of management." *Journal of Applied Ecology* 33 (1): 161–72.

Lechner, A. M., D. Sprod, O. Carter, and E. C. Lefroy. 2017. "Characterising landscape connectivity for conservation planning using a dispersal guild approach." *Landscape Ecology* 32 (1): 99–113. doi: 10.1007/s10980-016-0431-5.

Lees, A. C., and C. A. Peres. 2008. "Conservation value of remnant riparian forest corridors of varying quality for Amazonian birds and mammals." *Conservation Biology* 22 (2): 439–49.

Leis, J. M. 2018. "Paradigm lost: Ocean acidification will overturn the concept of larval-fish biophysical dispersal." *Frontiers in Marine Science* 5: 47.

Leopold, A., 1933. *Game Management.* New York: Scribners.

Lepczyk, C. A., M. F. J. Aronson, K. L. Evans, M. A. Goddard, S. B. Lerman, and J. S. MacIvor. 2017. "Biodiversity in the city: Fundamental questions for understanding the ecology of urban green spaces for biodiversity conservation." *BioScience* 67 (9): 799–807.

Lester, S. E., and B. S. Halpern. 2008. "Biological responses in marine no-take reserves versus partially protected areas." *Marine Ecology Progress Series* 367: 49–56.

Levchenko, V. F., and V. A. Kotolupov. 2010. "Levels of organization of living systems: Cooperons." *Journal of Evolutionary Biochemistry and Physiology* 46 (6): 631–41.

Levey, D. J. 2005. "Effects of landscape corridors on seed dispersal by birds." *Science* 309 (5731): 146–48. doi: 10.1126/science.1111479.

Levins, R. 1969. "Some demographic and genetic consequences of environmental heterogeneity for biological control." *Bulletin of the Entomological Society of America* 15 (3): 237–40. doi: 10.1093/besa/15.3.237.

Levins, R. A. 1970. "Extinction." *Lectures in Mathematical Life Sciences* 2: 75–107.

Lewison, R. L., L. B. Crowder, A. J. Read, and S. A. Freeman. 2004. "Understanding impacts of fisheries bycatch on marine megafauna." *Trends in Ecology & Evolution* 19 (11): 598–604.

Lewison, R., A. J. Hobday, S. Maxwell, E. Hazen, J. R. Hartog, D. C. Dunn, D. Briscoe, S. Fossette, C. E. O'keefe, and M. Barnes. 2015. "Dynamic ocean management: Identifying the critical ingredients of dynamic approaches to ocean resource management." *BioScience* 65 (5): 486–98.

Lexartza-Artza, I., and J. Wainwright. 2009. "Hydrological connectivity: Linking concepts with practical implications." *Catena* 79 (2): 146–52.

Li, Z., M. E. Hodgson, and W. Li. 2018. "A general-purpose framework for parallel processing of large-scale LiDAR data." *International Journal of Digital Earth* 11 (1): 26–47.

Lichstein, J. W., T. R. Simons, and K. E. Franzreb. 2002. "Landscape effects on breeding songbird abundance in managed forests." *Ecological Applications* 12 (3): 836–57. doi: 10.1890/1051-0761(2002)012[0836:leobsa]2.0.co;2.

Lidicker, W. Z., Jr. 1975. "The role of dispersal in the demography of small mammals." In *Small Mammals: Their Production and Population Dynamics*, edited by F. B. Golley, K. Petrusewicz, and L. Ryszkowski, 103–28. London: Cambridge University Press.

———. 1978. "Regulation of numbers in small mammal populations: Historical reflections and a synthesis." In *Populations of Small Mammals under Natural Conditions*, edited by D. P. Snyder, 122–41. Pittsburgh: University of Pittsburgh, Pymatuning Laboratory of Ecology, Special Publications.

———. 1985. "Population structuring as a factor in understanding microtine cycles." *Acta Zoologica Fennica* 173: 23–27.

———. 1988. "Solving the enigma of microtine 'cycles.'" *Journal of Mammalogy* 69 (2): 225–35. doi: 10.2307/1381374.

———. 1994. "A spatially explicit approach to vole population processes." *Polish Ecological Studies* 20 (3–4).

———. 1995a. *Landscape Approaches in Mammalian Ecology and Conservation*. Minneapolis: University of Minnesota Press.

———. 1995b. "The landscape concept: Something old, something new." In *Landscape Approaches in Mammalian Ecology and Conservation*, edited by W. Z. Lidicker Jr., 3–19. Minneapolis: University of Minnesota Press.

———. 1999. "Responses of mammals to habitat edges: An overview." *Landscape Ecology* 14 (4): 333–43.

———. 2000. "A food web/landscape interaction model for microtine rodent density cycles." *Oikos* 91 (3): 435–45. doi: 10.1034/j.1600–0706.2000.910304.x.

———. 2002. "From dispersal to landscapes: Progress in the understanding of population dynamics." *Acta Theriologica* 47 (S1): 23–37. doi: 10.1007/bf03192478.

———. 2007. Landscape ecology: Whence came this creature? Wiley Online Library.

———. 2008. "Levels of organization in biology: On the nature and nomenclature of ecology's fourth level." *Biological Reviews* 83 (1): 71–78.

———. 2010. "The Allee effect: Its history and future importance." *Open Ecology Journal* 3: 71–82.

———. 2015. "Genetic and spatial structuring of the California vole (*Microtus californicus*) through a multiannual density peak and decline." *Journal of Mammalogy* 96 (6): 1142–51.

Lidicker, W. Z., Jr., and W. D. Koenig. 1996. "Responses of terrestrial vertebrates to habitat edges and corridors." In *Metapopulations and Wildlife Conservation*, edited by D. R. McCullough, 85–109. Washington, DC: Island Press.

Lidicker, W. Z., Jr., and J. A. Peterson. 1999. "Responses of small mammals to habitat edges." In *Landscape Ecology of Small Mammals*, edited by G. W. Barrett and J. D. Peles, 211–27. New York: Springer-Verlag.

Lidicker, W. Z., Jr., and N. C. Stenseth. 1992. "To disperse or not to disperse: Who does it and why?" In *Animal Dispersal: Small Mammals as a Model*, edited by N. C. Stenseth, W. Z. Lidicker Jr., 21–36. London: Chapman and Hall.

Lidicker, W. Z., Jr., J. O. Wolff, L. N. Lidicker, and M. H. Smith. 1992. "Utilization of a habitat mosaic by cotton rats during a population decline." *Landscape Ecology* 6 (4): 259–68. doi: 10.1007/bf00129704.

Lindborg, R., and O. Eriksson. 2004. "Historical landscape connectivity affects present plant species diversity." *Ecology* 85 (7): 1840–45. doi: 10.1890/04-0367.

Lindenmayer, D. B., R. B. Cunningham, C. F. Donnelly, B. E. Triggs, and M. Belvedere. 1994. "Factors influencing the occurrence of mammals in retained linear strips (wildlife corridors) and contiguous stands of montane ash forest in the Central Highlands of Victoria, southeastern Australia." *Forest Ecology and Management* 67 (1–3): 113–33.

Lindenmayer, D. B., and J. F. Franklin. 2002. *Conserving Forest Biodiversity: A Comparative Multiscaled Approach*. Washington, DC: Island Press.

Linkie, M., G. Chapron, D. J. Martyr, J. Holden, and N. Leader-Williams. 2006. "Assessing the viability of tiger subpopulations in a fragmented landscape." *Journal of Applied Ecology* 43 (3): 576–86.

Liro, A., and J. Szacki. 1994. "Movements of small mammals along two ecological corridors in suburban Warsaw." *Polish Ecological Studies* 20 (3–4).

Littlefield, C. E., B. H. McRae, J. L. Michalak, J. J. Lawler, and C. Carroll. 2017. "Connecting today's climates to future climate analogs to facilitate movement of species under climate change." *Conservation Biology*. doi: 10.1111/cobi.12938.

Loarie, S. R., P. B. Duffy, H. Hamilton, G. P. Asner, C. B. Field, and D. D. Ackerly. 2009. "The velocity of climate change." *Nature* 462 (7276): 1052–57. doi: 10.1038/nature08649.

Lohr, J. N., P. David, and C. R. Haag. 2014. "Reduced lifespan and increased ageing driven by genetic drift in small populations." *Evolution* 68 (9): 2494–2508.

Lookingbill, T., and D. Urban. 2004. "An empirical approach towards improved spatial estimates of soil moisture for vegetation analysis." *Landscape Ecology* 19 (4): 417–33. doi: 10 .1023/b:land.0000030451.29571.8b.

López-Duarte, P. C., H. S. Carson, G. S. Cook, F. J. Fodrie, B. J. Becker, C. DiBacco, and L. A. Levin. 2012. "What controls connectivity? An empirical, multi-species approach." *Integrative and Comparative Biology* 52: 511–24.

Luschi, P., G. C. Hays, and F. Papi. 2003. "A review of long-distance movements by marine turtles, and the possible role of ocean currents." *Oikos* 103 (2): 293–302.

Lytle, D. A., and N. L. Poff. 2004. "Adaptation to natural flow regimes." *Trends in Ecology & Evolution* 19 (2): 94–100.

MacArthur, R., and E. O. Wilson. 1967. *The Theory of Island Biogeography*. Princeton, NJ: Princeton University Press.

Machtans, C. S., M. Villard, and S. J. Hannon. 1996. "Use of riparian buffer strips as movement corridors by forest birds." *Conservation Biology* 10 (5): 1366–79. doi: 10.1046/j.1523 –1739.1996.10051366.x.

Mackey, B. G., J. E. M. Watson, G. Hope, and S. Gilmore. 2008. "Climate change, biodiversity conservation, and the role of protected areas: An Australian perspective." *Biodiversity* 9 (3–4): 11–18.

Magura, T., G. L. Lövei, and B. Tóthmérész. 2017. "Edge responses are different in edges under natural versus anthropogenic influence: A meta-analysis using ground beetles." *Ecology and Evolution* 7 (3): 1009–17.

Maher, S. P., T. L. Morelli, M. Hershey, A. L. Flint, L. E. Flint, C. Moritz, and S. R. Beissinger. 2017. "Erosion of refugia in the Sierra Nevada meadows network with climate change." *Ecosphere* 8 (4). doi: 10.1002/ecs2.1673.

Malo, J. E., F. Suarez, and A. Diez. 2004. "Can we mitigate animal-vehicle accidents using predictive models?" *Journal of Applied Ecology* 41 (4): 701–10. doi: 10.1111/j.0021-8901 .2004.00929.x.

Marrotte, R. R., A. Gonzalez, and V. Millien. 2017. "Functional connectivity of the white-footed mouse in Southern Quebec, Canada." *Landscape Ecology* 32 (10): 1987–98.

Massolo, A., and A. Meriggi. 1998. "Factors affecting habitat occupancy by wolves in northern

Apennines (northern Italy): A model of habitat suitability." *Ecography* 21 (2): 97–107. doi: 10.1111/j.1600-0587.1998.tb00663.x.

Mateo-Sanchez, M. C., N. Balkenhol, S. Cushman, T. Perez, A. Dominguez, and S. Saura. 2015. "A comparative framework to infer landscape effects on population genetic structure: Are habitat suitability models effective in explaining gene flow?" *Landscape Ecology* 30 (8): 1405–20. doi: 10.1007/s10980-015-0194-4.

Maxwell, S. M., E. L. Hazen, S. J. Bograd, B. S. Halpern, G. A. Breed, B. Nickel, N. M. Teutschel, L. B. Crowder, S. Benson, and P. H. Dutton. 2013. "Cumulative human impacts on marine predators." *Nature Communications* 4: 2688.

Maxwell, S. M., E. L. Hazen, R. L. Lewison, D. C. Dunn, H. Bailey, S. J. Bograd, D. K. Briscoe, S. Fossette, A. J. Hobday, and M. Bennett. 2015. "Dynamic ocean management: Defining and conceptualizing real-time management of the ocean." *Marine Policy* 58: 42–50.

McCauley, D. E. 1993. "Genetic consequences of extinction and recolonization in fragmented habitats." In *Biotic Interactions and Global Change*, edited by P. M. Kareiva, J. G. Kingsolver, and R. B. Huey, 217–33. Sunderland, MA: Sinauer.

McCauley, D. J., M. L. Pinsky, S. R. Palumbi, J. A. Estes, F. H. Joyce, and R. R. Warner. 2015. "Marine defaunation: Animal loss in the global ocean." *Science* 347 (6219): 1255641.

McCloskey, J. M., and H. Spalding. 1989. "A reconnaissance-level inventory of the amount of wilderness remaining in the world." *Ambio* 18 (4): 221–27.

McCullough, D. R., J. K. Fischer, and J. D. Ballou. 1996. "From bottleneck to metapopulation: Recovery of the tule elk in California." In *Metapopulations and Wildlife Conservation*, edited by D. R. McCullough, 375–403. Washington, DC: Island Press.

McDonald, T. 2004. "The Wilderness Society's WildCountry Program. An interview with its National Coordinator, Virginia Young." *Ecological Management and Restoration* 5 (2): 87–97. doi: 10.1111/j.1442-8903.2004.00183.x.

McDonald, T., G. D. Gann, J. Jonson, and K. W. Dixon. 2016. "International standards for the practice of ecological restoration—including principles and key concepts." Society for Ecological Restoration, Washington, DC.

McGuire, J. L., J. J. Lawler, B. H. McRae, T. A. Nuñez, and D. M. Theobald. 2016. "Achieving climate connectivity in a fragmented landscape." *Proceedings of the National Academy of Sciences of the United States of America* 113 (26): 7195–7200. doi: 10.1073/pnas.160 2817113.

McKelvey, K. S., K. B. Aubry, and Y. K. Ortega. 2000. "History and distribution of lynx in the contiguous United States." In *Ecology and Conservation of Lynx in the United States* by L. F. Ruggiero, K. B. Aubry, S. W. Buskirk et al., 207–59. Boulder: University Press of Colorado.

McKelvey, K. S., J. P. Copeland, M. K. Schwartz, J. S. Littell, K. B. Aubry, J. R. Squires, S. A. Parks, M. M. Elsner, and G. S. Mauger. 2011. "Climate change predicted to shift wolverine distributions, connectivity, and dispersal corridors." *Ecological Applications* 21 (8): 2882–97.

McKenna, M. F., J. Calambokidis, E. M. Oleson, D. W. Laist, and J. A. Goldbogen. 2015. "Simultaneous tracking of blue whales and large ships demonstrates limited behavioral responses for avoiding collision." *Endangered Species Research* 27 (3): 219–32.

McKenna, M. F., S. L. Katz, C. Condit, and S. Walbridge. 2012. "Response of commercial ships to a voluntary speed reduction measure: Are voluntary strategies adequate for mitigating ship-strike risk?" *Coastal Management* 40 (6): 634–50.

McKenna, M. F., D. Ross, S. M. Wiggins, and J. A. Hildebrand. 2012. "Underwater radiated noise from modern commercial ships." *Journal of the Acoustical Society of America* 131 (1): 92–103.

McKeon, C. S., M. X. Weber, S. E. Alter, N. E. Seavy, E. D. Crandall, D. J. Barshis, E. D. Fechter-Leggett, and K. L. L. Oleson. 2016. "Melting barriers to faunal exchange across ocean basins." *Global Change Biology* 22 (2): 465–73.

McLachlan, J. S., J. S. Clark, and P. S. Manos. 2005. "Molecular indicators of tree migration capacity under rapid climate change." *Ecology* 86 (8): 2088–98.

McLeod, E., and R. V. Salm. 2006. *Managing Mangroves for Resilience to Climate Change.* World Conservation Union (IUCN).

McLeod, E., R. Salm, A. Green, and J. Almany. 2009. "Designing marine protected area networks to address the impacts of climate change." *Frontiers in Ecology and the Environment* 7 (7): 362–70.

McMillion, S. 2004. Bears in town seeking food. *Bozeman* [Montana] *Daily Chronicle*, February 15.

McRae, B. H., B. G. Dickson, T. H. Keitt, and V. B. Shah. 2008. "Using circuit theory to model connectivity in ecology, evolution, and conservation." *Ecology* 89 (10): 2712–24.

McRae, B. H., S. A. Hall, P. Beier, and D. M. Theobald. 2012. "Where to restore ecological connectivity? Detecting barriers and quantifying restoration benefits." *PLoS ONE* 7 (12). doi: 10.1371/journal.pone.0052604.

McRae, B. H., K. Popper, A. Jones, M. Schindel, S. Buttrick, K. Hall, R. S. Unnasch, and J. Platt. 2016. "Conserving nature's stage: Mapping omnidirectional connectivity for resilient terrestrial landscapes in the Pacific Northwest." The Nature Conservancy, Portland Oregon. Available online at: http://nature.org/resilienceNW. Accessed 6 November 2017.

Meentemeyer, R. K., S. E. Haas, and T. Václavík. 2012. "Landscape epidemiology of emerging infectious diseases in natural and human-altered ecosystems." *Annual Review of Phytopathology* 50: 379–402.

Meier, E. S., H. Lischke, D. R. Schmatz, and N. E. Zimmermann. 2012. "Climate, competition and connectivity affect future migration and ranges of European trees." *Global Ecology and Biogeography* 21 (2): 164–78. doi: 10.1111/j.1466-8238.2011.00669.x.

Merenlender, A. M. 2000. "Mapping vineyard expansion provides information on agriculture and the environment." *California Agriculture* 54 (3): 7–12. doi: 10.3733/ca.v054n03p7.

Merenlender, A. M., C. Kremen, M. Rakotondratsima, and A. Weiss. 1998. "Monitoring impacts of natural resource extraction on lemurs of the Masoala Peninsula, Madagascar." *Conservation Ecology* 2 (2). doi: 10.5751/es-00064-020205.

Merenlender, A. M., and M. K. Matella. 2013. "Maintaining and restoring hydrologic habitat connectivity in Mediterranean streams: An integrated modeling framework." *Hydrobiologia* 719 (1): 509–25.

Merkle, J. A., K. L. Monteith, E. O. Aikens, M. M. Hayes, K. R. Hersey, A. D. Middleton, B. A. Oates, H. Sawyer, B. M. Scurlock, and M. J. Kauffman. 2016. "Large herbivores surf waves of green-up during spring." *Proceedings of the Royal Society B* 283 (1833): 20160456.

Merriam, G., and A. Lanoue. 1990. "Corridor use by small mammals: Field measurement for three experimental types of *Peromyscus leucopus*." *Landscape Ecology* 4 (2–3): 123–31. doi: 10.1007/bf00132856.

Mesquita, R. C. G., P. Delamônica, and W. F. Laurance. 1999. "Effect of surrounding vegetation on edge-related tree mortality in Amazonian forest fragments." *Biological Conservation* 91 (2–3): 129–34. doi: 10.1016/s0006-3207(99)00086-5.

Mestre, F., B. B. Risk, A. Mira, P. Beja, and R. Pita. 2017. "A metapopulation approach to predict species range shifts under different climate change and landscape connectivity scenarios." *Ecological Modelling* 359: 406–14. doi: 10.1016/j.ecolmodel.2017.06.013.

Meyer-Gutbrod, E. L., C .H. Greene, and K. T. A. Davies. 2018. "Marine species range shifts necessitate advanced policy planning: The case of the North Atlantic right whale." *Oceanography* 31: https://doi.org/10.5670/oceanog.2018.209.

Meyerson, L. A., D. V. Viola, and R. N. Brown. 2010. "Hybridization of invasive *Phragmites australis* with a native subspecies in North America." *Biological Invasions* 12 (1): 103–11.

Midgley, G. F., I. D. Davies, C. H. Albert, R. Altwegg, L. Hannah, G. O. Hughes, L. R. O'halloran, C. Seo, J. H. Thorne, and W. Thuiller. 2010. "BioMove—An integrated platform simulating the dynamic response of species to environmental change." *Ecography* 33 (3): 612–16.

Milder, J. C. 2007. "A framework for understanding conservation development and its ecological implications." *AIBS Bulletin* 57 (9): 757–68.

Millenium Ecosystem Assessment. 2005. *Ecosystems and Human Well-Being: Synthesis*. Washington, DC: Island Press.

Miller, K., and K. Hyun. 2011. Ecological corridors: Legal framework for the Baekdu Daegan mountain system *(South Korea)*. IUCN Case Studies, part of a project on IUCN Guidelines for Protected Areas Legislation (Lausche, 2011). Gland, Switzerland: IUCN.

Millett, K. 2004. "Birds on a cool green roof." *Chicago Wilderness* 7: 6–9.

Mills, K. E., A. J. Pershing, C. J. Brown, Y. Chen, F. S. Chiang, D. S. Holland, S. Lehuta, J. A. Nye, J. C. Sun, and A. C. Thomas. 2013. "Fisheries management in a changing climate: Lessons from the 2012 ocean heat wave in the Northwest Atlantic." *Oceanography* 26 (2): 191–95.

Mills, L. S. 1996. "Fragmentation of a natural area: Dynamics of isolation for small mammals on forest remnants." In *National Parks and Protected Areas: Their Role in Environmental Protection*, edited by G. Wright, 199–218. Cambridge, England: Blackwell Science.

Mills, L. S., and F. W. Allendorf. 1996. "The one-migrant-per-generation rule in conservation and management." *Conservation Biology* 10 (6): 1509–18. doi: 10.1046/j.1523-1739.1996.10061509.x.

Minor, E. S., and D. L. Urban. 2008. "A graph-theory framework for evaluating landscape connectivity and conservation planning." *Conservation Biology* 22 (2): 297–307.

Mitra, A. 2013. *Sensitivity of Mangrove Ecosystem to Changing Climate*, Vol. 62. New York: Springer.

Moilanen, A., K. A. Wilson, and H. Possingham. 2009. *Spatial Conservation Prioritization: Quantitative Methods and Computational Tools*. Oxford: Oxford University Press.

Mokany, K., T. D. Harwood, and S. Ferrier. 2013. "Comparing habitat configuration strategies for retaining biodiversity under climate change." *Journal of Applied Ecology* 50 (2): 519–27. doi: 10.1111/1365-2664.12038.

Mokany, K., G. J. Jordan, T. D. Harwood, P. A. Harrison, G. Keppel, L. Gilfedder, O. Carter, and S. Ferrier. 2017. "Past, present and future refugia for Tasmania's palaeoendemic flora." *Journal of Biogeography* 44 (7): 1537–46. doi: 10.1111/jbi.12927.

Molinos, J. G., B. S. Halpern, D. S. Schoeman, C. J. Brown, W. Kiessling, P. J. Moore, J. M. Pandolfi, E. S. Poloczanska, A. J. Richardson, and M. T. Burrows. 2016. "Climate velocity and the future global redistribution of marine biodiversity." *Nature Climate Change* 6: 83. doi: 10.1038/nclimate2769.

Mooney, H. A., and R. J. Hobbs. 2000. *Invasive Species in a Changing World*. Washington, DC: Island Press.

Moore, T. J., J. V. Redfern, M. Carver, S. Hastings, J. D. Adams, and G. K. Silber. 2018. "Exploring ship traffic variability off California." *Ocean & Coastal Management* 163: 515–27.

Mora, C., D. P. Tittensor, S. Adl, A. G. B. Simpson, and B. Worm. 2011. "How many species are there on Earth and in the ocean?" *PLoS Biology* 9 (8): e1001127.

Morelli, T. L., C. Daly, S. Z. Dobrowski, D. M. Dulen, J. L. Ebersole, S. T. Jackson, J. D. Lundquist, C. I. Millar, S. P. Maher, and W. B. Monahan. 2016. "Managing climate change refugia for climate adaptation." *PLoS ONE* 11 (8): e0159909.

Morelli, T. L., S. P. Maher, M. C. W. Lim, C. Kastely, L. M. Eastman, L. E. Flint, A. L. Flint, S. R. Beissinger, and C. Moritz. 2017. "Climate change refugia and habitat connectivity promote species persistence. "*Climate Change Responses* 4 (1): 8.

Morgan, S. G. 2014. "Behaviorally Mediated Larval Transport in Upwelling Systems." *Advances in Oceanography* 2014. 1–7. doi: 10.1155/2014/364214.

Morgan, S. G., and J. L. Fisher. 2010. "Larval behavior regulates nearshore retention and offshore migration in an upwelling shadow and along the open coast." *Marine Ecology Progress Series* 404: 109–26.

Moritz, C., J. L. Patton, C. J. Conroy, J. L. Parra, G. C. White, and S. R. Beissinger. 2008. "Impact of a century of climate change on small-mammal communities in Yosemite National Park, USA." *Science* 322 (5899): 261–64.

Morley, J. W., R. L. Selden, R. J. Latour, T. L. Frölicher, R. J. Seagraves, and M. L. Pinsky. 2018. "Projecting shifts in thermal habitat for 686 species on the North American continental shelf." *PLoS ONE* 13 (5): e0196127.

Morrison, S. A., and W. M. Boyce. 2009. "Conserving connectivity: Some lessons from mountain lions in Southern California." *Conservation Biology* 23 (2): 275–85. doi: 10.11 11/j.1523-1739.2008.01079.x.

Moussa, R., and C. Bocquillon. 2009. "On the use of the diffusive wave for modelling extreme flood events with overbank flow in the floodplain." *Journal of Hydrology* 374 (1–2): 116–35.

Mumby, P. J. 2006. "Connectivity of reef fish between mangroves and coral reefs: Algorithms for the design of marine reserves at seascape scales." *Biological Conservation* 128 (2): 215–22.

Mumby, P. J., I. A. Elliott, C. M. Eakin, W. Skirving, C. B. Paris, H. J. Edwards, S. Enríquez, R. Iglesias-Prieto, L. M. Cherubin, and J. R. Stevens. 2011. "Reserve design for uncertain responses of coral reefs to climate change." *Ecology Letters* 14 (2): 132–40.

Mumme, R. L., S. J. Schoech, G. E. Woolfenden, and J. W. Fitzpatrick. 2000. "Life and death in the fast lane: Demographic consequences of road mortality in the Florida scrub-jay." *Conservation Biology* 14 (2): 501–12. doi: 10.1046/j.1523-1739.2000.98370.x.

Murcia, C. 1995. "Edge effects in fragmented forests: Implications for conservation." *Trends in Ecology & Evolution* 10 (2): 58–62. doi: 10.1016/s0169-5347(00)88977-6.

Murphy, M. T. 2001. "Source-sink dynamics of a declining eastern kingbird population and the value of sink habitats." *Conservation Biology* 15 (3): 737–48. doi: 10.1046/j.1523 -1739.2001.015003737.x.

Musick, J. A., and C. J. Limpus. 1997. "Habitat utilization and migration in juvenile sea turtles." *Biology of Sea Turtles* 1: 137–63.

Naeem, S. 1998. "Species redundancy and ecosystem reliability." *Conservation Biology* 12: 39–45.

Naeem, S., L. J. Thompson, S. P. Lawler, J. H. Lawton, and R. M. Woodfin. 1994. "Declining biodiversity can alter the performance of ecosystems." *Nature* 368 (6473): 734–37. doi: 10.1038/368734a0.

Nagelkerken, I., C. M. Roberts, G. Van Der Velde, M. Dorenbosch, M. C. van Riel, E. Co-cheret De La Morinière, and P. H. Nienhuis. 2002. "How important are mangroves and seagrass beds for coral-reef fish? The nursery hypothesis tested on an island scale." *Marine Ecology Progress Series* 244: 299–305.

Natural Resources Conservation District. 1999. "Conservation corridor planning at the land-scape level: Managing for wildlife." *National Biology Handbook*. Washington, DC: US Department of Agriculture, Natural Resources Conservation Service.

Newburn, D., S. Reed, P. Berck, and A. Merenlender. 2005. "Economics and land-use change in prioritizing private land conservation." *Conservation Biology* 19 (5): 1411–20. doi: 10.1111/j.1523-1739.2005.00199.x.

Newmark, W. D. 1987. "A land-bridge island perspective on mammalian extinctions in west-ern North American parks." *Nature* 325 (6103): 430–32. doi: 10.1038/325430a0.

———. 1995. "Extinction of mammal populations in western North American national parks." *Conservation Biology* 9 (3): 512–26. doi: 10.1046/j.1523-1739.1995.09030512.x.

Newmark, W. D., C. N. Jenkins, S. L. Pimm, P. B. McNeally, and J. M. Halley. 2017. "Targeted habitat restoration can reduce extinction rates in fragmented forests." *Proceedings of the National Academy of Sciences* 114: 9635–40.

Ng, S. J., J. W. Dole, R. M. Sauvajot, S. P. D. Riley, and T. J. Valone. 2004. "Use of high-way undercrossings by wildlife in southern California." *Biological Conservation* 115 (3): 499–507. doi: 10.1016/s0006-3207(03)00166-6.

Nicholls, C. I., M. Parrella, and M. A. Altieri. 2001. "The effects of a vegetational corridor on the abundance and dispersal of insect biodiversity within a northern California organic vineyard." *Landscape Ecology* 16 (2): 133–46. doi: 10.1023/a:1011128222867.

Niskanen, A., M. Luoto, H. Vare, and R. K. Heikkinen. 2017. "Models of Arctic-alpine refu-gia highlight importance of climate and local topography." *Polar Biology* 40 (3): 489–502. doi: 10.1007/s00300-016-1973-3.

Norton, D. A., R. J. Hobbs, and L. Atkins. 1995. "Fragmentation, disturbance, and plant distribution: Mistletoes in woodland remnants in the western Australian wheatbelt." *Conservation Biology* 9 (2): 426–38. doi: 10.1046/j.1523-1739.1995.9020426.x.

Norton, T. W., and H. A. Nix. 1991. "Application of biological modelling and GIS to identify regional wildlife corridors." In *Nature Conservation 2: The Role of Corridors*, edited by D. A. Saunders and R. J. Hobbs, 19–26. Chipping Norton, New South Wales, Australia: Surrey Beatty.

Noss, R. F. 1987. "Corridors in real landscapes: A reply to Simberloff and Cox." *Conservation Biology* 1 (2): 159–64. doi: 10.1111/j.1523-1739.1987.tb00024.x.

———. 2003. "A checklist for wildlands network designs." *Conservation Biology* 17 (5): 1270–75. doi: 10.1046/j.1523-1739.2003.02489.x.

Noss, R. F., and B. Csuti. 1997. "Habitat fragmentation." In *Principles of Conservation Biology*, edited by G. K. Meffe and C. R. Carroll, 269–304. Sunderland, MA: Sinauer.

Noss, R. F., A. P. Dobson, R. Baldwin, P. Beier, C. R. Davis, D. A. Dellasala, J. Francis, H. Locke, K. Nowak, and R. Lopez. 2012. "Bolder thinking for conservation." *Conserva-tion Biology* 26: 1–4.

NPK Consultants. 2011. Ribeira das Jardas Linear Park in Cacém, Portugal.

Nuñez, T. A., J. J. Lawler, B. H. McRae, D. J. Pierce, M. B. Krosby, D. M. Kavanagh, P. H.

Singleton, and J. J. Tewksbury. 2013. "Connectivity planning to address climate change." *Conservation Biology* 27 (2): 407–16. doi: 10.1111/cobi.12014.

Obenour, D. R., D. Scavia, N. N. Rabalais, R. E. Turner, and A. M. Michalak. 2013. "Retrospective analysis of midsummer hypoxic area and volume in the northern Gulf of Mexico, 1985–2011." *Environmental Science & Technology* 47 (17): 9808–15.

Oberndorfer, E., J. Lundholm, B. Bass, R. R. Coffman, H. Doshi, N. Dunnett, S. Gaffin, Manfred Köhler, K. K. Y. Liu, and B. Rowe. 2007. "Green roofs as urban ecosystems: Ecological structures, functions, and services." *BioScience* 57 (10): 823–33.

Olds, A. D., R. M. Connolly, K. A. Pitt, and P. S. Maxwell. 2012. "Primacy of seascape connectivity effects in structuring coral reef fish assemblages." *Marine Ecology Progress Series* 462: 191–203. doi: 10.3354/meps09849.

Olds, A. D., R. M. Connolly, K. A. Pitt, S. J. Pittman, P. S. Maxwell, C. M. Huijbers, B. R. Moore, S. Albert, D. Rissik, and R. C. Babcock. 2016. "Quantifying the conservation value of seascape connectivity: A global synthesis." *Global Ecology and Biogeography* 25 (1): 3–15.

Olds, A. D., K. A. Pitt, P. S. Maxwell, and R. M. Connolly. 2012. "Synergistic effects of reserves and connectivity on ecological resilience." *Journal of Applied Ecology* 49 (6): 1195–1203.

Olson, D., M. O'Connell, Y. Fang, J. Burger, and R. Rayburn. 2009. "Managing for climate change within protected area landscapes." *Natural Areas Journal* 29 (4): 394–99.

Olsson, M., P. Widen, and J. Larkin. 2008. "Effectiveness of a highway overpass to promote landscape connectivity and movement of moose and roe deer in Sweden." *Landscape and Urban Planning* 85:133–39.

Opdam, P., R. Foppen, R. Reijnen, and A. Schotman. 1995. "The landscape ecological approach in bird conservation: Integrating the metapopulation concept into spatial planning." *Ibis* 137 (s1).

Opedal, Ø. H., W. S. Armbruster, and B. J. Graae. 2015. "Linking small-scale topography with microclimate, plant species diversity and intra-specific trait variation in an alpine landscape." *Plant Ecology & Diversity* 8 (3): 305–15.

Orrock, J. L., and E. I. Damschen. 2005. "Corridors cause differential seed predation." *Ecological Applications* 15 (3): 793–98. doi: 10.1890/04-1129.

Ortega, Y. K., and D. E. Capen. 1999. "Effects of forest roads on habitat quality for ovenbirds in a forested landscape." *The Auk* 116 (4): 937–46. doi: 10.2307/4089673.

Ostfeld, R. S. 1992. "Small-mammal herbivores in a patchy environment: Individual strategies and population responses." In *Effects of Resource Distribution on Animal–Plant Interactions*, edited by M. D. Hunter, T. Ohgushi, and P. W. Price. Amsterdam: Elsevier.

Oza, G. M. 1983. "Deteriorating habitat and prospects of the Asiatic lion." *Environmental Conservation* 10 (04): 349. doi: 10.1017/s0376892900013096.

Pachauri, R. K., M. R. Allen, V. R. Barros, J. Broome, W. Cramer, R. Christ, J. A. Church, L. Clarke, Q. Dahe, and P. Dasgupta. 2014. *Climate Change 2014: Synthesis Report.* Contribution of Working Groups I, II, and III to the fifth assessment report of the Intergovernmental Panel on Climate Change. IPCC.

Palmer, M. A., J. B. Zedler, and D. A. Falk. 2016. *Foundations of Restoration Ecology.* Washington, DC: Island Press.

Palumbi, S. R. 2003. "Population genetics, demographic connectivity, and the design of marine reserves." *Ecological Applications* 13 (sp1): 146–58.

Panetta, F. D. 1991. "Negative values of corridors." In *Nature Conservation 2: The Role of*

Corridors, edited by D. A. Saunders and R. J. Hobbs, 410. Chipping Norton, New South Wales, Australia: Surrey Beatty.

Panetta, F. D., and A. J. M. Hopkins. 1991. "Weeds in corridors: Invasion and management." In *Nature Conservation 2: The Role of Corridors*, edited by D. A. Saunders and R. J. Hobbs, 341–52. Chipping Norton, New South Wales, Australia: Surrey Beatty.

Paradis, E., S. R. Baillie, W. J. Sutherland, and R. D. Gregory. 1999. "Dispersal and spatial scale affect synchrony in spatial population dynamics." *Ecology Letters* 2 (2): 114–20.

Parks Canada. 2017. 10 quick facts about highway wildlife crossings in the park. Banff National Park, www.pc.gc.ca/en/pn-np/ab/banff/info/gestion-management/enviro/trans port/tch-rtc/passages-crossings/faq/10. Accessed 28 April 2018.

Parmesan, C. 2006. "Ecological and evolutionary responses to recent climate change." *Annual Review of Ecology, Evolution, and Systematics* 37: 637–69.

Parodi, J. J., L. Giambastiani, N. E. Seavy, I. M. Thalmayer, E. Lasky, and T. Gardali. 2014. "A How-to Guide and Metadata for the Riparian Restoration Design Database for Marin and Sonoma Counties." Point Blue Conservation Science, www.pointblue.org /restorationtools.

Pasternack, G. B., C. L. Wang, and J. E. Merz. 2004. "Application of a 2D hydrodynamic model to design of reach-scale spawning gravel replenishment on the Mokelumne River, California." *River Research and Applications* 20 (2): 205–25.

Paton, P. W. C. 1994. "The effect of edge on avian nest success: How strong is the evidence?" *Conservation Biology* 8 (1): 17–26. doi: 10.1046/j.1523-1739.1994.08010017.x.

Pearson, R. G., and T. P. Dawson. 2005. "Long-distance plant dispersal and habitat fragmentation: Identifying conservation targets for spatial landscape planning under climate change." *Biological Conservation* 123 (3): 389–401. doi: 10.1016/j.biocon.2004.12 .006.

Pecl, G. T., M. B. Araújo, J. D. Bell, J. Blanchard, T. C. Bonebrake, I. Chen, T. D. Clark, R. K. Colwell, F. Danielsen, and B. Evengård. 2017. "Biodiversity redistribution under climate change: Impacts on ecosystems and human well-being." *Science* 355 (6332): eaai9214.

Pellatt, M. G., S. J. Goring, K. M. Bodtker, and A. J. Cannon. 2012. "Using a down-scaled bioclimate envelope model to determine long-term temporal connectivity of Garry oak (*Quercus garryana*) habitat in western North America: Implications for protected area planning." *Environmental Management* 49 (4): 802–15. doi: 10.1007/s00267-012 -9815-8.

Peluso, N. L. 1993. "Coercing conservation?" *Global Environmental Change* 3 (2): 199–217. doi: 10.1016/0959-3780(93)90006-7.

Pence, G. Q. K., M. A. Botha, and J. K. Turpie. 2003. "Evaluating combinations of on- and off-reserve conservation strategies for the Agulhas Plain, South Africa: A financial perspective." *Biological Conservation* 112 (1–2): 253–73. doi: 10.1016/s0006-3207(02)00413-5.

Peng, L. L. H., and C. Y. Jim. 2013. "Green-roof effects on neighborhood microclimate and human thermal sensation." *Energies* 6 (2): 598–618.

Perault, D. R., and M. V. Lomolino. 2000. "Corridors and mammal community structure across a fragmented, old-growth forest landscape." *Ecological Monographs* 70 (3): 401–22. doi: 10.1890/0012-9615(2000)070[0401:camcsa]2.0.co;2.

Pérez-García, N., J. H. Thorne, and F. Domínguez-Lozano. 2017. "The mid-distance dispersal optimum, evidence from a mixed-model climate vulnerability analysis of an edaphic endemic shrub." *Diversity and Distributions* 23 (7): 771–82.

Pérez-Méndez, N., P. Jordano, and A. Valido. 2018. "Persisting in defaunated landscapes: Reduced plant population connectivity after seed dispersal collapse." *Journal of Ecology* 106 (3): 936–47.

Perry, A. L., P. J. Low, J. R. Ellis, and J. D. Reynolds. 2005. "Climate change and distribution shifts in marine fishes." *Science* 308 (5730): 1912–15.

Peterson, J. A. 1996. "Gray-tailed vole population responses to inbreeding and environmental stress." PhD dissertation, University of California, Berkeley.

Phillips, S. J., P. Williams, G. Midgley, and A. Archer. 2008. "Optimizing dispersal corridors for the cape proteaceae using network flow." *Ecological Applications* 18 (5): 1200–11. doi: 10.1890/07-0507.1.

Pinsky, M. L., B. Worm, M. J. Fogarty, J. L. Sarmiento, and S. A. Levin. 2013. "Marine taxa track local climate velocities." *Science* 341 (6151): 1239–42.

Planes, S., G. P. Jones, and S. R. Thorrold. 2009. "Larval dispersal connects fish populations in a network of marine protected areas." *Proceedings of the National Academy of Sciences*:pnas. 0808007106.

Plotnick, R. E., and M. L. McKinney. 1993. "Ecosystem organization and extinction dynamics." *PALAIOS* 8 (2): 202. doi: 10.2307/3515172.

Plowright, R. K., P. Foley, H. E. Field, A. P. Dobson, J. E. Foley, P. Eby, and P. Daszak. 2011. "Urban habituation, ecological connectivity and epidemic dampening: The emergence of Hendra virus from flying foxes (*Pteropus* spp.)." *Proceedings of the Royal Society of London B: Biological Sciences* 278 (1725): 3703–12.

Poff, N. L., B. D. Richter, A. H. Arthington, S. E. Bunn, R. J. Naiman, E. Kendy, M. Acreman, C. Apse, B. P. Bledsoe, and M. C. Freeman. 2010. "The ecological limits of hydrologic alteration (ELOHA): A new framework for developing regional environmental flow standards." *Freshwater Biology* 55 (1): 147–70.

Poff, N. L., and J. K. H. Zimmerman. 2010. "Ecological responses to altered flow regimes: A literature review to inform the science and management of environmental flows." *Freshwater Biology* 55 (1): 194–205.

Poloczanska, E. S., C. J. Brown, W. J. Sydeman, W. Kiessling, D. S. Schoeman, P. J. Moore, K. Brander, J. F. Bruno, L. B. Buckley, and M. T. Burrows. 2013. "Global imprint of climate change on marine life." *Nature Climate Change* 3 (10): 919.

Polovina, J. J., J. P. Dunne, P. A. Woodworth, and E. A. Howell. 2011. "Projected expansion of the subtropical biome and contraction of the temperate and equatorial upwelling biomes in the North Pacific under global warming." *ICES Journal of Marine Science* 68 (6): 986–95.

Polovina, J. J., E. A. Howell, and M. Abecassis. 2008. "Ocean's least productive waters are expanding." *Geophysical Research Letters* 35 (3).

Polovina, J. J., E. Howell, D. R. Kobayashi, and M. P. Seki. 2001. "The transition zone chlorophyll front, a dynamic global feature defining migration and forage habitat for marine resources." *Progress in Oceanography* 49 (1–4): 469–83.

———. 2017. "The Transition Zone Chlorophyll Front updated: Advances from a decade of research." *Progress in Oceanography* 150: 79–85.

Pongracz, J. D., D. Paetkau, M. Branigan, and E. Richardson. 2017. "Recent hybridization between a polar bear and grizzly bears in the Canadian Arctic." *Arctic* 70 (2): 151–60.

Popescu, A. 2016. Bridge seen as key to mountain lion survival near L.A. https://www.marketplace.org/2016/01/01/sustainability/bridge-seen-as-key-to-mountain-lion-survival-near-la. Accessed 3 July 2018. *Marketplace.*

Poulsen, J. R., and C. J. Clark. 2004. "Densities, distributions, and seasonal movements of gorillas and chimpanzees in swamp forest in northern Congo." *International Journal of Primatology* 25 (2): 285–306. doi: 10.1023/b:ijop.0000019153.50161.58.

Power, M. E., D. Tilman, J. A. Estes, B. A. Menge, W. J. Bond, L. S. Mills, G. Daily, J. C. Castilla, J. Lubchenco, and R. T. Paine. 1996. "Challenges in the quest for keystones." *BioScience* 46 (8): 609–20. doi: 10.2307/1312990.

Pressey, R. L., and K. H. Taffs. 2001. "Scheduling conservation action in production landscapes: Priority areas in western New South Wales defined by irreplaceability and vulnerability to vegetation loss." *Biological Conservation* 100 (3): 355–76. doi: 10.1016/s0006-3207(01)00039-8.

Prevedello, J. A., and M. V. Vieira. 2010. "Does the type of matrix matter? A quantitative review of the evidence." *Biodiversity and Conservation* 19 (5): 1205–23.

Prevedello, J. A., N. J. Gotelli, and J. P. Metzger. 2016. "A stochastic model for landscape patterns of biodiversity." *Ecological Monographs* 86 (4): 462–79.

Price, O. F., J. C. Z. Woinarski, and D. Robinson. 1999. "Very large area requirements for frugivorous birds in monsoon rainforests of the Northern Territory, Australia." *Biological Conservation* 91 (2–3): 169–80. doi: 10.1016/s0006-3207(99)00081-6.

Primm, S. A., and T. W. Clark. 1996. "Making sense of the policy process for carnivore conservation." *Conservation Biology* 10 (4): 1036–45. doi: 10.1046/j.1523-1739.1996.1004 1036.x.

Pringle, C. M. 2001. "Hydrologic connectivity and the management of biological reserves: A global perspective." *Ecological Applications* 11 (4): 981–98. doi: 10.2307/306 1006.

Pringle, C. M. 2003. "What is hydrologic connectivity and why is it ecologically important?" *Hydrological Processes* 17 (13): 2685–89.

Prugh, L. R., K. E. Hodges, A. R. E. Sinclair, and J. S. Brashares. 2008. "Effect of habitat area and isolation on fragmented animal populations." *Proceedings of the National Academy of Sciences* 105 (52): 20770–75.

Prugh, L. R., C. J. Stoner, C. W. Epps, W. T. Bean, W. J. Ripple, A. S. Laliberte, and J. S. Brashares. 2009. "The rise of the mesopredator." *BioScience* 59 (9): 779–91.

Pulsford, I., G. Worboys, G. Howling, and T. Barrett. 2012. "Great eastern ranges, Australia." In *Climate and Conservation: Landscape and Seascape Science, Planning, and Action*, edited by J. A. Hilty, C. C. Chester, and M. S. Cross, 202–16. Washington, DC: Island Press/Center for Resource Economics.

Putman, R. J. 1997. "Deer and road traffic accidents: Options for management." *Journal of Environmental Management* 51 (1): 43–57.

Pyšek, P., and D. M. Richardson. 2010. "Invasive species, environmental change and management, and health." *Annual Review of Environment and Resources* 35: 25–55.

Quammen, D. 2012. *Spillover: Animal Infections and the Next Human Pandemic*. New York: W.W. Norton.

Rabalais, N. N., R. E. Turner, and D. Scavia. 2002. "Beyond science into policy: Gulf of Mexico hypoxia and the Mississippi River. Nutrient policy development for the Mississippi River watershed reflects the accumulated scientific evidence that the increase in nitrogen loading is the primary factor in the worsening of hypoxia in the northern Gulf of Mexico." *AIBS Bulletin* 52 (2): 129–42.

Rahel, F. J. 2013. "Intentional fragmentation as a management strategy in aquatic systems." *BioScience* 63 (5): 362–72.

Rail, J., M. Darveau, A. Desrochers, and J. Huot. 1997. "Territorial responses of boreal forest birds to habitat gaps." *Condor* 99 (4): 976. doi: 10.2307/1370150.

Ranta, E., V. Kaitala, J. Lindström, and H. Linden. 1995. "Synchrony in population dynamics." *Proceedings of the Royal Society of London B* 262 (1364): 113–18.

Ražen, N., A. Brugnoli, C. Castagna, C. Groff, P. Kaczensky, F. Kljun, F. Knauer, I. Kos, M. Krofel, and R. Luštrik. 2016. "Long-distance dispersal connects Dinaric-Balkan and Alpine grey wolf (*Canis lupus*) populations." *European Journal of Wildlife Research* 62 (1): 137–42.

Razgour, O. 2015. "Beyond species distribution modeling: A landscape genetics approach to investigating range shifts under future climate change." *Ecological Informatics* 30: 250–56. doi: 10.1016/j.ecoinf.2015.05.007.

Rebelo, A. G., and W. R. Siegfried. 1992. "Where should nature reserves be located in the Cape Floristic Region, South Africa? Models for the spatial configuration of a reserve network aimed at maximizing the protection of floral diversity." *Conservation Biology* 6 (2): 243–52. doi: 10.1046/j.1523-1739.1992.620243.x.

Redfern, J. V., L. T. Hatch, C. Caldow, M. L. DeAngelis, J. Gedamke, S. Hastings, L. Henderson, M. F McKenna, T. J. Moore, and M. B. Porter. 2017. "Assessing the risk of chronic shipping noise to baleen whales off Southern California, USA." *Endangered Species Research* 32: 153–67.

Redfern, J. V., M. F. McKenna, T. J. Moore, J. Calambokidis, M. L. Deangelis, E. A. Becker, J. Barlow, K. A. Forney, P. C. Fiedler, and S. J. Chivers. 2013. "Assessing the risk of ships striking large whales in marine spatial planning." *Conservation Biology* 27 (2): 292–302.

Reed, D. H. 2005. "Relationship between population size and fitness." *Conservation Biology* 19 (2): 563–68. doi: 10.1111/j.1523-1739.2005.00444.x.

Reed, S. E., J. A. Hilty, and D. M. Theobald. 2014. "Guidelines and incentives for conservation development in local land-use regulations." *Conservation Biology* 28 (1): 258–68.

Reed, S. E., and A. M. Merenlender. 2008. "Quiet, nonconsumptive recreation reduces protected area effectiveness." *Conservation Letters* 1 (3): 146–54.

Reilly, M., M. Tobler, D. Sonderegger, and P. Beier. 2017. "Spatial and temporal response of wildlife to recreational activities in the San Francisco Bay ecoregion." *Biological Conservation* 207: 117–26.

Renton, M., S. Childs, R. Standish, and N. Shackelford. 2013. "Plant migration and persistence under climate change in fragmented landscapes: Does it depend on the key point of vulnerability within the lifecycle?" *Ecological Modelling* 249: 50–58. doi: 10.1016/j.ecolmodel.2012.07.005.

Renton, M., N. Shackelford, and R. J. Standish. 2012. "Habitat restoration will help some functional plant types persist under climate change in fragmented landscapes." *Global Change Biology* 18 (6): 2057–70. doi: 10.1111/j.1365-2486.2012.02677.x.

Resasco, J., N. M. Haddad, J. L. Orrock, D. Shoemaker, L. A. Brudvig, E. I. Damschen, J. J. Tewksbury, and D. J. Levey. 2014. "Landscape corridors can increase invasion by an exotic species and reduce diversity of native species." *Ecology* 95 (8):2033–39.

Reunanen, P., M. Mönkkönen, and A. Nikula. 2000. "Managing boreal forest landscapes for flying squirrels." *Conservation Biology* 14 (1): 218–26. doi: 10.1046/j.1523-1739.2000.98387.x.

Rhymer, J. M., and D. Simberloff. 1996. "Extinction by hybridization and introgression." *Annual Review of Ecology and Systematics* 27 (1): 83–109. doi: 10.1146/annurev.ecolsys.27.1.83.

Rich, A. C., D. S. Dobkin, and L. J. Niles. 1994. "Defining forest fragmentation by corridor width: The influence of narrow forest-dividing corridors on forest-nesting birds in southern New Jersey." *Conservation Biology* 8 (4): 1109–21.

Richard, Y., and D. P. Armstrong. 2010. "Cost distance modelling of landscape connectivity and gap-crossing ability using radio-tracking data." *Journal of Applied Ecology* 47 (3): 603–10. doi: 10.1111/j.1365-2664.2010.01806.x.

Ricketts, T. H. 2004. "Tropical forest fragments enhance pollinator activity in nearby coffee crops." *Conservation Biology* 18 (5): 1262–71. doi: 10.1111/j.1523-1739.2004.00227.x.

Ricketts, T. H., G. C. Daily, P. R. Ehrlich, and J. P. Fay. 2001. "Countryside biogeography of moths in a fragmented landscape: Biodiversity in native and agricultural habitats." *Conservation Biology* 15 (2): 378–88. doi: 10.1046/j.1523-1739.2001.015002378.x.

Ricketts, T. H., G. C. Daily, P. R. Ehrlich, and C. D. Michener. 2004. "Economic value of tropical forest to coffee production." *Proceedings of the National Academy of Sciences* 101 (34): 12579–82. doi: 10.1073/pnas.0405147101.

Ries, L., R. J. Fletcher, J. Battin, and T. D. Sisk. 2004. "Ecological responses to habitat edges: Mechanisms, models, and variability explained." *Annual Review of Ecology, Evolution, and Systematics* 35 (1): 491–522. doi: 10.1146/annurev.ecolsys.35.112202.130148.

Riley, S. P. D., L. E. Serieys, J. P. Pollinger, J. A. Sikich, L. Dalbeck, R. K. Wayne, and H. B. Ernest. 2014. "Individual behaviors dominate the dynamics of an urban mountain lion population isolated by roads." *Current Biology* 24: 1989–94.

Riley, S. P. D., H. B. Shaffer, S. R. Voss, and B. M. Fitzpatrick. 2003. "Hybridization between a rare, native tiger salamander (*Ambystoma californiense*) and its introduced congener." *Ecological Applications* 13: 1263–75.

Ripple, W. J., J. A. Estes, R. L. Beschta, C. C. Wilmers, E. G. Ritchie, M. Hebblewhite, J. Berger, B. Elmhagen, M. Letnic, and M. P. Nelson. 2014. "Status and ecological effects of the world's largest carnivores." *Science* 343 (6167): 1241484.

Ripple, W. J., C. Wolf, T. M. Newsome, M. Galetti, M. Alamgir, E. Crist, M. I. Mahmoud, W. F. Laurance, and 15,364 scientist signatories from 184 countries. 2017. "World scientists' warning to humanity: A second notice." *BioScience* 67 (12): 1026–28.

Rissman, A. R., and A. M. Merenlender. 2008. "The conservation contributions of conservation easements: Analysis of the San Francisco Bay Area protected lands spatial database." *Ecology and Society* 13 (1).

Roberts, D. R., and A. Hamann. 2016. "Climate refugia and migration requirements in complex landscapes." *Ecography* 39 (12): 1238–46.

Robillard, C. M., L. E. Coristine, R. N. Soares, and J. T. Kerr. 2015. "Facilitating climate-change-induced range shifts across continental land-use barriers." *Conservation Biology* 29 (6): 1586–95. doi: 10.1111/cobi.12556.

Rockwood, R. C., J. Calambokidis, and J. Jahncke. 2017. "High mortality of blue, humpback and fin whales from modeling of vessel collisions on the US West Coast suggests population impacts and insufficient protection." *PLoS ONE* 12 (8): e0183052.

Rodenhouse, N. L., G. W. Barrett, D. M. Zimmerman, and J. C. Kemp. 1992. "Effects of uncultivated corridors on arthropod abundances and crop yields in soybean agroecosystems." *Agriculture, Ecosystems & Environment* 38 (3): 179–91. doi: 10.1016/0167-8809(92)90143-y.

Rodriguez, A., G. Crema, and M. Delibes. 1996. "Use of non-wildlife passages across a high speed railway by terrestrial vertebrates." *Journal of Applied Ecology* 33 (6): 1527. doi: 10.2307/2404791.

Roehrdanz, P. R., and L. Hannah. 2016. "Climate change, California wine, and wildlife habitat." *Journal of Wine Economics* 11 (1): 69–87.

Roff, D. 2003. "Evolutionary danger for rainforest species." *Science* 301 (5629): 58–59. doi: 10.1126/science.1087384.

Rogers, C. M., M. J. Taitt, J. N. M. Smith, and G. Jongejan. 1997. "Nest predation and cowbird parasitism create a demographic sink in wetland-breeding song sparrows." *Condor* 99 (3): 622–33.

Roll, J., R. J. Mitchell, R. J. Cabin, and D. L. Marshall. 1997. "Reproductive success increases with local density of conspecifics in a desert mustard (*Lesquerella fendleri*)." *Conservation Biology* 11 (3): 738–46. doi: 10.1046/j.1523-1739.1997.96013.x.

Rolland, R. M., W. A. McLellan, M. J. Moore, C. A. Harms, E. A. Burgess, and K. E. Hunt. 2017. "Fecal glucocorticoids and anthropogenic injury and mortality in North Atlantic right whales *Eubalaena glacialis*." *Endangered Species Research* 34: 417–29.

Rolstad, J. 1991. "Consequences of forest fragmentation for the dynamics of bird populations: Conceptual issues and the evidence." *Biological Journal of the Linnean Society* 42 (1–2): 149–163. doi: 10.1111/j.1095-8312.1991.tb00557.x.

Rolston, H. 1988. *Environmental Ethics: Duties to and Values in the Natural World*. Philadelphia: Temple University Press.

Root, J. J., R. B. Puskas, J. W. Fischer, C. B. Swope, M. A. Neubaum, S. A. Reeder, and A. J. Piaggio. 2009. "Landscape genetics of raccoons (*Procyon lotor*) associated with ridges and valleys of Pennsylvania: Implications for oral rabies vaccination programs." *Vector-Borne and Zoonotic Diseases* 9 (6): 583–88.

Rösch, V., T. Tscharntke, C. Scherber, and P. Batáry. 2015. "Biodiversity conservation across taxa and landscapes requires many small as well as single large habitat fragments." *Oecologia* 179 (1): 209–22.

Rose, N. A., and P. J. Burton. 2009. "Using bioclimatic envelopes to identify temporal corridors in support of conservation planning in a changing climate." *Forest Ecology and Management* 258: S64–S74. doi: 10.1016/j.foreco.2009.07.053.

Rosen, S. 1974. "Hedonic prices and implicit markets: Product differentiation in pure competition." *Journal of Political Economy* 82 (1): 34–55. doi: 10.1086/260169.

Rouget, M., R. M. Cowling, A. T. Lombard, A. T. Knight, and I. H. K. Graham. 2006. "Designing large-scale conservation corridors for pattern and process." *Conservation Biology* 20 (2): 549–61. doi: 10.1111/j.1523-1739.2006.00297.x.

Rouget, M., R. M. Cowling, R. L. Pressey, and D. M. Richardson. 2003. "Identifying spatial components of ecological and evolutionary processes for regional conservation planning in the Cape Floristic Region, South Africa." *Diversity and Distributions* 9 (3): 191–210. doi: 10.1046/j.1472-4642.2003.00025.x.

Rowe, D. B. 2011. "Green roofs as a means of pollution abatement." *Environmental Pollution* 159 (8–9): 2100–2110.

Roy, M. L., and C. Le Pichon. 2017. "Modelling functional fish habitat connectivity in rivers: A case study for prioritizing restoration actions targeting brown trout." *Aquatic Conservation: Marine and Freshwater Ecosystems* 27: 927–37.

Rudman, S. M., and D. Schluter. 2016. "Ecological impacts of reverse speciation in threespine stickleback." *Current Biology* 26 (4): 490–95.

Rudnick, D. A., S. J. Ryan, P. Beier, S. A. Cushman, F. Dieffenbach, C. W. Epps, L. R. Gerber et al. 2012. "The role of landscape connectivity in planning and implementing conservation and restoration priorities." *Issues in Ecology* 16: 1–20.

Ryszkowski, L., N. R. French, and A. Kedziora, eds. 1996. *Dynamics of an Agricultural Landscape*. PoznaÐ: Polish Academy of Sciences, Zaklad BadaÐ Ðrodowiska Rolniczego i Les'nego [Research Center for Agricultural and Forest Environment].

Saarman, E., M. Gleason, J. Ugoretz, S. Airamé, M. Carr, E. Fox, A. Frimodig, T. Mason, and J. Vasques. 2013. "The role of science in supporting marine protected area network planning and design in California." *Ocean & Coastal Management* 74: 45–56.

Sala, E., S. Giakoumi, and handling editor, L. Pendleton. 2017. "No-take marine reserves are the most effective protected areas in the ocean." *ICES Journal of Marine Science* 75 (3): 1166–68.

Sandel, B., L. Arge, B. Dalsgaard, R. G. Davies, K. J. Gaston, W. J. Sutherland, and J. Svenning. 2011. "The influence of Late Quaternary climate-change velocity on species endemism." *Science* 334 (6056): 660–64.

Sanderson, E. W., M. Jaiteh, M. A. Levy, K. H. Redford, A. V. Wannebo, and G. Woolmer. 2002. "The human footprint and the last of the wild." *BioScience* 52 (10): 891. doi: 10.1641/0006-3568(2002)052[0891:thfatl]2.0.co;2.

Sanderson, E. W., J. Walston, and J. G. Robinson. 2018. "From bottleneck to breakthrough: Urbanization and the future of biodiversity conservation." *BioScience* 68 (6): 412–26.

Santos, A., C. Branquinho, P. Gonçalves, and M. Santos. 2015. Lisbon, Portugal: Case Study City Portrait. Part of a GREEN SURGE study on urban green infrastructure planning and governance in 20 European cities. https://greensurge.eu/products/case-studies/Case_Study_Portrait_Lisbon.pdf. Accessed 20 June 2018: Reis Fundação da Faculdade de Ciências de Lisboa (FFCUL), Portugal, Feb. 5, 2015, version 1.0.

Saraiva, G. 2018. "The Ribeira das Jardas stream: An urban floodplain in Lisbon." In *Managing Flood Risk, Innovative Approaches from Big Floodplain Rivers and Urban Streams*, edited by A. Serra-Llobet, G. M. Kondolf, K. Schafer, and S. Nicholson, 126–31. London: Palgrave Macmillan.

Saraiva, M. G., G. M. Kondolf, P. Simoes, K. Morgado, and J. Alves. 2002. *River Conservation and Restoration Potential in the Metropolitan Area of Lisbon, Portugal*. Fourth International Workshop on Sustainable Land Use Planning: Collaborative planning for the metropolitan landscape. Bellingham: Western Washington University.

Saunders, D. A., and C. P. de Rebeira. 1991. "Values of corridors to avian populations in a fragmented landscape." In *Nature Conservation 2: The Role of Corridors*, edited by D. A. Saunders and R. J. Hobbs, 221–40. Chipping Norton, New South Wales, Australia: Surrey Beatty.

Save LA Cougars. 2018. Donate to help us #savelacougars. https://savelacougars.org/:savela cougars.org.

Sawaya, M. A., A. P. Clevenger, and S. Kalinowski. 2013. "Wildlife crossing structures connect *Ursid* populations in Banff National Park." *Conservation Biology* 27: 721–30.

Sawyer, S. C., C. W. Epps, and J. S. Brashares. 2011. "Placing linkages among fragmented habitats: Do least-cost models reflect how animals use landscapes?" *Journal of Applied Ecology* 48 (3): 668–78. doi: 10.1111/j.1365-2664.2011.01970.x.

Scales, K. L., E. L. Hazen, M. G. Jacox, F. Castruccio, S. M. Maxwell, R. L. Lewison, and S. J. Bograd. 2018. "Fisheries bycatch risk to marine megafauna is intensified in Lagrangian coherent structures." *Proceedings of the National Academy of Sciences*: 201801270.

Scales, K. L., P. I. Miller, L. A. Hawkes, S. N. Ingram, D. W. Sims, and S. C. Votier. 2014. "On the front line: Frontal zones as priority at-sea conservation areas for mobile marine vertebrates." *Journal of Applied Ecology* 51 (6): 1575–83.

Schill, S. R., G. T. Raber, J. J. Roberts, E. A. Treml, J. Brenner, and P. N. Halpin. 2015. "No reef is an island: Integrating coral reef connectivity data into the design of regional-scale marine protected area networks." *PLoS ONE* 10 (12): e0144199.

Schloss, C. A., T. A. Nuñez, and J. J. Lawler. 2012. "Dispersal will limit ability of mammals to track climate change in the Western Hemisphere." *Proceedings of the National Academy of Sciences* 109 (22): 8606–11. doi: 10.1073/pnas.1116791109.

Schlotterbeck, C. 2012. "The coal canyon story." *Ecological Restoration* 30 (4): 290–93. doi: 10.3368/er.30.4.290.

Schmidt, N. M., and P. M. Jensen. 2003. "Changes in mammalian body length over 175 years: Adaptations to a fragmented landscape?" *Conservation Ecology* 7 (2). doi: 10.5751/es-00520-070206.

Schmiegelow, F. K. A., S. G. Cumming, S. Harrison, S. Leroux, K. Lisgo, R. Noss, and B. Olsen. 2006. "Conservation beyond crisis management: A conservation-matrix model." Edmonton: Canadian BEACONs Project Discussion Paper 1.

Schmiegelow, F. K. A., and M. Mönkkönen. 2002. "Habitat loss and fragmentation in dynamic landscapes: Avian perspectives from the boreal forest." *Ecological Applications* 12 (2): 375. doi: 10.2307/3060949.

Schmitz, O. J., J. J. Lawler, P. Beier, C. Groves, G. Knight, D. A. Boyce, J. Bulluck et al. 2015. "Conserving biodiversity: Practical guidance about climate change adaptation approaches in support of land-use planning." *Natural Areas Journal* 35 (1): 190–203.

Schultz, C. B. 1995. "Corridors, islands and stepping stones: The role of dispersal behavior in designing reserves for a rare Oregon butterfly." *Bulletin of the Ecological Society of America* 76: 240.

Schumaker, N. H., A. Brookes, J. R. Dunk, B. Woodbridge, J. A. Heinrichs, J. J. Lawler, C. Carroll, and D. LaPlante. 2014. "Mapping sources, sinks, and connectivity using a simulation model of northern spotted owls." *Landscape Ecology* 29 (4): 579–92.

Schuster, R., and P. Arcese. 2015. "Effects of disputes and easement violations on the cost-effectiveness of land conservation." *PeerJ* 3: e1185.

Schwartz, C. C., P. H. Gude, L. Landenburger, M. A. Haroldson, and S. Podruzny. 2012. "Impacts of rural development on Yellowstone wildlife: Linking grizzly bear *Ursus arctos* demographics with projected residential growth." *Wildlife Biology* 18 (3): 246–57.

Seal, U. S., E. T. Thorne, M. A. Bogan, and S. H. Anderson. 1989. *Conservation Biology and the Black-Footed Ferret*. New Haven, CT: Yale University Press.

Seddon, P. J., C. J. Griffiths, P. S. Soorae, and D. P. Armstrong. 2014. "Reversing defaunation: Restoring species in a changing world." *Science* 345 (6195): 406–12.

Sefa Dei, G. J. 2000. "Rethinking the role of indigenous knowledges in the academy." *International Journal of Inclusive Education* 4 (2): 111–32.

Selonen, V., and I. K. Hanski. 2003. "Movements of the flying squirrel *Pteromys volans* in corridors and in matrix habitat." *Ecography* 26 (5): 641–51. doi: 10.1034/j.1600-0587.2003.03548.x.

Semlitsch, R. D., and J. R. Bodie. 2003. "Biological criteria for buffer zones around wetlands and riparian habitats for amphibians and reptiles." *Conservation Biology* 17 (5): 1219–28. doi: 10.1046/j.1523-1739.2003.02177.x.

Shaffer, M. L. 1981. "Minimum population sizes for species conservation." *BioScience* 31 (2): 131–34. doi: 10.2307/1308256.

Shafer, S. L., P. J. Bartlein, E. M. Gray, and R. T. Pelltier. 2015. "Projected future vegetation changes for the northwest United States and southwest Canada at a fine spatial resolution using a dynamic global vegetation model." *PLoS ONE* 10 (10): e0138759.

Shafroth, P. B., A. C. Wilcox, D. A. Lytle, J. T. Hickey, D. C. Andersen, V. B. Beauchamp, A. Hautzinger, L. E. McMullen, and A. Warner. 2010. "Ecosystem effects of environmental flows: Modelling and experimental floods in a dryland river." *Freshwater Biology* 55 (1): 68–85.

Shanks, A. L. 2009. "Pelagic larval duration and dispersal distance revisited." *Biological Bulletin* 216 (3): 373–85.

Shanks, A. L., and G. L. Eckert. 2005. "Population persistence of California Current fishes and benthic crustaceans: A marine drift paradox." *Ecological Monographs* 75 (4): 505–24.

Shanks, A. L., B. A. Grantham, and M. H. Carr. 2003. "Propagule dispersal distance and the size and spacing of marine reserves." *Ecological Applications* 13 (sp1): 159–69.

Shannon, G., M. F. McKenna, L. M. Angeloni, K. R. Crooks, K. M. Fristrup, E. Brown, K. A. Warner, M. D. Nelson, C. White, and J. Briggs. 2016. "A synthesis of two decades of research documenting the effects of noise on wildlife." *Biological Reviews* 91 (4): 982–1005.

Shanu, S., J. Idiculla, Q. Qureshi, Y. Jhala, and S. Bhattacharya. 2016. "A graph theoretic approach for modelling wildlife corridors." arXiv preprint arXiv:1603.01939.

Sheppe, W. 1972. "The annual cycle of small mammal populations on a Zambian floodplain." *Journal of Mammalogy* 53 (3): 445–60. doi: 10.2307/1379036.

Shirk, A. J., D. O. Wallin, S. A. Cushman, C. G. Rice, and K. I. Warheit. 2010. "Inferring landscape effects on gene flow: A new model selection framework." *Molecular Ecology* 19 (17): 3603–19.

Sieving, K. E., T. A. Contreras, and K. L. Maute. 2004. "Heterospecific facilitation of forest-boundary crossing by mobbing understory birds in north-central Florida." *The Auk* 121 (3): 738. doi: 10.1642/0004-8038(2004)121[0738:hfofcb]2.0.co;2.

Sieving, K. E., M. F. Willson, and T. L. De Santo. 2000. "Defining corridor functions for endemic birds in fragmented south-temperate rainforest." *Conservation Biology* 14 (4): 1120–32. doi: 10.1046/j.1523-1739.2000.98417.x.

Silva, J. B., M. G. Saraiva, I. Loupa-Ramos, F. Bernardo, and F. Monteiro. 2004. Classification of the aesthetic value of Jardas stream. Deliverable 4-3, URBEM (Urban River Basin Enhancement Methods) project, EVK-CT-2002-00082 CESUR, IST. Lisbon, Portugal: Technical University of Lisbon.

Simberloff, D., and L. G. Abele. 1976. "Island biogeography theory and conservation practice." *Science* 191 (4224): 285–86. doi: 10.1126/science.191.4224.285.

Simberloff, D., and J. Cox. 1987. "Consequences and costs of conservation corridors." *Conservation Biology* 1 (1): 63–71. doi: 10.1111/j.1523-1739.1987.tb00010.x.

Simberloff, D., J. A. Farr, J. Cox, and D. W. Mehlman. 1992. "Movement corridors: Conservation bargains or poor investments?" *Conservation Biology* 6 (4): 493–504. doi: 10.1046/j.1523-1739.1992.06040493.x.

Singleton, P. H., and J. F. Lehmkuhl. 1999. "Assessing wildlife habitat connectivity in the Interstate 90 Snoqualmie Pass corridor, Washington." Proceedings of the Third International Conference on Wildlife Ecology and Transportation.

Sizer, N., and E. V. J. Tanner. 1999. "Responses of woody plant seedlings to edge formation in a lowland tropical rainforest, Amazonia." *Biological Conservation* 91 (2–3): 135–42. doi: 10.1016/s0006-3207(99)00076-2.

Snep, R. P. H., M. F. W. DeVries, and P. Opdam. 2011. "Conservation where people work: A role for business districts and industrial areas in enhancing endangered butterfly populations?" *Landscape and Urban Planning* 103 (1): 94–101.

Snyder, S., J. S. F. Peter, D. T. Lynne, Y. Xu, and S. Kohin. 2017. "Crossing the line: Tunas actively exploit submesoscale fronts to enhance foraging success." *Limnology and Oceanography Letters* 2 (5): 187–94. doi: 10.1002/lol2.10049.

Sommers, A. P., C. C. Price, C. D. Urbigkit, and E. M. Peterson. 2010. "Quantifying economic impacts of large-carnivore depredation on bovine calves." *Journal of Wildlife Management* 74 (7): 1425–34.

Söndgerath, D., and B. Schröder. 2002. "Population dynamics and habitat connectivity affecting the spatial spread of populations—a simulation study." *Landscape Ecology* 17 (1): 57–70.

Sonoma County Agricultural Preservation and Open Space District. 2000. *Acquisition Plan: A Blueprint for Agricultural and Open Space Preservation*. Santa Rosa, CA.

Sonoma County Agricultural Preservation and Open Space District. 2015. Sonoma Developmental Center Draft Resource Assessment.

Sonoma Land Trust. 2014a. Sonoma Valley wildlife corridor project: Management and monitoring strategy. Santa Rosa, California: Sonoma Land Trust.

Sonoma Land Trust. 2014b. (with Pathways for Wildlife). Sonoma Valley wildlife corridor road underpass use report, 2013–2014.

Soulé, M. E., and M. E. Gilpin. 1991. "The theory of wildlife corridor capability." In *Nature Conservation 2: The Role of Corridors*, edited by D. A. Saunders and R. J. Hobbs, 3–8. Chipping Norton, New South Wales, Australia: Surrey Beatty.

Spackman, S. C., and J. W. Hughes. 1995. "Assessment of minimum stream corridor width for biological conservation: Species richness and distribution along mid-order streams in Vermont, USA." *Biological Conservation* 71 (3): 325–32. doi: 10.1016/0006-3207(94)00055-u.

Spear, S. F., N. Balkenhol, M. J. Fortin, B. H. McRae, and K. Scribner. 2010. "Use of resistance surfaces for landscape genetic studies: Considerations for parameterization and analysis." *Molecular Ecology* 19 (17): 3576–91. doi: 10.1111/j.1365-294X.2010.04657.x.

Spencer, W. D., R. H. Barrett, and W. J. Zielinski. 1983. "Marten habitat preferences in the northern Sierra Nevada." *Journal of Wildlife Management* 47 (4): 1181. doi: 10.2307/380 8189.

St. Clair, C. C., M. Bélisle, A. Desrochers, and S. Hannon. 1998. "Winter responses of forest birds to habitat corridors and gaps." *Conservation Ecology* 2 (2). doi: 10.5751/es-000 68-020213.

Stacey, P. B., M. L. Taper, and V. A. Johnson. 1997. "Migration within metapopulations: The impact upon local population dynamics." In *Metapopulation Biology: Ecology, Genetics and Evolution*, edited by I. Hanski and M. E. Gilpin, 267–91. San Diego, CA: Academic Press.

Stamps, J. A., M. Buechner, and V. V. Krishnan. 1987. "The effects of edge permeability and habitat geometry on emigration from patches of habitat." *American Naturalist* 129 (4): 533–52. doi: 10.1086/284656.

Staudt, A., A. K. Leidner, J. Howard, K. A. Brauman, J. S. Dukes, L. J. Hansen, C. Paukert, J. Sabo, and L. A. Solórzano. 2013. "The added complications of climate change: Understanding and managing biodiversity and ecosystems." *Frontiers in Ecology and the Environment* 11 (9): 494–501.

Steffen, W., K. Richardson, J. Rockström, S. E. Cornell, I. Fetzer, E. M. Bennett, R. Biggs, S. R. Carpenter, W. De Vries, and C. A. de Wit. 2015. "Planetary boundaries: Guiding human development on a changing planet." *Science* 347 (6223): 1259855.

Stenseth, N. C., and W. Z. Lidicker Jr. 1992. "The study of dispersal: A conceptual guide." In *Animal Dispersal: Small Mammals as a Model*, edited by N. C. Stenseth and W. Z. Lidicker Jr., 5–20. London: Chapman and Hall.

Stephens, S. E., D. N. Koons, J. J. Rotella, and D. W. Willey. 2004. "Effects of habitat fragmentation on avian nesting success: A review of the evidence at multiple spatial scales." *Biological Conservation* 115 (1): 101–10. doi: 10.1016/s0006-3207(03)00098-3.

Stewart, J. 2017. Bridges for animals to safely cross freeways are popping up around the world. My Modern Met. https://mymodernmet.com/wildlife-crossings/. Accessed 28 April 2018.

Stith, B. M., J. W. Fitzpatrick, G. E. Woolfenden, and B. Pranty. 1996. "Classification and conservation of metapopulations: A case study of the Florida scrub jay." In *Metapopulations and Wildlife Conservation*, edited by D. R. McCullough, 187–215. Washington, DC: Island Press.

Strange, E. 2012. "Compensation for cooling water intake entrainment effects and the habitat production forgone method." Prepared for the California Energy Commission by Stratus Consulting, Inc.

Stromberg, J. C. 1997. "Growth and survivorship of Fremont cottonwood, Goodding willow, and salt cedar seedlings after large floods in central Arizona." *Great Basin Naturalist* 57 (3):198–208.

Stumpf, K. J., T. C. Theimer, M. A. Mcleod, and T. J. Koronkiewicz. 2012. "Distance from riparian edge reduces brood parasitism of southwestern willow flycatchers, whereas parasitism increases nest predation risk." *Journal of Wildlife Management* 76 (2): 269–77.

Sullivan, L. L., B. L. Johnson, L. A. Brudvig, and N. M. Haddad. 2011. "Can dispersal mode predict corridor effects on plant parasites?" *Ecology* 92: 1559–64.

Suryan, R. M., J. A. Santora, and W. J. Sydeman. 2012. "New approach for using remotely sensed chlorophyll a to identify seabird hotspots." *Marine Ecology Progress Series* 451: v213–25.

Synes, N. W., K. Watts, S. C. F. Palmer, G. Bocedi, K. A. Barton, P. E. Osborne, and J. M. J. Travis. 2015. "A multi-species modelling approach to examine the impact of alternative climate change adaptation strategies on range shifting ability in a fragmented landscape." *Ecological Informatics* 30: 222–29. doi: 10.1016/j.ecoinf.2015.06.004.

Szacki, J., and A. Liro. 1991. "Movements of small mammals in the heterogeneous landscape." *Landscape Ecology* 5 (4): 219–24. doi: 10.1007/bf00141436.

Tabarelli, M., W. Mantovani, and C. A. Peres. 1999. "Effects of habitat fragmentation on plant guild structure in the montane Atlantic forest of southeastern Brazil." *Biological Conservation* 91 (2–3): 119–27. doi: 10.1016/s0006-3207(99)00085-3.

Taha, H. 1997. "Modeling the impacts of large-scale albedo changes on ozone air quality in the South Coast Air Basin." *Atmospheric Environment* 31: 1667–76.

Tanzania. 2018. The Wildlife Conservation Act (Cap. 283). Supplement No. 10. Subsidiary Legislation. 16 March 2018, to the *Gazette of the United Republic of Tanzania*, no. 11, vol. 99.

Tardy, O., A. Massé, F. Pelletier, and D. Fortin. 2018. "Interplay between contact risk, conspecific density, and landscape connectivity: An individual-based modeling framework." *Ecological Modelling* 373: 25–38.

Taylor, A. D. 1991. "Studying metapopulation effects in predator-prey systems." *Biological Journal of the Linnean Society* 42 (1–2): 305–23. doi: 10.1111/j.1095-8312.1991.tb00 565.x.

Templeton, A. R. 1986. "Coadaptation and outbreeding depression." In *Conservation Biology: The Science of Scarcity and Diversity*, edited by M. E. Soulé, 105–16. Sunderland, MA: Sinauer.

Tetzlaff, D., C. Soulsby, P. J. Bacon, A. F. Youngson, C. Gibbins, and I. A. Malcolm. 2007. "Connectivity between landscapes and riverscapes—a unifying theme in integrating hydrology and ecology in catchment science?" *Hydrological Processes* 21 (10): 1385–89.

Tewksbury, J. J., D. J. Levey, N. M. Haddad, S. Sargent, J. L. Orrock, A. Weldon, B. J. Danielson, J. Brinkerhoff, E. I. Damschen, and P. Townsend. 2002. "Corridors affect plants, animals, and their interactions in fragmented landscapes." *Proceedings of the National Academy of Sciences* 99 (20): 12923–26. doi: 10.1073/pnas.202242699.

Theobald, D. M., S. E. Reed, K. Fields, and M. Soulé. 2012. "Connecting natural landscapes using a landscape permeability model to prioritize conservation activities in the United States." *Conservation Letters* 5 (2): 123–33. doi: 10.1111/j.1755-263X.2011.00218.x.

Thomas, C. D. 2011. "Translocation of species, climate change, and the end of trying to recreate past ecological communities." *Trends in Ecology & Evolution* 26 (5): 216–21.

Thomas, N., R. Lucas, P. Bunting, A. Hardy, A. Rosenqvist, and M. Simard. 2017. "Distribution and drivers of global mangrove forest change, 1996–2010." *PLoS ONE* 12 (6): e0179302.

Thorne, J. H., D. Cameron, and J. F. Quinn. 2006. "A conservation design for the central coast of California and the evaluation of mountain lion as an umbrella species." *Natural Areas Journal* 26 (2): 137–48.

Thorne, J. H., H. Choe, P. A. Stine, J. C. Chambers, A. Holguin, A. C. Kerr, and M. W. Schwartz. 2017. "Climate change vulnerability assessment of forests in the Southwest USA." *Climatic Change*. doi: 10.1007/s10584-017-2010-4.

Thorne, K., G. MacDonald, G. Guntenspergen, R. Ambrose, K. Buffington, B. Dugger, C. Freeman, C. Janousek, L. Brown, and J. Rosencranz. 2018. "US Pacific coastal wetland resilience and vulnerability to sea-level rise." *Science Advances* 4 (2): eaao3270.

Thorne, L. H., M. G. Conners, E. L. Hazen, S. J. Bograd, M. Antolos, D. P. Costa, and S. A. Shaffer. "Effects of El Niño-driven changes in wind patterns on North Pacific albatrosses." *Journal of The Royal Society Interface* 13: 20160196.

Thorne, L. H., E. L. Hazen, S. J. Bograd, D. G. Foley, M. G. Conners, M. A. Kappes, H. M. Kim, D. P. Costa, Y. Tremblay, and S. A. Shaffer. 2015. "Foraging behavior links climate variability and reproduction in North Pacific albatrosses." *Movement Ecology* 3 (1): 27.

Thornton, D., K. Zeller, C. Rondinini, L. Boitani, K. Crooks, C. Burdett, A. Rabinowitz, and H. Quigley. 2016. "Assessing the umbrella value of a range-wide conservation network for jaguars (*Panthera onca*)." *Ecological Applications* 26 (4): 1112–24.

Thuiller, W., C. Albert, M. B. Araujo, P. M. Berry, M. Cabeza, A. Guisan, T. Hickler et al. 2008. "Predicting global change impacts on plant species' distributions: Future challenges." *Perspectives in Plant Ecology Evolution and Systematics* 9 (3–4): 137–152. doi: 10.1016/j.ppees.2007.09.004.

Tiemann, S., and R. Siebert. 2008. Ecological networks implemented by participatory approaches as a response to landscape fragmentation—A review of German literature. Clermont-Ferrand, France: 8th European IFSA Symposium.

Tilman, D. 1990. "Constraints and tradeoffs: Toward a predictive theory of competition and succession." *Oikos* 58 (1): 3. doi: 10.2307/3565355.

——. 1994. "Competition and biodiversity in spatially structured habitats." *Ecology* 75 (1): 2–16. doi: 10.2307/1939377.

Tilman, D., and P. Kareiva, eds. 1997. *Spatial Ecology: The Role of Space in Population Dynamics and Interspecific Interactions.* Princeton, NJ: Princeton University Press.

Tilman, D., C. L. Lehman, and P. Kareiva. 1997. "Population dynamics in spatial habitats." In *Spatial Ecology: The Role of Space in Population Dynamics and Interspecific Interactions*, edited by D. Tilman and P. Kareiva, 3–20. Princeton: Princeton University Press.

Tinker, D. B., W. H. Romme, and D. G. Despain. 2003. "Historic range of variability in landscape structure in subalpine forests of the Greater Yellowstone Area, USA." *Landscape Ecology* 18 (4): 427–39. doi: 10.1023/a:1026156900092.

Tjørve, E. 2010. "How to resolve the SLOSS debate: Lessons from species-diversity models." *Journal of Theoretical Biology* 264 (2): 604–12.

Toronto Pollinator Protection Strategy. 2017. https://www.toronto.ca/wp-content/uploads/2017/11/9819-Toronto-Pollinator-Strategy-Booklet-Draft-Priorities-and-Actions-2017.pdf. Accessed 19 June 2018.

Toronto Ravine Strategy. 2017. https://www.toronto.ca/wp-content/uploads/2017/10/9183-TorontoRavineStrategy.pdf.

Torres, A., J. A. G. Jaeger, and J. C. Alonso. 2016. "Assessing large-scale wildlife responses to human infrastructure development." *Proceedings of the National Academy of Sciences* 113 (30): 8472–77.

Town of Canmore, Town of Banff, and Alberta Environment and Parks. 2018. Human–wildlife coexistence: Recommendations for improving human–wildlife coexistence in the Bow Valley. ISBN 978-1-4601-4005-5, http://banff.ca/DocumentCenter/View/5520. Accessed 25 June 2018.

Townsend, P. A., and K. L. Masters. 2015. "Lattice-work corridors for climate change: A conceptual framework for biodiversity conservation and social-ecological resilience in a tropical elevational gradient." *Ecology and Society* 20 (2). doi: 10.5751/es-07324-200201.

Tracey, J. A., S. N. Bevins, S. VandeWoude, and K. R. Crooks. 2014. "An agent-based movement model to assess the impact of landscape fragmentation on disease transmission." *Ecosphere* 5 (9). doi: 10.1890/es13-00376.1.

Tracey, J. A., J. Zhu, E. Boydston, L. Lyren, R. N. Fisher, and K. R. Crooks. 2013. "Mapping behavioral landscapes for animal movement: A finite mixture modeling approach." *Ecological Applications* 23 (3): 654–69.

Travis, S. E., J. E. Marburger, S. Windels, and B. Kubátová. 2010. "Hybridization dynamics of invasive cattail (*Typhaceae*) stands in the Western Great Lakes Region of North America: A molecular analysis." *Journal of Ecology* 98 (1): 7–16.

Treadway, C. 2017. Berkeley Hills: Tilden Park road will close to protect amorous newts. *East Bay Times.* https://www.eastbaytimes.com/2017/10/27/berkeley-hills-tilden-park-road-will-close-to-protect-amorous-newts/. Accessed 3 July 2018: Bay Area News Group.

Treml, E. A., J. J. Roberts, Y. Chao, P. N. Halpin, H. P. Possingham, and C. Riginos. 2012. "Reproductive output and duration of the pelagic larval stage determine seascape-wide connectivity of marine populations." *Integrative and Comparative Biology* 52 (4): 525–37. doi: 10.1093/icb/ics101.

Treves, A., L. Naughton-Treves, E. K. Harper, D. J. Mladenoff, R. A. Rose, T. A. Sickley, and A. P. Wydeven. 2004. "Predicting human–carnivore conflict: A spatial model derived from 25 years of data on wolf predation on livestock." *Conservation Biology* 18 (1): 114–25. doi: 10.1111/j.1523-1739.2004.00189.x.

Treves, A., R. B. Wallace, L. Naughton-Treves, and A. Morales. 2006. "Co-managing human–wildlife conflicts: A review." *Human Dimensions of Wildlife* 11 (6): 383–96.

Trombulak, S. D., and R. F. Baldwin, eds. 2010. *Landscape-Scale Conservation Planning.* New York: Springer.

Tucker, M. A., K. Böhning-Gaese, W. F. Fagan, J. M. Fryxell, B. Van Moorter, S. C. Alberts, A. H. Ali, A. M. Allen, N. Attias, and T. Avgar. 2018. "Moving in the Anthropocene: Global reductions in terrestrial mammalian movements." *Science* 359 (6374): 466–69.

Tucker, Nigel I. J. 2000. "Linkage restoration: Interpreting fragmentation theory for the design of a rainforest linkage in the humid Wet Tropics of north-eastern Queensland." *Ecological Management and Restoration* 1 (1): 35–41. doi: 10.1046/j.1442-8903.2000.00006.x.

Turchin, P. 2003. *Complex Population Dynamics: A Theoretical/Empirical Synthesis.* Princeton: Princeton University Press.

Turner, M. G. 1989. "Landscape ecology: The effect of pattern on process." *Annual Review of Ecology and Systematics* 20: 171–97.

———. 2005. "Landscape ecology: What is the state of the science?" *Annual Review of Ecology, Evolution, and Systematics* 36: 319–44.

Turpie, J. K., C. Marais, and J. N. Blignaut. 2008. "The working for water programme: Evolution of a payments for ecosystem services mechanism that addresses both poverty and ecosystem service delivery in South Africa." *Ecological Economics* 65 (4): 788–98.

Tuxbury, S. M., and M. Salmon. 2005. "Competitive interactions between artificial lighting and natural cues during seafinding by hatchling marine turtles." *Biological Conservation* 121 (2): 311–16. doi: 10.1016/j.biocon.2004.04.022.

Tynan, C. T. 1998. "Ecological importance of the southern boundary of the Antarctic Circumpolar Current." *Nature* 392 (6677): 708.

Urban, D., and T. Keitt. 2001. "Landscape connectivity: A graph-theoretic perspective." *Ecology* 82 (5): 1205–18.

Urban, D. L., R. V. O'Neill, H. H., Shugart Jr. 1987. "Landscape Ecology: A hierarchical perspective can help understand spatial patterns." *BioScience* 37: 119–27.

USDA/USDI. 1994. Standards and guidelines for management of habitat for late-successional and old-growth forest related species within the range of the northern spotted owl. Washington, DC: US Department of Agriculture, Bureau of Land Management, and US Department of Interior, Forest Service.

Valhalla Wilderness Society. 2017. Newsletter No. 60. https://www.vws.org/wp-content/up loads/2018/02/VWSNewsL-2017FINALcolor-ilovepdf-compressed.pdf.

Valiela, I., J. L. Bowen, and J. K. York. 2001. "Mangrove forests: One of the world's threatened major tropical environments: At least 35% of the area of mangrove forests has been lost in the past two decades, losses that exceed those for tropical rain forests and coral reefs, two other well-known threatened environments." *AIBS Bulletin* 51 (10): 807–15.

Van Bohemen, H. 2002. "Infrastructure, ecology and art." *Landscape and Urban Planning* 59 (4): 187–201. doi: 10.1016/s0169-2046(02)00010-5.

Van Bohemen, H. D. 1996. Mitigation and compensation of habitat fragmentation caused by roads: Strategy, objectives, and practical measures. Transportation Research Record, no. 1475. Delft, Netherlands: Ministry of Transport, Public Works, and Water Management.

Van der Hoop, J. M., A. S. M. Vanderlaan, T. V. N. Cole, A. G. Henry, L. Hall, B. Mase-Guthrie, T. Wimmer, and M. J. Moore. 2015. "Vessel strikes to large whales before and after the 2008 Ship Strike Rule." *Conservation Letters* 8 (1): 24–32.

Van der Ree, R., and A. F. Bennett. 2003. "Home range of the squirrel glider (*Petaurus norfolcensis*) in a network of remnant linear habitats." *Journal of Zoology* 259 (4): 327–36. doi: 10.1017/s0952836902003229.

Van der Ree, R., A. F. Bennett, and D. C. Gilmore. 2004. "Gap-crossing by gliding marsupials: Thresholds for use of isolated woodland patches in an agricultural landscape." *Biological Conservation* 115 (2): 241–49. doi: 10.1016/s0006-3207(03)00142-3.

Van der Ree, R., D. J. Smith, and C. Grilo. 2015. *Handbook of Road Ecology*. Hoboken: John Wiley.

Van Horne, B. 1983. "Density as a Misleading Indicator of Habitat Quality." *Journal of Wildlife Management* 47 (4): 893–901.

Van Meter, K. J., P. Van Cappellen, and N. B. Basu. 2018. "Legacy nitrogen may prevent achievement of water quality goals in the Gulf of Mexico." *Science*: eaar4462.

Van Schalkwyk, J., J. S. Pryke, and M. J. Samways. 2017. "Wide corridors with much environmental heterogeneity best conserve high dung beetle and ant diversity." *Biodiversity and Conservation* 26 (5): 1243–56.

Van Valen, L. M. 1973. "Pattern and the balance of nature." *Evolutionary Theory* 1: 31–49.

Vannote, R. L., G. W. Minshall, K. W. Cummins, J. R. Sedell, and C. E. Cushing. 1980. "The river continuum concept." *Canadian Journal of Fisheries and Aquatic Sciences* 37 (1): 130–37.

Velasquez, L. S. 2012. Greenroof and greenwall projects database. Available from http://www.greenroofs.com/projects. Accessed 5 July 2018.

Venter, O., E. W. Sanderson, A. Magrach, J. R. Allan, J. Beher, K. R. Jones, H. P. Possingham, W. F. Laurance, P. Wood, and B. M. Fekete. 2016. "Sixteen years of change in the global terrestrial human footprint and implications for biodiversity conservation." *Nature Communications* 7: 12558.

Villarreal, E. L., and L. Bengtsson. 2005. "Response of a Sedum green-roof to individual rain events." *Ecological Engineering* 25 (1): 1–7.

Vos, C. C., P. Berry, P. Opdam, H. Baveco, B. Nijhof, J. O'Hanley, C. Bell, and H. Kuipers. 2008. "Adapting landscapes to climate change: Examples of climate-proof ecosystem networks and priority adaptation zones." *Journal of Applied Ecology* 45 (6): 1722–31. doi: 10.1111/j.1365-2664.2008.01569.x.

Vucetich, J. A., J. T. Bruskotter, and M. P. Nelson. 2015. "Evaluating whether nature's intrinsic value is an axiom of or anathema to conservation." *Conservation Biology* 29 (2): 321–32.

Wade, A. A., K. S. McKelvey, and M. K. Schwartz. 2015. "Resistance-surface-based wildlife conservation connectivity modeling: Summary of efforts in the United States and guide for practitioners." *General Technical Report*. RMRS–GTR–333. Fort Collins, CO: US Department of Agriculture, Forest Service, Rocky Mountain Research Station. 93.

Wade, T. G., K. Riitters, J. D. Wickham, and K. B. Jones. 2003. "Distribution and causes of global forest fragmentation." *Conservation Ecology* 7 (2). doi: 10.5751/es-00530-070207.

Waits, L. P., and C. W. Epps. 2015. "Population genetics and wildlife habitat." In *Wildlife Habitat Conservation: Concepts, Challenges, and Solutions*, edited by M. L. Morrison and H. A. Mathewson, 63. Baltimore: Johns Hopkins University Press.

Walker, B., and D. Salt. 2006. *Resilience Thinking: Sustaining Ecosystems and People in a Changing World*. Washington, DC: Island Press.

Walker, B. H. 1992. "Biodiversity and ecological redundancy." *Conservation Biology* 6 (1): 18–23. doi: 10.1046/j.1523-1739.1992.610018.x.

Walker, R., and L. Craighead. 1997. "Analyzing wildlife movement corridors in Montana using GIS." In *ESRI User Conference Proceedings*. California. http://gis.esri.com/library/userconf/proc97/proc97/abstract/a116.htm.

Wallace, J. B. 1997. "Multiple trophic levels of a forest stream linked to terrestrial litter inputs." *Science* 277 (5322): 102–104. doi: 10.1126/science.277.5322.102.

Wang, J. 2004. "Application of the one-migrant-per-generation rule to conservation and management." *Conservation Biology* 18 (2): 332–43. doi: 10.1111/j.1523-1739.2004.00440.x.

Ward, J. V., K. Tockner, D. B. Arscott, and C. Claret. 2002. "Riverine landscape diversity." *Freshwater Biology* 47 (4): 517–39.

Watson, J. E. M., N. Dudley, D. B. Segan, and M. Hockings. 2014. "The performance and potential of protected areas." *Nature* 515 (7525): 67.

Watson, J. E. M., M. Rao, K. Ai-Li, and X. Yan. 2012. "Climate change adaptation planning for biodiversity conservation: A review." *Advances in Climate Change Research* 3 (1): 1–11.

Weaver, J. 2015. Vital lands, sacred lands: Innovative conservation of wildlife and cultural values, Badger–Two Medicine area, Montana. WCS Canada Working Paper, 44. ISSN 1530-4426.

Weimerskirch, H., T. Guionnet, J. Martin, S. A. Shaffer, and D. P. Costa. 2000. "Fast and fuel efficient? Optimal use of wind by flying albatrosses." *Proceedings of the Royal Society of London. Series B: Biological Sciences* 267 (1455): 1869–74. doi: 10.1098/rspb.2000.1223.

Weiss, S. B. 1999. "Cars, cows, and checkerspot butterflies: Nitrogen deposition and management of nutrient-poor grasslands for a threatened species." *Conservation Biology* 13 (6): 1476–86. doi: 10.1046/j.1523-1739.1999.98468.x.

West, D. R., J. S. Briggs, W. R. Jacobi, and J. F. Negrón. 2014. "Mountain pine beetle–caused mortality over eight years in two pine hosts in mixed-conifer stands of the southern Rocky Mountains." *Forest Ecology and Management* 334: 321–30.

Westemeier, R. L. 1998. "Tracking the long-term decline and recovery of an isolated population." *Science* 282 (5394): 1695–98. doi: 10.1126/science.282.5394.1695.

Whiteley, A. R., S. W. Fitzpatrick, W. C. Funk, and D. A. Tallmon. 2015. "Genetic rescue to the rescue." *Trends in Ecology & Evolution* 30 (1): 42–49.

Whittinghill, L. J., D. B. Rowe, R. Schutzki, and B. M. Cregg. 2014. "Quantifying carbon sequestration of various green roof and ornamental landscape systems." *Landscape and Urban Planning* 123: 41–48.

Whittington, J., M. Hebblewhite, N. J. DeCesare, L. Neufeld, M. Bradley, J. Wilmshurst, and M. Musiani. 2011. "Caribou encounters with wolves increase near roads and trails: A time-to-event approach." *Journal of Applied Ecology* 48 (6): 1535–42.

Wiens, J. A. 1992. "What is landscape ecology, really?" *Landscape Ecology* 7 (3): 149–50.

———. 2007. *Foundation Papers in Landscape Ecology*. New York: Columbia University Press.

Wikramanayake, E., M. McKnight, E. Dinerstein, A. Joshi, B. Gurung, and D. Smith. 2004. "Designing a conservation landscape for tigers in human-dominated environments." *Conservation Biology* 18 (3): 839–44.

Wilcove, D., M. Bean, R. Bonnie, and M. McMillan. 1996. *Rebuilding the Arc: Toward a More Effective Endangered Species Act for Private Land*. Washington, DC: Environmental Defense Fund.

Wilcox, B. A. 1978. "Supersaturated island faunas: A species-age relationship for lizards on post-Pleistocene land-bridge islands." *Science* 199 (4332): 996–98. doi: 10.1126/science .199.4332.996.

Wiley, D. N., M. Thompson, R. M. Pace III, and J. Levenson. 2011. "Modeling speed restrictions to mitigate lethal collisions between ships and whales in the Stellwagen Bank National Marine Sanctuary, USA." *Biological Conservation* 144 (9): 2377–81.

Williams, J. C., C. S. ReVelle, and S. A. Levin. 2004. "Using mathematical optimization models to design nature reserves." *Frontiers in Ecology and the Environment* 2 (2): 98. doi: 10 .2307/3868216.

Williams, N. S. G., J. Lundholm, and J. S. MacIvor. 2014. "Do green roofs help urban biodiversity conservation?" *Journal of Applied Ecology* 51 (6): 1643–49.

Williams, P., L. Hannah, S. Andelman, G. Midgley, M. Araujo, G. Hughes, L. Manne, E. Martinez–Meyer, and R. Pearson. 2005. "Planning for climate change: Identifying minimum-dispersal corridors for the Cape proteaceae." *Conservation Biology* 19 (4): 1063–74. doi: 10.1111/j.1523-1739.2005.00080.x.

Wilson, D. S. 1992. "Complex interactions in metacommunities, with implications for biodiversity and higher levels of selection." *Ecology* 73 (6): 1984–2000. doi: 10.2307/1941449.

Wilson, E. O. 1988. "The current state of biological diversity." In *Biodiversity*, edited by E. O. Wilson and F. M. Peter, 3–18. Washington, DC: National Academy Press.

———. 2016. *Half-Earth: Our Planet's Fight for Life*. New York: W.W. Norton.

Wilson, J. R., S. Lomonico, D. Bradley, L. Sievanen, T. Dempsey, M. Bell, S. McAfee, C. Costello, C. Szuwalski, and H. McGonigal. 2018. "Adaptive comanagement to achieve climate-ready fisheries." *Conservation Letters*: e12452.

Wilson, K., R. L. Pressey, A. Newton, M. Burgman, H. Possingham, and C. Weston. 2005. "Measuring and incorporating vulnerability into conservation planning." *Environmental Management* 35 (5): 527–43. doi: 10.1007/s00267-004-0095-9.

Wojcik, V. A., and S. Buchmann. 2012. "Pollinator conservation and management on electrical transmission and roadside rights-of-way: A review." *Journal of Pollination Ecology* 7.

Wolff, J. O. 1999. "Behavioral model systems." In *Landscape Ecology of Small Mammals*, edited by G. W. Barrett and J. D. Peles, 11–40. New York: Springer Verlag.

Wolff, J. O., and W. Z. Lidicker Jr. 1980. "Population ecology of the taiga vole, *Microtus xanthognathus*, in interior Alaska." *Canadian Journal of Zoology* 58 (10): 1800–1812. doi: 10.1139/z80-247.

Wolff, J. O., E. M. Schauber, and W. D. Edge. 1997. "Effects of habitat loss and fragmentation on the behavior and demography of gray-tailed voles." *Conservation Biology* 11 (4): 945–56. doi: 10.1046/j.1523-1739.1997.96136.x.

Wood, P. A., and M. J. Samways. 1991. "Landscape element pattern and continuity of butterfly flight paths in an ecologically landscaped botanic garden, Natal, South Africa." *Biological Conservation* 58 (2): 149–66. doi: 10.1016/0006-3207(91)90117-r.

Woodson, C. B., and S. Y. Litvin. 2015. "Ocean fronts drive marine fishery production and biogeochemical cycling." *Proceedings of the National Academy of Sciences* 112 (6): 1710–15.

Woodson, C. B., M. A. McManus, J. A. Tyburczy, J. A. Barth, L. Washburn, J. E. Caselle, M. H. Carr, D. P. Malone, P. T. Raimondi, and B. A. Menge. 2012. "Coastal fronts set recruitment and connectivity patterns across multiple taxa." *Limnology and Oceanography* 57 (2): 582–96.

Worboys, G. L., W. L. Francis, and M. Lockwood (eds.). 2010. *Connectivity Conservation Management: A Global Guide*. London: Earthscan.

Worboys, G. L., and M. Lockwood. 2010. "Connectivity conservation management framework and key tasks." In *Connectivity Conservation Management: A Global Guide*, edited by G. L. Worboys, W. L. Francis, and M. Lockwood. London: Earthscan.

Worboys, G. L., M. Lockwood, A. Kothari, S. Feary, and I. Pulsford. 2015. *Protected Area Governance and Management*. Canberra: ANU Press.

World Bank Group. 2017. Terrestrial protected areas (% of total land area). https://data .worldbank.org/indicator/ER.LND.PTLD.ZS?end=2014&start=1990&view=chart& year=2014. Accessed 23 April 2018.

Wright, J. P., C. G. Jones, and A. S. Flecker. 2002. "An ecosystem engineer, the beaver, increases species richness at the landscape scale." *Oecologia* 132 (1): 96–101.

Wu, J. 2013. "Key concepts and research topics in landscape ecology revisited: 30 years after the Allerton Park workshop." *Landscape Ecology* 28 (1): 1–11.

Wu, Y. J., S. G. DuBay, R. K. Colwell, J. H. Ran, and F. M. Lei. 2017. "Mobile hotspots and refugia of avian diversity in the mountains of south-west China under past and contemporary global climate change." *Journal of Biogeography* 44 (3): 615–26. doi: 10.1111 /jbi.12862.

Wuebbles, D. J., D. W. Fahey, K. A. Hibbard, B. DeAngelo, B. S. Doherty, K. Hayhoe, R. Horton et al. 2017. "Executive summary." In *Climate Science Special Report: Fourth National Climate Assessment*, Vol. 1., edited by D. J. Wuebbles, D. W. Fahey, K. A. Hibbard, D. J. Dokken, B. C. Stewart, and T. K. Maycock, 12–34. Washington, DC: US Global Change Research Program.

Wyrtki, K. 1975. "El Niño—the dynamic response of the equatorial Pacific Ocean to atmospheric forcing." *Journal of Physical Oceanography* 5 (4): 572–84.

Yang, J., Q. Yu, and P. Gong. 2008. "Quantifying air pollution removal by green roofs in Chicago." *Atmospheric Environment* 42 (31): 7266–73.

Ye, X., A. K. Skidmore, and T. Wang. 2013. "Within-patch habitat quality determines the resilience of specialist species in fragmented landscapes." *Landscape Ecology* 28 (1): 135–47.

Young, M. A., L. W. Wedding, and M. H. Carr. Forthcoming. "Applying landscape ecology to evaluate the design of marine protected area networks." In *Seascape Ecology: Taking Landscape Ecology into the Sea*, edited by S. Pittman. Chichester, West Sussex, UK: John Wiley.

Zanette, L., and B. Jenkins. 2000. "Nesting success and nest predators in forest fragments: A study using real and artificial nests." *The Auk* 117 (2): 445. doi: 10.1642/0004-8038 (2000)117[0445:nsanpi]2.0.co;2.

Zeller, K. A., K. McGarigal, and A. R. Whiteley. 2012. "Estimating landscape resistance to movement: A review." *Landscape Ecology* 27 (6): 777–97. doi: 10.1007/s10980-012-9737-0.

Zeller, K. A., S. Nijhawan, R. Salom-Pérez, S. H. Potosme, and J. E. Hines. 2011. "Integrating occupancy modeling and interview data for corridor identification: A case study for jaguars in Nicaragua." *Biological Conservation* 144 (2): 892–901.

Zhu, K., C. W. Woodall, and J. S. Clark. 2012. "Failure to migrate: Lack of tree range expansion in response to climate change." *Global Change Biology* 18 (3): 1042–52. doi: 10.1111/j.1365-2486.2011.02571.x.

Zube, E. H. 1995. "Greenways and the US National Park system." *Landscape and Urban Planning* 33 (1–3): 17–25. doi: 10.1016/0169-2046(94)02011-4.

An expert on wildlife corridors, DR. JODI HILTY is the president and chief scientist of the Yellowstone to Yukon Conservation Initiative. For over twenty years she has worked to advance conservation by leading research projects, engaging in community-based conservation, and promoting policy and management changes. In the last fifteen years, she has focused her work on North America: among other accomplishments her team successfully established the first federally designated wildlife corridor in the United States, the Path of the Pronghorn in Wyoming, and led the science that served to inform the expansion of Nahanni National Park Reserve in the Northwest Territories to more than 6.5 times its former size. Dr. Hilty also has been coeditor or lead author on three books, most recently, *Climate and Conservation: Landscape and Seascape Science, Planning, and Action* (2012). She fills a range of advisory roles and is vice chair for North America of the International Union for Conservation of Nature (IUCN) connectivity committee.

DR. ANNIKA KEELEY is a wildlife ecologist with expertise in landscape connectivity science. She earned her PhD at the School of Forestry, Northern Arizona University. She has published several scientific research articles on aspects of wildlife corridors, ecosystem science, and animal behavior. She is currently a postdoctoral scholar at the University of California, Berkeley, systematically reviewing the literature on the intersection of connectivity and climate change science, as well as exploring the challenges and opportunities to implementing connectivity.

DR. WILLIAM LIDICKER is an ecologist, conservation biologist, and vertebrate biologist. With more than sixty years in academia, he brings extraordinary scholarship to bear on our subject. His research interests include population dynamics, social behavior, population genetics, community ecology, mammalian systematics, evolution, and landscape ecology. During his tenure at the University of California, Berkeley, he has published extensively and taught courses encompassing these and other interests. A recent paper by Dr. Lidicker concerns how to make conservation more effective. He has served as president of the American Society of Mammalogists and the International Federation of Mammalogists and chaired IUCN committees. He also was elected a Foreign Member of the Polish Academy of Sciences. Thus, he has an extensive worldwide network of contacts. He is currently professor of Integrative Biology and Curator of Mammals (Museum of Vertebrate Zoology) Emeritus.

DR. ADINA MERENLENDER is a Cooperative Extension Specialist at University of California, Berkeley, in the department of Environmental Science, Policy, and Management, and is an internationally recognized conservation biologist working on environmental problem solving at the landscape scale. She has published over 100 scientific research articles focused on underlying relationships between land use and biodiversity. She has trained graduate students as well as worked with decision makers to address the forces that influence loss of biodiversity, including the use of spatially explicit decision support systems. She recently founded and directs the UC California Naturalist program to foster a committed corps of volunteer naturalists and citizen scientists trained and ready to take an active role in natural resource conservation, education, and restoration. See more at http://ucanr.org/sites/merenlender.

Invited Authors for Chapter 9

DR. MARK CARR is a professor of marine ecology in the department of Ecology and Evolutionary Biology at the University of California, Santa Cruz. He received his PhD from the University of California, Santa Barbara, in 1991. Mark's research focuses on coastal marine ecosystems and the ecology of marine fishes. His research informs marine ecosystem-based management, including the design and evaluation of marine protected area (MPA) networks.

DR. ELLIOTT HAZEN is a research ecologist at the NOAA Southwest Fisheries Science Center in Monterey, California. He received his PhD from Duke University in 2008. He combines technologies, including fisheries acoustics, bio-logging tags, and oceanographic data with spatial statistics in his research. His studies aim to tease apart ocean, climate, and trophic effects on marine ecosystems.

INDEX

A

adaptation corridors, 128

adaptive evolution, 10

African Wildlife Foundation (AWF), 264–265

agreements, conservation, 248–251

agricultural systems, 113–114

Aichi Biodiversity Targets, 6

Alabama beach mice (*Peromyscus polionotus ammobates*), 85

Allee Effect, 40

Alligator Alley (Florida), 103

Amazon River basin, 143, 148

Amboseli National Park (Kenya), 263–266, 265f

American martens (*Martes americana*), 83

amphibians, 84, 178, 240–241, 242b, 243f. *See also Specific animals*

antagonistic relationships, 124

Antarctic Circumpolar Current, 220

anti-regulating forces, 39–40, 42f

aquatic environments, 21, 188. *See also* Freshwater environments; Marine environment

ArcGIS, 166

area centroids, 169

Asiatic lions (*Panthera leo leo*), 14, 60

assessment of connectivity, 186–193

assisted evolution, 122

assisted migration, 11, 13, 122

Atlantic bluefin tuna (*Thunnus thynnus*), 220, 221f, 222

AWF. *See* African Wildlife Foundation

B

Bachman's sparrows (*Aimophila aestivalis*), 109

Baekdu Daegan Mountain System Act, 244–245

balanced polymorphisms, 45

Banff National Park (Canada), 1, 100, 110–111, 111f, 132f, 147b, 162, 270

bank voles (*Clethrionomys glareolus*), 85

barriers, 17, 70, 125–127, 129–130

bats, 154–155

BDFFP. *See* Biological Dynamics of Forest Fragments Project

beach mice (*Peromyscus polionotus*), 74

behavioral factors, , 123–124

Belding's ground squirrels (*Urocitellus beldingi*), 200

Bhutan, 268–269, 269f

bighorn sheep (*Ovis canadensis*), 112

biodiversity: Bhutan and, 269; connectivity for, 99; corridors and, 103b, 104; habitat loss, fragmentation and, 7; human-induced fragmentation and, 58–63; metacommunities and, 48; negative effects of linkages and, 160; protected areas and, 13; restoration and, 253; riparian areas and, 141

Biological Dynamics of Forest Fragments Project (BDFFP), 59–60

birds: corridor dimensions and, 137; edge zones and, 78–79, 148; genetics, species extinction and, 66; green roofs and, 244b; habitat fragmentation and, 65, 65b; habitat quality and, 135–136; habitat requirements and, 120; hybridization and, 158; matrix, fragmentation and, 68; riparian areas and, 101, 141; risks from roads and, 99–100. *See also Specific birds*

birth rates, population growth and, 36–37

black bears (*Ursus americanus*), 74, 102, 103

black lion tamarins (*Leon topithecus chrysopygus*), 253

black-footed ferrets (*Mustela nigripes*), 66

black-tailed prairie dogs (*Cynomys ludovicianus*), 152

339

blind snakes (*Leptotyphlops dulcis*), 72
bobolinks (*Dolichonyx oryzivorus*), 78
bog turtles (*Glyptemys muhlenbergii*), 121
Boston Emerald Necklace, 98
Bow Valley (Alberta), 147b
Bowhead whales (*Balaena mysticetus*), 223–225
Braulio Carrillo National Park (Costa Rica), 92
brood parasitism, 124, 155, 156f
brown-headed cowbirds (*Molothrus ater*), 124, 155, 156f, 252
buffers, riparian corridors and, 205
butterflies, 44, 64, 133, 202

C
Cabinet Purcell Mountains, 270
California Current, 222, 229
California Gap Analysis program, 166
California Undercurrent, 230
California voles (*Microtus californicus*), 61, 86–87
camera traps, 170, 187
Canada lynx (*Lynx canadensis*), 131
Canmore, Alberta, Canada, 147b
Cap-and-Trade programs, 209
carbon-stock corridors, 209–210
Caribbean Challenge Initiative (CCI), 266–267, 267f
caribou (*Rangifer tarandus*), 153
carnivores: corridor quality and, 149; design objectives and, 125; as umbrella species, 120. *See also Specific animals*
cascading effects, 64
CCI. *See* Caribbean Challenge Initiative
cheetahs (*Acinonyx jubatus*), 110
chipmunks (*Tamias striatus*), 129
circuit theory, 174–175, 174b, 175f, 193–194
climate analogs, 198–199
climate change: benefits of resilience and, 11, 185–186; coastal marine environment and, 234; corridor assessment and, 188, 189f; corridors and, 111–112; design objectives and, 127–129; designing connectivity for, 205–210, 214–215; Great Eastern Ranges Corridor and, 108;

hybridization, introgression and, 159; oceans and, 225–226, 234; overview of, 9–13, 195–196; reduced genetic variation and, 46; refugia and, 201f, 210; species distribution modeling and, 210–213
Climate Linkage Mapper, 206
climate refugia, 199–200, 210, 236
climate space: climate analogs and, 198–199; climate velocity and, 196–198, 197f; range dynamics and, 200–202; range expansion and, 202–204, 204t; refugia and, 199–200
climate velocity, 196–198, 197f
climate-biota mismatches, 127
coarse-filter approaches, 12–13, 172–173
coastal marine environment: connectivity and corridors in, 226–232; threats to species, ecosystems, and connectivity of, 233–234
collaborations: establishing, 164–165; importance of, 239–240; large-scale, 241–242; prioritization and, 185–186
colonists, defined, 32
colonization, island biogeography and, 21–23
colonizing fronts, 201
communities, 47. *See also* Metacommunities
community relaxation, 23, 24f
commuting corridors, defined, 18b
conflicting ecological objectives, 160–161
conflicts, human-wildlife, 14, 63
connectivity, 18b, 56, 59–61, 91. *See also* Ecological connectivity
connectivity planning: addressing scale and, 165–167; behavioral factors and, 123–124; changing approach to, 11; climate change and, 205–210; collaborations and, 164–165; conflicting ecological objectives and, 160–161; continuous corridors and, 129–132; corridor dimensions and, 136–139; dispersal and, 122; ecological networks for conservation and, 144–145, 144f; focal species and, 117–120; generalists vs. specialists and, 123; habitat quality and, 134–136; habitat requirements and, 120–121; human activity and, 124–125; hydrologic habi-

tats and, 142–143; identifying corridors for conservation and restoration and, 167–178; landscape configuration and, 139–141; overview of, 116–117; physical limitations and, 125–127; prioritization and, 178–185; riparian corridors and, 141–142; stepping-stone connectivity and, 132–134; topography, microclimate and, 127–129

conservation agreements, 248–251

conservation easements, 162, 249–250

conservation network planning, 209–210

Conservation Reserve Program (USDA), 250

Conserving Nature's Stage, 208, 209

continuous corridors, 129–132

corridors: assessing, 186–193; benefits to humans, 112–115; biodiversity and, *See* biodiversity; biological benefits of, 108–112; climate change and, *See* climate change; complexities of, 103b, 104–108; definitions and descriptions of, 18b, 90–93; design of, *See* connectivity planning; dimensions of, 136–139; ecological connectivity and, 16–17; ecological networks and, 15, 106–108, 144–145, 144f; as filters, 123, 148–151; for individual species conservation, 102–104; matrix and, 70; overview of, 14–16, 17, 18b, 90–93, 91f; planning and, *See* connectivity planning; potential problems with, *See* pitfalls/disadvantages; riparian, 101–102; site selection and, 167–178, 168b; types of, 18b, 93–101, 128–132, 143, 173, 206–210

cotton rats (*Sigmodon hispidus*), 80

Coxen's Fig Parrot Rainforest Restoration Project (Australia), 253

Crater Lakes National Park (Australia), 92–93

culverts, 126, 189f. *See also* Underpasses

D

dams, 143, 152

deer (*Odocoileus virginianus*, *Cervus elaphus*), 76–77, 110

demes, metapopulations and, 25, 31b

demographic processes: of extinction,

36–40, 37f; island biogeography and, 22; marine connectivity and, 232; metapopulations and, 30; negative with linkages, 155–157

demographic stochasticity, 39

design objectives. *See* Connectivity planning

dimensions of corridors, 136–139

directionality, 127–128

disadvantages. *See* Pitfalls/disadvantages

diseases. *See* Parasites; Pathogens

dispersal: connectivity and, 109; corridors and, 104–105, 105f; demography of extinction and, 36–37; design objectives and, 122; matrix and, 69–73; metapopulations and, 25–27, 31b; overview of, 30–32; phoresy and, 32, 72–73, 155; range shifts and, 201; reasons for, 35–36, 38t; synchronicity and, 38; terminology of, 32–33

dispersal corridors, defined, 18b

dispersal distance, defined, 33, 34f

dispersers, defined, 32

disseminules, defined, 32, 33

domestic animals effects, 77, 163

Donaghy's Corridor project, 92–93

Douglas fir (*Pseudotsuga menziesii*) forests, 80

drones, 165–166

drought refugia, 210

dual-function landscapes, 69, 245–248

Ducks Unlimited, 250

E

easements, 162, 249–250

eastern kingbirds (*Tyrannus tyrannus*), 85

ecoducts, 110

ecological connectivity, 14–17, 18b

ecological framework: dispersal and, 30–36, 38t; extinction demography and, 36–40, 37f; genetic structuring and, 40–44; island biogeography and, 19–24; landscape and ecoscape concepts and, 52–54; longer-term perspective and, 44–46; metacommunity theory and, 47–52, 50b, 50f; metapopulation process and, 30, 31b; metapopulation theory and, 24–30, 31b; overview of, 19

ecological land units, 173, 207–208, 208f
ecological networks, 15, 106–108, 144–145, 144f, 275–276
ecological resilience. *See* Resilience
ecological response models, 192
ecologically/biologically significant areas (EBSA), 220
economic issues, 161–163, 182–183
ecoscapes, 53, 53t
ecosystem 52 [definition]
ecosystem services, 8, 51, 112, 113
ecotonal edge effects, 80–81, 81f
ecotones, defined, 74
edge effects: theory 59; overview 73-84; birds 65; species loss 67; avoid 128, 139; impacts 148
edges: design objectives and, 123; encroachment and, 75, 75f; fragmentation and, 73–82; graph theory and, 179–181; negative effects of, 148, 152; range shifts and, 200–202
education programs, 246–247.
El Niño Southern Oscillation (ENSO), 221, 229
elephants, 263–266
emergent diseases, 154
emigrants, defined, 32
emigration rates, 36–37, 87–89
encroachment, 75, 75f
enduring features, 173, 207–208, 208f
ENSO. *See* El Niño Southern Oscillation
epistatic complexes, 45
equatorial upwelling zone, 220
equilibrium, island biogeography and, 20–21, 21f
equilibrium densities, anti-regulating forces and, 42f
estuaries, 234
eucalyptus (*Eucalyptus salmonophloia*), 64, 133
European common voles (*Microtus arvalis*), 86–87
European Green Belt, 97f
European mink (*Mustela lutreola*), 182
Everglades National Park (Florida), 100
evolution, 10, 22, 33–34, 122
exotic species, 76, 120, 135, 151–153, 251–252

expanding edges, 201
expert opinion models, 170
ex-situ refugia, 200, 210
extinction of species: demography of, 36–40, 37f; habitat loss, fragmentation and, 7; island biogeography and, 20, 21f; metapopulations and, 25–27; outbreeding depression and, 45–46
extinction debt, 72

F

fecal surveys, 170
fencerows as corridors, 95, 135
fencing, 126–127, 126f, 131, 162–163, 270
Fender's blue butterfly (*Icaricia icarioides fenderi*), 133
feral animals, 77
fertilizers, edge zones and, 77
filters, corridors as, 123, 148–151
fine-filter approaches, 12–13, 170–172, 210–213
Fitzgerald Biosphere Project, 246
flagship species, 117, 118, 241
Flathead Indian Reservation (Montana), 110, 270
flickers (*Colaptes auratus*), 158
floodplains, 141
Florida scrub jays (*Aphelocoma coerulescens*), 28–29, 29f, 78
flying foxes (*Pteropus* spp.), 122, 154–155
flying squirrels (*Pteromys volans*), 68
focal species, 117–120, 125, 187
foraging niches, 218
forward velocity, 198
fragmentation: consequences of human-induced, 58–63, 65b; dispersal and, 33–34; edge effects and, 73–89; genetic considerations, extinctions and, 66–67; matrix and, 67–73, 83–89; metacommunities and, 48; natural vs. human-induced, 55–58; overview of, 55; species composition of patches and, 63–65; speed and pattern of change and, 57–58
freshwater environments, 116, 188–192, 190f
frogs, 178
functional connectivity: corridors for enhancing, 103–104; overview of, 18b,

94–95; in riverscapes, 143–144; stepping-stone connectivity, 132–134
future research and directions, 277–279, 280–281b

G
Gadgarra State Forest (Australia), 92–93
gap: forest 78; sagebrush 107; water 124; built environment 129; size 130; crossing 131, 150;
gatekeeper butterflies (*Pyronia tithonus*), 133
gene flow, metapopulations and, 44
generalists, 123
genetic diversity: island biogeography and, 22; ; marine protected areas and, 236; oceans and, 223; range shifts and, 202; species extinction, fragmentation and, 66
genetic factors: habitat suitability models and, 170, 171; species extinction, fragmentation and, 66–67
genetic structuring: connectivity and, 109; metapopulations and, 30, 31b, 43–44; overview of, 40–41; small populations and, 41–43, 42f; species filtering and, 149–151
geodiversity, 208, 209
geomorphology, 227
geophysical settings, 173, 207–208, 208f
geotaxis, 231
Gir Forest National Park (India), 14
GIS (geographic information system) 167; kernel 169; methods 173; explanatory variables 184;tracking 188; science 190; caveats 193;
global footprint map, 7, 8f
global positioning systems (GPS), 105, 150, 170, 171, 188
global warming. *See* Climate change
GPS. *See* Global positioning systems
Grand Teton National Park, 106, 107f
graph theory, 179–181, 180f
gray wolves (*Canis lupus*), 82, 105, 105f, 125, 153
gray-tailed voles (*Microtus canicaudus*), 130
Great Basin of North America, 23
Great Eastern Ranges Corridor (Australia), 106, 108

greater prairie chickens (*Tympanuchus cupido pinnatus*), 66
Greater Yellowstone Ecosystem: dispersal and, 105; scale and, 166; Yellowstone National Park and, 5, 57f; Yellowstone to Yukon Conservation Initiative and, 108, 131, 269–272, 271f, 272f
green roofs, 241, 244b, 245f
greenways/greenbelts, 98–99, 113
grizzly bears (*Ursus arctos*): fragmentation and, 62f, 63, 63f; humans and, 1, 4, 125, 147b; prioritization and, 183; protected areas and, 14; range shifts and, 1, 159; roads and, 77–78, 110, 130; Yellowstone to Yukon Conservation Initiative and, 270–272, 271f
grosbeaks (*Pheuticus* spp.), 158
Gross National Happiness, 269
Gulf of Mexico dead zone, 233
Gulf Stream, 219, 229
gyres, 230

H
habitat connectivity, defined, 18b
habitats: communities and, 47; corridors and quality of, 134–136; design objectives and, 120–121, 130; human-induced loss of, 7–9; marine connectivity and, 232; modeling suitability of, 170–172, 171t, 172f
Hanski, Ilkka, 44, 46
hardness. *See* Permeability
heat islands, 184
HEC-RAS program, 191
hedgerows, 96f, 113–114, 137
hedonic approach, 183
Hendra virus, 154–155
heterosis, 42–43
honeybees, 43
human activity, design objectives and, 124–125
human-influenced land use, 67. *See also* Matrix
human-wildlife conflict, 14, 63
humpback whales, 218–219, 219f
hybrid zones, 45
hybridization, 158–159, 159f, 223

hydraulic modeling, 190–192
hydrologic connectivity, 142–143, 188–192, 190f

I

identification of corridors, 167–178, 168b
immigrants, defined, 32
immigration rates, population growth and, 36–37
implementation, framework for, 239f
inbreeding, 45
inbreeding depression, 41, 42–43, 109
India, 14, 60
indicator species, 117
Indigenous Protected Areas, 247
individual-based models, 174b, 178
in-situ refugia, 200
integrated modeling frameworks, 188–192, 190f
introgression, 158–159
invasions, 151–155, 251–252
inverse density dependent forces, 39–40, 42f
island biogeography, 19–24, 21f, 48
isolation, 23, 49, 51–52

J

jaguars (*Panthera onca*), 118

K

kelp forests, 64
Kenya, 263–266, 265f
kernel analysis, 169, 194
keystone species, 117
Kilimanjaro Transboundary Landscape (Kenya), 263–266, 265f

L

La Niña, 221, 222–223
La Selva Biological Station (Costa Rica), 92
land facet corridors, 173, 207–208, 208f
Land Trust Alliance, 249
land trusts, 249
land-bridge islands, 23
Landcare, Australia (organization), 246
landscape configuration, 139–141
landscape connectivity, 18b
landscape ecology, 52–53

landscape permeability. *See* Permeability
land-use change, 183–185, 276
lantana (*Lantana camara*), 135
larval dispersal, 226–227, 228f, 230–231, 230f
lattice-work corridors, 206, 207f
laws, 243–244
leading edges, 201
lease agreements, 250–251, 264–266
least-cost path analysis, 173–174, 174b, 175f, 193
legal mechanisms, 243–244
length, corridor quality and, 136–139
Leopold, Aldo, 74
Levins, Richard, 25–27
Levins's classic metapopulations, 28, 28f
LiDAR data, 166
life-history traits, larval dispersal and, 231
light, 74, 77–78, 125, 217–218
linkage, defined, 18b
linkage groups, 45
lions (*Panthera leo*), 157
Lisbon, Portugal, 255–256, 257f
livestock, 163
living fences, 127
lizards, 23, 24f, 64
Los Angeles, California, 1–2
Lumholtz tree kangaroos (*Dendrolagus lumholtzi*), 252

M

Maasai pastoralists, 263–264
MacArthur, R. A., 20
Maine Wildlands Network, 108
mainland-island metapopulations, 27, 28f
management agreements, 250
mangrove deforestation, 233–234
marine environment: Caribbean Challenge Initiative and, 266–267; human-induced change and, 7; island biogeography and, 22–23; overview of, 216–217; protected areas in, 13–14; range shifts and, 201–202; stepping-stone connectivity and, 134f. *See also* Coastal marine environment; Pelagic marine environment
marine protected areas (MPA), 217, 226, 235–237, 266–267, 267f

marlin (*Makaira mazara*), 222–223
matrix: corridor utility and, 140; defined, 4, 18b, 67; differential use of, 72; fragmentation and, 67–73; metapopulations and, 31b; movements and, 69–73; resistance estimation and, 172–173; as resource, 83–84; as secondary habitat, 84–87, 86f; as sink or stopper, 87–89, 87f, 88f
matrix edge effects, 79–81, 81f
Mayacamas Mountains, California, 260, 262f
McKenna, Catherine, 247
meadow voles (*Microtus pennsylvanicus*), 83
Meadoway (Toronto, Ontario), 258, 259f
Mesoamerican Biological Corridor. *See* Paseo Pantera Project
metacommunities: overview of, 47–52, 50b, 50f; pathogen and parasite movement in, 155; stepping-stone connectivity and, 133–134
metapopulations: coastal marine environment and, 235; conceptual history of, 24–30, 31b; demography of extinction and, 36–40; genetic structuring and, 43–44; planning and, 117; processes of, 30, 31b; types of, 27–28, 28f
mice: beach mice (*Peromyscus polionotus*), 74; house mice (*Mus musculus*), 25, 83; salt marsh harvest mice (*Reithrodontomys raviventris*), 83; white-footed mice (*Peromyscus leucopus*), 95, 135, 136
microclimate, 75–76, 128–129, 200, 207
migration: assisted, 11, 13, 122; corridors and, 18, 106, 107f; defined, 32; North Pacific Transition Zone and, 220, 221f; ocean connectivity and, 218–219; patch-matrix interactions and, 84
migratory: birds 95, 156, 244; pollinators 115; ungulates 149; marine 218;
MIKE-11 program, 191
mistletoe (*Amyema miquelii*), 64, 122, 124
mistletoe birds (*Dicaeum hirundinaceum*), 122, 124
mitigation: carbon 10; species 55; transportation 150, 270; measures 162-163; climate 209, 216; power plants 233;
mobility, fragmentation and, 62–63

modeling: approaches to, 173–178, 174b; changing approach to, 11; climate change and, 10; of connectivity and extinctions, 15–16; corridor filtering and, 151; future research and, 278; hydraulic, 190–192, 190f; of larval dispersal, 231–232; limitations of, 193–194; matrix, fragmentation and, 68–69; of matrix permeability, 79–80; of metapopulation analysis, 46; new approaches for, 164; of pathogen and parasite movement, 155; of permeability, 167; of range shifts, 203–204, 210–213; resistance and, 169–173
monitoring, design objectives and, 132
Morro do Diablo State Park (Brazil), 252–253, 254f
mountain caribou (*Rangifer tarandus caribou*), 77
mountain lions (*Puma concolor*): design objectives and, 118–119, 125, 129–130; fragmentation and, 61; permeability and, 3f; range shifts and, 1–2, 76–77; roads and, 78, 110
mountain pine beetles (*Dendroctonus ponderosae*), 11
moving window analysis, 169
movement: facilitate 185; barriers to 186, 194; animal data 188; social 194; marine patterns 216-217,219
MPA. *See* Marine protected areas
muskox (*Ovibos moschatus*), 40
mutualism, 122, 124, 150

N

National Wildlife Corridors Plan, 194
native species, invasions of, 153–155
naturalness, resistance and, 172–173
naturalness-based corridors, 206–207
negative consequences. *See* Pitfalls/disadvantages
nest parasitism, 124, 155, 156f
Netherlands, 101
newts (*Taricha* sp.), 242b, 243f
node-less analysis, 168, 169
nodes, 168–169, 179–181
noise, marine environments and, 224
nonequilibrium metapopulations, 28, 28f

nonnative species. *See* Exotic species
North Pacific Transition Zone (NPTZ),
220, 221f, 222
Norway, 81, 82f
Norwegian lemmings (*Lemmus lemmus*), 83
NPTZ. *See* North Pacific Transition Zone

O
objectives, defining, 167–168
occupancy data, 168
oceans. *See* Marine environment
Olmsted, Frederick Law, 98
omniscape approach, 174b, 175, 176–177f
open populations, 227
open-space systems. *See* Greenways/
   greenbelts
orioles (*Icterus* spp.), 158
otters (*Lontra canadensis*), 104
outbreeding depression, 45–46, 159–160
outreach campaigns, 240–241, 242b
ovenbirds (*Seiurus aurocapillus*), 78
overpasses, 100–101, 110, 132f, 150,
   162–163

P
paleoclimate models, 212
panthers. *See* mountain lion (*Panther concolor*)
parasites, 154–155, 201
partnerships. *See* Collaborations
Paseo Pantera Project (Central America), 106
passive restoration, 251
pastoralism, 263
patch size, 56, 58–59, 60b
pathogens, 25, 49, 152, 154–155, 201
pattern of change, fragmentation and,
   56–58, 57f
pelagic larval duration, 231
pelagic marine environment: connectivity
   and corridors in, 217–219; major corridors of, 219–221, 221f; overview of,
   217; threats to and conservation
   approaches for, 222–226
permeability: corridors as biotic filters and,
   148–150; defined, 18b; edge zones and,
   79; matrix and, 70–72, 70f; matrix as
   sink or stopper and, 87–89; modeling
   of, 167.

Perpetual Ocean visualization, 230
persistence-like index, 182
pesticides, 77, 233
pests, corridors and, 114
Peterson Creek corridor (Australia), 252
phenotypic plasticity, 10
phoresy, 32, 72–73, 155
photic zone, 217
phototaxis, 231
physical environment, 47–48, 125–127,
   186–187
physiological bias, 22
phytoplankton blooms, 218
pied imperial pigeons (*Ducula bicolor*), 61
Pinhook Swamp (Florida), 103
pink salmon (*Onchorhynchus gorbuscha*), 46
pioneer plant species, 77
pitfalls/disadvantages: corridors as biotic
   filters, 148–151; demographic impacts,
   155–157; economic, 161–163; edge effect impacts, 148; introduction of pathogens and parasites, 154–155; invasions
   of exotic species, 151–153; invasions of
   native species, 153–155; negative genetic
   effects, 160–161; overview of, 146–148;
   social behavior and, 157–158
plague, 152
planning. *See* Connectivity planning
plasticity, 10
policy mechanisms, 243–244
pollinators: coffee plantations and, 83,
   115; corridors and, 98, 112, 114–115;
   dispersal and, 33; electricity transmission
   corridors and, 258
pollution, 11, 233, 236, 252
population growth, 7, 42f, 201
population viability analyses, 181–182
predators, 64, 76, 153, 218–219, 252.
   See also Specific predators
presaturation dispersal, 35–36, 37f
prioritization: climate resilience and, 185–
   186; conflicting objectives and, 160–162;
   economic issues and, 182–183; graph
   theory and, 179–181; land-use change
   and, 183–185; overview of, 178–179;
   species persistence and, 181–182; steps
   for, 168b

private lands, 162, 248–251

pronghorn antelope (*Antilocapra americana*), 106, 107f, 125–126

protected areas: Bhutan and, 268–269; changing approach to, 11; limitations to, 13–14; naturalness-based corridors and, 207; reconnecting, 14–16; as stepping stones, 133. *See also* Marine protected areas; Wilderness areas

public, communicating with, 241

Q

qualitative genetic factors, 38t

quantitative genetic factors, 38t

Quaternary Period, 9–10

R

railways, 131

range dynamics, 200–202

range expansion, 202–204, 204t

range shifts, 10, 127–129, 210–213

rasters, 169–170

ratio of optimal to marginal habitat patch area (ROMPA), 85–87, 86f

RCP scenarios, 213

rear edges, 201

recreation, connectivity and, 112–113

recruitment, range shifts and, 201

red deer (*Cervus elaphus*), 110

Red Desert (Wyoming), 106, 107f

red-backed voles (*Clethrionomys californicus*), 61, 123

Reducing Emissions from Deforestation and Forest Degradation (REDD+), 209

redundancy, 8–9, 132

refugia, 199–200, 210, 236

regulating forces, 42f

Representative Concentration Pathways (RCP) scenarios, 213

reproduction, 201, 231

resilience, 8–9, 11, 15, 185–186

resistance: assessing movement and, 169; least-cost analysis and, 173–174, 175f; matrix structure and, 172–173; overview of, 169–170; species-specific estimates of, 170–172, 171t

resistance layers, 169–170

resistant kernel approach, 174b, 175, 178

resolution, 165–166

resource, matrix as, 83–84

resource extraction, 11

resource selection functions, 171t

restoration, 11, 251–253

reverse velocity, 198

rheotaxis, 231

Ribeira das Jardas (Portugal), 255–256, 257f

rice rats (*Oryzomys palustris*), 83–84

richness, island biogeography and, 20–21, 21f

ringtail opossum (*Hemibelideus lemuroides*), 35

riparian areas: climate-wise connectivity and, 205–206, 214; connectivity and, 101–102, 102f; corridor design objectives and, 141–142; invasive species and, 152

risk assessments, 11

river continuum concept, 143

river corridor principles, 143

rivers, 7, 143, 233

roadkill 2, 78, 99, 131, 187, 260, 302

roads: closure of, 242b; collisions and, 78, 162–163; design objectives and, 131; edge zones and, 77–78; mitigating risks of, 99–101; Yellowstone to Yukon Conservation Initiative and, 270

roadside corridors, 95

ROMPA. *See* Ratio of optimal to marginal habitat patch area

root voles (*Microtus oeconomus*), 124

rose-crowned fruit doves (*Ptilinopus regina*), 61

Russian River (California), 12f

S

Saguaro National Park, 78

Saint-Venant equations, 191

salmon (*Onchorhynchus* spp.), 45–46, 74

salt marsh harvest mice (*Reithrodontomys raviventris*), 83

San Francisco Bay Area Critical Linkage map, California, 119f

Santa Rita-Tumacacori planning area
(Arizona), 208f
satellite data, 165–166
saturation dispersal, 35, 37f
scale: addressing, 165–167; climate change
and, 12; connectivity and, 16, 106; corridors and, 92, 103b, 104;; ocean connectivity and, 225–226, 230–231, 230f;
perception of patchiness and, 153–154;
planning and, 116–117, 274–275
SCAPOSD. *See* Sonoma County Agricultural Preservation and Open Space
District
scarlet rose finches (*Carpodacus erythrinus*),
78–79
screech owls (*Otus asio*), 72, 79
sea otters (*Enhydra lutris*), 64
sea surface height, 218
seascape connectivity, defined, 18b
seasonal forcing, 38, 49–51, 50b
seasonal migration corridors, defined, 18b
seeds, 11–12, 33, 34f, 115
selection functions, 171t
shelterbelts, 95, 114
shipping lanes (marine), 222, 223–225, 224f
Sierra Madre Occidental Biological Corridor, 108
simulation models, 182
sink, matrix as, 87–89
sink populations, 84–85
Sky Island Wildland Network, U.S. Southwest, 108
Slaty Creek Underpass Wildlife Project
(Australia), 246–247
SLOSS (single large or several small)
debate, 60b
small populations, 38–43
small skippers (*Thymelicus sylvestris*), 133
snowpack, 113–114, 210
social systems: corridor assessment and,
186; design objectives and, 123–124;
dispersal and, 34–35; island biogeography and, 22; matrix, movement and, 73;
negative with linkages, 157–158
song sparrows (*Melospiza melodia*), 66
Sonoma County Agricultural Preservation

and Open Space District (SCAPOSD),
California, 113
Sonoma Land Trust, 262
Sonoma Valley Wildlife Corridor, California
(SVWC), 2, 260–262, 262f
source populations, defined, 84
source-sink axis, 84–85
source-sink metapopulations, 27, 28f
spatial scale, 165
specialist species, 117, 123
species composition, changing in patches,
63–65
species distribution modeling, 210–213
species number (richness), 20–21, 21f
species persistence, 109, 181–182
species relaxation, 64
species size, island biogeography and, 22
species-specific connectivity, 170–172,
195–196, 210–213
spores, dispersal and, 33
spotted owls (*Strix occidentalis caurina*),
102, 122
Stebbins, Robert C., 242b
stepping-stone connectivity: corridor, 91;
patches, 123; leading to hybridization
158; for butterflies 202; limitations
203; green roofs and, 241, 244b, 245f;
oceans and, 227, 232-235; overview of,
132–134
stewardship, 245–248, 279
stochastic influences, small populations and,
39, 41
stopper, matrix as, 87–89
stratified random sampling, 187
streams, 95, 101–102, 102f, 188–192,
190f, 209
structural connectivity: climate change and,
195, 205–210; coarse-filter approaches
and, 172–173; defined, 18b; overview of,
93–95; in riverscapes, 143–144; species
distribution modeling and, 210–213
substratum type, 227, 229f, 231
sugar gliders (*Petaurus norfolcensis*), 121
survivorship, 21–23, 104–105
SVWC. *See* Sonoma Valley Wildlife Corridor
sylvatic plague (*Yersinia pestis*), 152

synchronicity, 30, 37–38
synergistic effects, 10, 46

T

taiga voles (*Microtus xanthognathus*), 73, 80
Targhee National Forest (USA), 57, 57f
temperature frontal boundaries, oceanic, 218
temperature gradients, 205, 218
termini, identification of, 169
terrestrial ecosystems, 23, 167–178
Theory of Island Biogeography, The (MacArthur and Wilson), 20
thermotaxis, 231
thresholds, anti-regulating forces and, 39–40
tigers (Panthera tigris), 180f, 182
Tilden Regional Park (California), 242b, 243f
topography: climate refugia and, 199–200, 201f; climate velocity index and, 198; design objectives and, 127–129; naturalness-based corridors and, 207; vulnerability to climate change and, 210
Toronto, Ontario, 258, 259f
track surveys, 170
trailing edges, 201
Trans-Canada Highway, 110–111, 132f, 162, 270
Transition Zone Chlorophyll Front (TZCF), 220
trophic levels, 49, 50f
trusts, 249
tufted titmouse (*Baeolophus bicolor*), 79
tule elk (*Cervus elaphus nannodes*), 40
tundra voles (*Microtus oeconomus*), 130
tunnels. *See* Underpasses
turtles, 126
Two Countries One Forest, 92
TZCF. *See* Transition Zone Chlorophyll Front

U

umbrella species, 117, 118–120
underpasses: as corridors, 96f; economic issues and, 162–163; for individual species conservation, 102–103; roads and, 100, 110, 111f, 131; species filtering and, 150. *See also* Culverts
unimodal distributions, 34f
United Nations Convention for the Law of the Seas, 220
upwelling, 230
urban areas, 113, 255–259, 257f
urban sprawl, 113

V

Vagility of species, 21–22, 72
vegetation: community classifications and, 187; corridor dimensions and, 137; design objectives and, 122; effects of fragmentation on over time, 44; habitat quality and, 135–136; microclimate alterations and, 75–76; pioneer species, edge zones and, 77
vehicle collisions, 78, 162–163
velocity, climate, 196–198, 197f
vertical migration, 217–218
voles (*Microtus* spp.): corridor continuity and, 130; corridor quality and, 150; density dependence and, 124; dispersal and, 73; edge effects and, 80; genetic variation and, 61; inbreeding, insecticide and, 46; matrix and, 71, 85–87; riparian corridors and, 142; as specialists, 123
Vreeland, Christy, 2
vulnerable species, 117

W

water clover (*Marsilea quadrifoliar*), 182
water column structure, 217, 227
watershed connectivity, 143
western toads (*Bufo boreas*), 93
wetlands, 121, 166, 192, 234
whales, 218–219, 219f, 223–225
white-footed mice (*Peromyscus leucopus*), 135, 136
whooping cranes (*Grus americana*), 133
width, corridor quality and, 137–139, 149, 155, 214–215
wilderness areas, statistics on, 7
wildfires, 260
Wildlands Network, 108

willow flycatchers (*Empidonax traillii extimus*), 155, 156f
Wilson, David S., 47
Wilson, E. O., 20
windbreaks, 97f
wolverines (*Gulo gulo*), 105, 183, 210
wolves, 125, 153
wood thrushes (*Hylocichla mustelina*), 79
woodpeckers, 121

Working for Water program, South Africa, 252
working landscapes, 69, 245–248

Y
Yellowstone National Park, 14, 57, 57f. *See also* Greater Yellowstone Ecosystem
Yellowstone to Yukon (Y2Y) Conservation Initiative, 108, 131, 269–272, 271f, 272f